*Generations of Reason*

# Generations of Reason

A FAMILY'S SEARCH
FOR MEANING IN
POST-NEWTONIAN ENGLAND

Joan L. Richards

Yale UNIVERSITY PRESS

New Haven & London

Published with assistance from the Annie Burr Lewis Fund.
Published with assistance from the foundation established in memory of Amasa
Stone Mather of the Class of 1907, Yale College.

Yale University Press books may be purchased in quantity for educational,
business, or promotional use. For information, please e-mail sales.press@yale.edu
(U.S. office) or sales@yaleup.co.uk (U.K. office).

Set in Adobe Garamond type by Tseng Information Systems, Inc.
Printed in the United States of America.

Library of Congress Control Number: 2021931524
ISBN 978-0-300-25549-2 (hardcover : alk. paper)

A catalogue record for this book is available from the British Library.

This paper meets the requirements of ANSI/NISO z39.48-1992
(Permanence of Paper).

10 9 8 7 6 5 4 3 2 1

For Rick

# *Contents*

# *Acknowledgments*

This book may justly be seen as the culmination of a life devoted to understanding the place of mathematics in modern European cultural and intellectual history. Examining the English reception of non-Euclidean geometry in my first book, *Mathematical Visions,* I laid the groundwork for my career-long fascination with the peculiarities of Victorian mathematics. In my second book, *Angles of Reflection,* I began to explore the people who were pursuing mathematics. In particular, I found myself intrigued by the relationship between the mathematical and logical thinker Augustus De Morgan and his wife, Sophia, who was deeply involved in early Victorian spiritualism. I thought the juxtaposition of a logician and a spiritualist was fascinating, and so when I finished *Angles,* I set out to understand how these two managed to pursue such apparently divergent paths within the context of their early Victorian world.

Historians are shaped at least as much by their archival material as by their ideas, and what I found as I researched the De Morgans fundamentally altered the nature of my project. The first shock came from reading the papers of Sophia's father, the eighteenth-century political radical William Frend. I initially approached Frend as a background figure, but his letters clearly revealed that his thinking formed the essential foundation for the ways both of the De Morgans understood their world. And so my project expanded to include two generations, and I spent several years following the development of Frend's ideas through his long and complicated life.

Just as I was drawing my eighteenth-century sojourn to a close, archival forces again intervened, this time in the form of a gut-wrenching correspondence between two men in Frend's wife's family: her Anglican grandfather, Francis Blackburne, and her Unitarian uncle, Theophilus Lindsey. Their desperate efforts to reconcile their reasoned conclusions about matters of faith with their positions in the Anglican Church drove home the practical immediacy of the issues that faced those determined to pursue reason in eighteenth-century England. It also led me to see how essential Lindsey's decision to create an alternative to the Anglican Church was to Frend's life in the next generation. And so a project that began as an account of a marriage in science became *Generations of Reason,* a chronicle of the ways three generations of a remarkable family pursued reason in their lives.

I owe enormous debts of gratitude for those who supported me as this book took shape. For organizational ease I will divide them into three parts—research, conceptualization, and writing—although in reality these processes are never really separate. On the research side, the American Council of Learned Societies, the NSF, and the Guggenheim Foundation generously supported my work for the two academic years 2001–2003. During those years and beyond, a number of the libraries opened their collections to me. These include the large collection of De Morgan papers to be found in the archives of the Senate House Library of the University of London and the archives of University College London. Dr. Williams's Library in London contains the pivotal correspondence between Francis Blackburne and Theophilus Lindsey. The Cambridge University Library houses a considerable collection of William Frend papers, as well as correspondence between Augustus De Morgan and George Biddel Airy. The Archive of the Noel, Byron, and Lovelace Families in the Bodleian Library of Oxford University contains extensive correspondences between Lady Byron and William Frend and Sophia De Morgan, as well as letters between Ada Lovelace and Augustus De Morgan. Additional materials are to be found in the archives of Trinity College and Jesus College in Cambridge. I visited these archives in a series of quick trips to England spread over several years and am grateful to Phillip Benedict and Patricia Fara, who helped me find places to stay while in Cambridge and Oxford. I am equally grateful to the host of librarians and archivists, who expertly guided me through the material in their collections. My appreciation for their efforts has only increased in the past few months, when they have gone to great lengths to obtain the images I needed in the midst of the Covid-19 lockdown. Karen Attar and Dean Hanlon of the Senate House Library were particularly helpful, but I am also grateful to all of the determined and creative librarians who lie behind every image in this book.

The material I found in the archives profoundly affected my thinking about this project. My evolving understandings of its scope were supported by several intense periods as a fellow at the Max Planck Institute for the History of Science in Berlin in spring 2001, fall 2004, and August 2005. Following the thinking of people, as opposed to focusing on the development of ideas, has led me in many new and unexpected directions. The most significant challenge lay in understanding De Morgan's logical work. I knew very little logic when I began this project, and I simply could not have finished it without the help of Daniel Merrill and more particularly Michael Hobart, who devoted years to helping me make sense of De Morgan's approach to the subject. On a somewhat different level, I am also deeply grateful to all of those who listened and shared their expertise on early morning rows at the Narragansett Boat Club. Lynn McCracken helped me think about Lindsey's and Frend's theological issues, Ruth Berenson gave insights into Sophia's efforts to raise children, Kyna Leski helped me respond to visual evidence, while John Duke, Eric Goetz, and Amy Abbott were ready to think about virtually anything. Years of conversations among the birds, wind, and waves of the Seekonk River informed the thinking behind this book.

Then came the writing. Working with students has always been a wonderful stimulus to my thinking, but I found it very difficult to write while teaching. After several frustrating years trying to write between classes, I was rescued by Sachiko Kusakawa, who sponsored my application to spend 2010–11 as a visiting fellow commoner at Trinity College, Cambridge. It was a remarkable opportunity that not only allowed me to write but also constituted the first time I had ever spent more than five weeks in England. In that extraordinary year, conversations over meals with a variety of fellows of Trinity College, in the Colloquium of the Department of History and Philosophy of Science, and with Simon Schaffer and Anne and James Secord opened my understandings in innumerable ways.

Writing about people is very different from the kind of focused history of mathematics I was educated to do, and after returning to a full life of teaching, I realized I needed help. Over the course of three summers, Susan Dearing guided me through the process of turning an enormous sprawling manuscript into a book. It is impossible for me to overstate Susan's importance for this project. There is no word that she has not read at least twice, and every non-mathematician and logician who understands one of the technical passages can thank her for insisting that I be clear. I, for my part, can't thank her enough.

After three summers making the manuscript accessible, the next step was to convene a workshop whose members could professionally assess the book. I am infinitely grateful to Lorraine Daston, Janet Browne, Kevin Lambert, Lukas Rieppel, Tara Nummedal, and Susan Dearing, who gathered for an intense day of discussion in early April 2017. Four more months as a fellow of the Max Planck Institute in fall 2018 gave me time to incorporate their comments and insights into my thinking. Through it all, my colleagues in the Brown History Department—Nancy Jacobs, Holly Case, Kerry Smith, Cynthia Brokaw, Seth Rockman, Robert Self, Ethan Pollock—have always been ready to offer advice and support. More recently, I've been grateful to my editor, Jean Thomson Black, who recognized the value of what I was doing. Writing this book has been a long and complicated process, which I could not have completed without considerable help along the way.

And then there is my family. Having begun by recognizing that this book represents the culmination of a life's work, I will close by thanking my husband, Rick Richards, who has faithfully stood by my side for more than two-thirds of that life, and my children, Brady and Ned, who grew up in the midst of it. Without their constant presence none of the work I have done, including this book, would have been possible.

# GENERATIONS OF REASON

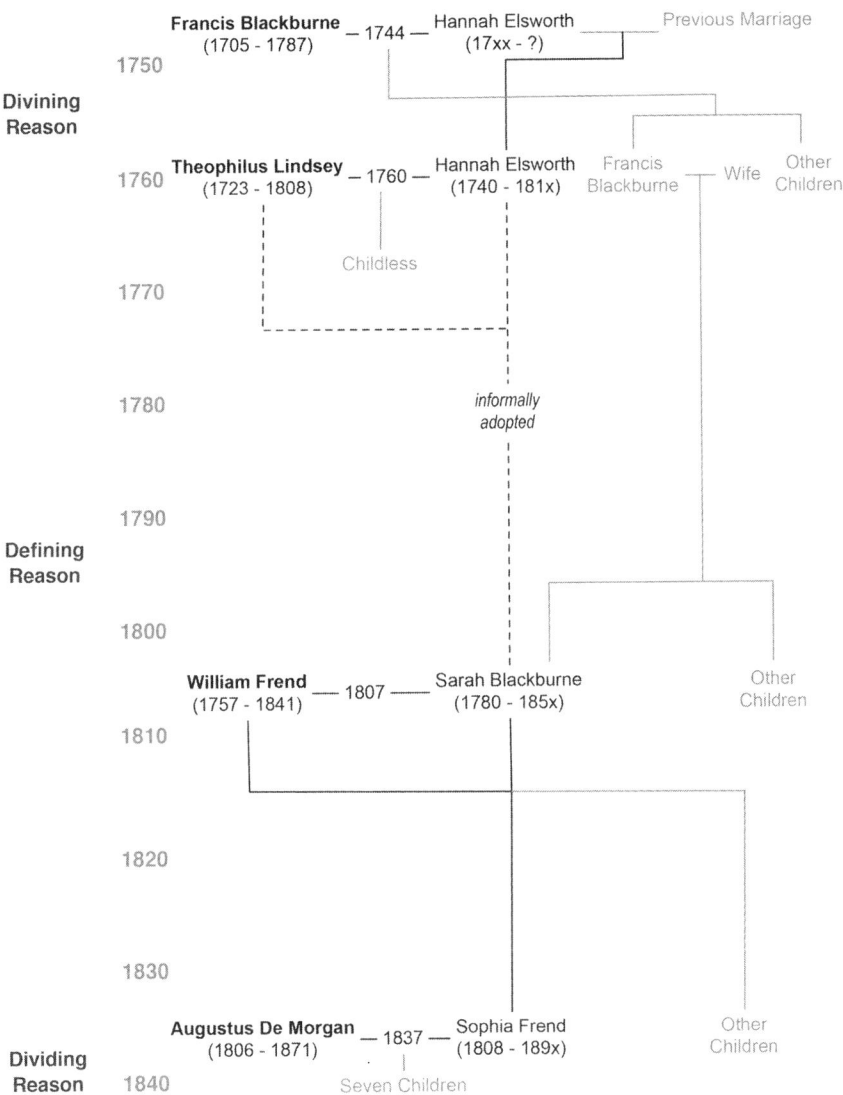

Divining
Reason

1750

Francis Blackburne — 1744 — Hannah Elsworth
(1705 - 1787)                    (17xx - ?)

Previous Marriage

1760    Theophilus Lindsey — 1760 — Hannah Elsworth    Francis    Wife    Other
        (1723 - 1808)                 (1740 - 181x)    Blackburne            Children

1770                                          Childless

1780                                          informally
                                              adopted

1790

Defining
Reason

1800

1810    William Frend — 1807 — Sarah Blackburne          Other
        (1757 - 1841)          (1780 - 185x)             Children

1820

1830                                                      Other
                                                          Children

        Augustus De Morgan — 1837 — Sophia Frend
        (1806 - 1871)              (1808 - 189x)
Dividing
Reason   1840    Seven Children

# *Introduction*

On the 3rd Augst at the Superintendent's Registrars Office, St. Pancrs, by the Rev'd T. Madge, Minister of Essex Street Chapel, Augustus De Morgan Esq. to Sophia Elizabeth, oldest daughter of William Frend Esq.

"My dear Lady Byron," the father of the bride, William Frend, wrote to his longtime friend and correspondent in 1837, "the above contains the important intelligence of the proceedings of this morning. The happy couple have a fine day for their excursion. . . . I am very much pleased at this new mode of marrying. . . . The idea of the Registrar's Office at first startled the females but by degrees they were reconciled to it."[1]

The wedding of Augustus De Morgan and Sophia Frend was a family event. Except for the couple and minister, the only people present were the parents and siblings of the bride and groom. Even Lady Byron, who was both a close friend of William Frend and a confidant of his daughter Sophia, participated only through the letters she received after the fact. That the wedding was located in the Registrar's Office was a powerful statement about what it meant to be a member of that family. The elderly man who beamed at the couple over his cane had spent decades fighting for the choice that the Marriage Act had extended to young couples less than a year before. Sophia's and Augustus's decision to be married in a secular space was for Frend the triumphal conclusion to a life spent struggling against the monopolistic power of England's established church.

Even as Frend exulted in the wedding's location, he recognized that others of the group did not. The most "startled" of the attending females was undoubtedly Augustus's deeply religious mother, who was quite upset. "Why are you not married in a church, as your father was before you?" she wailed, but her son would have none of it. Actually, he reminded her, she and his father had not been married in a church but rather in a colonial office in Ceylon by a minister "who was a model of riotous living and unclerical notoriety."[2] This was undoubtedly true, but by the time of her son's marriage, hard experience had effectively destroyed whatever spirit of adventure had propelled Elizabeth De Morgan into hers. But having spent most of his growing up negotiating her anxious fears about propriety, Augustus was not about to capitulate on the venue of his wedding.

The form of the service was another issue. Sophia rather liked the Anglican wedding service, though Augustus did not. The details of their negotiations are lost, but Sophia later reported, "My respect for his scruples made me willing to comply with his wish." In the event, the Reverend Madge married them following a script "differing from that in the prayer book only by the omission of the very small part to which we could not assent with our whole hearts."[3] That Augustus had his way with the service in St Pancras does not mean that he did so in everything, however. He viscerally hated leaving London, and three weeks honeymooning in Normandy left him totally depleted.

Frend may thus have been the only person who had no problems with Sophia's and Augustus's secular wedding. Part of his satisfaction lay in the recognition that even as his beloved daughter was moving out of his house and changing her name to De Morgan, she was not leaving his family. On the contrary, when Augustus married Sophia, he was joining a family of which Frend was the oldest living member.

The larger family, whose spirits suffused the proceedings on that sunny August day in 1837, was not the usual patrilineal one, held together by the flow of material and other goods from father to son. Instead, it was defined by a commitment to reason that was carefully passed from one generation to the next. In each generation, two unrelated men were drawn together by their shared understanding of the centrality of reason to their lives. Then, after ten years or more, these masculine friendships became familial when the younger man married one of the older man's young female relatives: a stepdaughter, a niece, a daughter. The family thus created is displayed in the family tree at the beginning of this introduction. The men whose friendships preceded their marriages are on the left side of the page, and the women who tied them together are listed to their right.

The family so depicted originated in the friendship between Francis Blackburne and Theophilus Lindsey, which flourished in the 1750s and became a family affair when Lindsey married Blackburne's stepdaughter, Hannah Elsworth, in 1760. The friendship between Lindsey and Frend began in the late 1780s, and culminated with Frend's marriage to Lindsey's niece, Sarah Blackburne, who was also the granddaughter of Francis Blackburne. Twenty years later, the pattern repeated when De Morgan and Frend formed a friendship that led to the wedding in St Pancras. That women formed the biological rope binding the family together means it did not carry a common name across generations. Nonetheless, all of the men, women, and children within it recognized its central importance to their identity.

The concept of reason that defined this family's outlook had been recognized as the essential capacity separating humans from animals since at least the time of Aristotle. In the eighteenth century, the dramatic success of Newtonian cosmology gave such immediacy to that capacity that the whole century has been referred to as the Age of Reason, or Enlightenment. The conviction that all humans could use their minds to reach valid conclusions about the world supported the thinking that erupted in the American and French Revolutions. This conviction played out differently in England, which remained a monarchy in which membership in the Anglican Church was a prerequisite for full participation in civil society. The strength of the established church has led some to deny that England had an Enlightenment at all, but that is to miss the development of reason within religion itself. Members of the family that gathered in spirit around the De Morgan wedding understood it as a divine gift and a sacred charge. For them, to reason with others was to carry out a conversation on a common ground hallowed by God's love for his reasoning creatures.

The family drew their understanding of reason from the writings of John Locke. For Locke, Isaac Newton's success in understanding the workings of the universe served as a staggering demonstration of the power of reason. The remarkable implication of his accomplishment was that when humans deployed their reason they could come to understand the movements of the sun, planets, and stars as absolutely and completely as they did Euclidean geometry. For Locke, the wonder of Newton's cosmology lay in the way it enabled him to align his thoughts with those of the God who created the universe.

The epistemological breakthroughs represented by Newton's cosmology led many continental thinkers to embrace reason, as opposed to scriptural revelation, as the best way to understand themselves and their position in the world. None of the members of the family of reason took this route, however. Even as they rebelled against what they saw as Anglican restrictions on reason, they all remained deeply committed to scriptural Christianity. In their firm religious commitment, the family was following a Locke who was all but invisible to continental thinkers. In his final work, *The Reasonableness of Christianity, As Delivered in the Scriptures,* Locke embedded human beings in a world defined by a loving relationship between God and the rational creatures God had created to reason alongside him. Locke argued that such a God, being both reasonable and merciful, would certainly communicate with his creatures in ways they could understand. For him, the teachings of Jesus as recorded in the four Gospels and Acts constituted that communication. Applying the power of reason to Jesus's words would bring people to an understanding of God's will that was as clear as Newton's understanding of the universe God had created.

The members of the family that gathered in body and spirit around the De Morgan wedding were all Lockean Christians, for whom the ability to reason took the form of a sacred trust. All of the men in the family encountered Locke's work as students at Anglican Cambridge. All of the women learned of reason under their parents' wings and passed it on to the children who flourished under theirs. All members of the family followed Psalm 15, which defined the truly godly person as one who "speaketh the truth in his heart" even though it be "to his own hurt."[4] In many situations, the truths they spoke were controversial enough to cost them friends, jobs, and social standing.

Their commitment to practicing reason steered the family through more than one hundred eventful years of English history. Their story begins in the middle of the eighteenth century, when Lindsey began struggling with his growing conviction that exercising reason required him to break away from the Anglican Church. In 1774 he finally resolved the conflict by forming England's first Unitarian Church in London. Over the course of the next fifteen years Lindsey and his congregation stood at the forefront of an English Enlightenment focused on eliminating religious discrimination in politics, while welcoming diversity into their church. When reason led William Frend out of the church in 1788, he became a satellite of Lindsey's movement in Cambridge.

In the decades after the French Revolution broke out in 1789, these Unitarian efforts to open their society to the full humanity of reason's power were cut to the ground. The English reaction to the political dramas played out in France over the course of the next twenty-five years was severe and the consequences of publicly speaking reason's truth could be harsh. Lindsey was retired by this time, but Frend, who was in his prime, was severely punished for his positions. He never wavered in his sacred duty to practice the reason he had been granted, but prudence required that he shift his focus from religion and politics to mathematics and astronomy, and his audience from the public to the domestic sphere.

Lindsey's and Frend's wives both remained steadfastly supportive in the many instances when their husbands defended unpopular positions. Living in a world in which all of their power and legitimacy came from men complicated the charge that they live reasoned lives. Hannah Lindsey was known, and often disliked, for loudly speaking her truth, while Sarah Frend, who was much more quiet, devoted herself to cultivating her children's reason at home. Hannah and Sarah were very different people, but both devoted their lives to furthering the cause of reason in the spheres they inhabited.

After the fall of Napoleon in 1815, the English slowly recovered enough confidence to open themselves to the kind of religious and social diversity that Lindsey had pursued in the 1780s. In 1828 Parliament repealed the Test and Corporation Acts, which had restricted the political rights of dissenters. The following year they lifted the specific restrictions placed on Catholics with the passage of the Catholic Emancipation Bill. Action on these two religious bills was a prelude to the Great Reform Bill of 1832, which ameliorated the blatant unfairness of representation in the English Parliament. These adjustments significantly altered the religious and political landscape of the previous century.

For Frend, the political and religious changes of the 1820s and 1830s represented a phoenix Enlightenment, in which ideals and commitments that had been brutally trampled in the 1790s burst forth again. However, much had changed since he and Lindsey had pushed for religious and political reform in the 1780s. In the year before Augustus and Sophia exchanged their vows in the Registrar's Office of St Pancras, a not-yet-eighteen-year-old Victoria had become the queen of England. As the young couple pledged themselves to one another, they were committing themselves to a path of reason that had been blazed by the frail man who beamed upon them from a seat in the front and the deeply caring woman who sat by his side. The challenge for this new generation was to carry the torch of reason lit in the insular eighteenth century into the expansive nineteenth-century world around them.

The place of religion in their lives was a major factor that divided the De Morgans from their predecessors. It is very difficult to imagine either Blackburne or Lindsey being comfortable with the couple's decision to be married in the St Pancras Registrar's Office. Frend was happy with their choice, but a closer look reveals that he disagreed with them as to its meaning. He saw the civil space as an inclusive one in which all religions could be practiced equally, but for Sophia and Augustus it was a neutral space essentially devoid of religion. The Victorian secular world they were moving into was to be essentially defined by its separation into a masculine public sphere of action and a female private one that sheltered religious practice.

Both of the De Morgans embraced this division. Augustus spent his days cultivating young men's reason by teaching them mathematics in the secular spaces of University College London. In the evenings he worked to strengthen the rational capacities of his countrymen in writings that kept them apprised of the latest developments in mathematics and astronomy. His accomplishments as arguably the foremost mathematical and logical thinker of his age were played out against the backdrop of these educational efforts.

At the same time, Sophia embraced her role as an early Victorian woman. She recognized that there were advantages to her position and spoke out in areas that were accepted as female. She was deeply involved in ameliorating the conditions of the poor, developing humane responses to the problems of mental health, caring for unwed mothers, and, at the very end of her life, protecting animals from vivisection. She supported Augustus's interests in mathematics and logic but was more immediately interested in applying her reason to an enormous variety of texts, from the Bible to Swedenborg to the utterances of the neighbors, servants, and children with whom she lived. In all of these endeavors, Sophia attested to the power of reason in her female world.

While the De Morgans carried on the family commitment to exercise reason, their forebears' eighteenth-century world was coming to seem static and parochial. One of the casualties of the tumultuous years of political revolution and reaction was the essentially cyclical view of history that supported the eighteenth-century Unitarians in their efforts to find the true meaning of Jesus's message in some kind of pure and distant past. At the same time, England's imperial expansion was raising awareness that different forms of truth were spread across the world, while the patterns of its emergence in different times and places were enabling new approaches to historical possibility. De Morgan devoted his life to adjusting mathematics and logic to fit the new understanding of historical progress. Sophia devoted hers to exploring new vistas of religious understanding that were being opened by the influx of different languages and spiritual traditions. Both worked to adjust the parameters of the reason they had inherited to fit their early Victorian world.

Sophia and Augustus devoted themselves to defending the family tradition in which the exercise of reason could lead to transcendent understandings validated by wondrous experiences of insight. Nonetheless, over the course of their lives, understandings of reason in England were evolving. One powerful force for change came from the continent, where the religious vision of Locke's *Reasonableness* never took hold. In addition, with the rise of industry, reason began to be valued for its effectiveness in solving problems rather than for its support of transcendent insight. The implications of these developments were at once subtle, hard to identify, and all-pervasive. Even as the De Morgans reveled in the progress that allowed them to expand their thinking beyond eighteenth-century boundaries, they were always at least equally engaged in defending the essential power of human reason against a myriad of forces that were undercutting it.

By the time the De Morgans were ready to pass the family torch to the next generation, their understanding of reason as the essential path to insight into the deepest meanings of the human experience was losing currency in the world around them. Today, the most well-known member of the family is their son, William Frend De Morgan, who is remembered for the captivating beauty of his ceramics. In a world increasingly divided between the arts and the sciences, this well-known bearer of two generations of the family name carried the humanistic values of his forebears' reason into the next generation as an artist rather than theologian or mathematician.

It can be maddeningly difficult for historians to capture the meaning of centrally important words and concepts because of the ways their import and implications slide. In the case of reason, the power to shape so many different lives arose from the ways it twisted a set of loosely related possibilities and implications into a powerfully dynamic whole.[5] Essentially the same analogy that supports approaching reason as a twisted rope of implications is also to be found at the heart of the yarns that we spin, the stories that we weave. From this perspective, the scene before the Unitarian minister in St Pancras may be understood as a detail in a rich tapestry that members of the family had been weaving for more than seventy years. The sense of the reason that they embraced is to be found in the human stories with which it is entangled. In *Generations of Reason,* a twisting, spinning thread of reason binds the characters together and holds their place in a story that stretches from the mid-eighteenth to the end of the nineteenth century.

I'm going to start at the beginning.

# PART ONE

# Divining Reason

# CHAPTER 1

## *Faith in Reason*

The historical record is silent about Sophia's mother's reactions to the De Morgan wedding. This is characteristic; the surviving sources reveal but little about the woman William Frend referred to simply and lovingly as Mrs. F. She may have been "startled" that Sophia decided to be married in a registrar's office, but she undoubtedly approved of the rest of the proceedings. Creating a personal service by tinkering with the Anglican rite had long been common practice among the Unitarians at the Essex Street Chapel, where she had attended services off and on since she was sixteen, and the Reverend Madge had been her trusted minister for more than twenty years.

Family was of primary importance to Sarah Frend. All of her other children—Frances, Alicia, Henry, and Alfred—were present for Sophia's wedding, and in the weeks leading up to it she was making sure her increasingly frail husband had adequate transportation, advising her children about their clothes, and consulting about everyone's proper place. At least equally important to the shape of the proceedings, however, was the family in which she had grown up. Sophia's mother was born Sarah Blackburne, and the story of reason that in 1837 paused in the parlor of the St Pancras Registrar's Office began with a friendship between her grandfather, Francis Blackburne, the archdeacon of Cleveland in Yorkshire, and Theophilus Lindsey, the rector of a small parish in Blackburne's diocese. When Lindsey married Blackburne's stepdaughter, Hannah Elsworth, in 1760, they became the first couple of the family of reason whose descendants gathered at the wedding of Sophia Frend and Augustus De Morgan.

The frontispiece to Sarah's grandfather's collected works was etched in 1780, the year she was born. In the image Francis Blackburne appears as a somewhat jowly, hawk-nosed clergyman in a simple white wig (figure 1). There is no sign of levity in his face; the eyes that look out of the page are clear and straightforward. The directness of the image is appropriate to the man. At the time it was made, Blackburne had for more than forty years served as the Anglican rector of St Mary's Church in the Yorkshire town of Richmond. He had also for more than thirty years been the archdeacon of Cleveland, but he was never comfortable with the trappings of that position. The simple black robes he chose for his portrait are fitting to a man who spent his life trying to turn the Anglican Church away from its controlling hierarchies.

Blackburne was born in the Richmond he was to serve as rector seventy-five years before this portrait was made. His town was on the western edge of the "rocky steeps" of the Yorkshire Dales, which were in the early eighteenth century cut off from the rest of England by a combination of mountains and all but impassable roads.[1] The square battlements of a twelfth-century Norman castle rose above the town's large curved market square. Tucked into the hill, on the level of the town, the square steeple of St Mary's Church echoed the castle's towers. Below it all, the River Swale rushed and broke in burbling cataracts. The hill and the river that made Richmond so beautiful created problems as well. Originally built as a market town, over the centuries its hilliness and the difficulty of navigating the Swale hindered competition with its neighbors. In the early eighteenth century, the population was holding steady at about three thousand, as it had for the previous four hundred years.

The population may have been stable, but Richmond was not stagnant, and when Blackburne was born there in 1705, the stocking trade was beginning to boom.[2] His grandfather, who for the convenience of this paragraph will be Francis Blackburne I, had made a great deal of money in the stocking trade and settled a goodly fortune on his son, Francis Blackburne II. However, Blackburne II drank himself to death by the age of twenty-nine, leaving behind a wife and three young children, the eldest of whom was Sarah's grandfather. In due time, his widowed mother remarried, this time to William Kirkby, a Lancashire man who conscientiously furthered the education of his Blackburne charges. In 1722, Kirkby sent seventeen-year-old Francis to Catharine's Hall (now St Catharine's) at Cambridge University.

When Blackburne went to Cambridge, he traveled from England's north-

Figure 1: Francis Blackburne. Frontispiece to Francis
Blackburne, *The Works Theological and Miscellaneous*
(Cambridge: Benjamin Flower, 1804).

ern periphery to the center of its intellectual establishment. The Catharine's
Hall he entered was one of sixteen colleges loosely gathered under the umbrella of Cambridge University. The two largest were St John's and Trinity,
whose turreted gatehouses and elaborately decorated buildings had been constructed in the sixteenth century as outward and visible signs that England's
Anglicanism was as powerful as the Catholicism it was superseding. By contrast, Catharine's Hall was the architectural product of the Puritan seventeenth century. The money ran out before its sheltering wall was built, so its

L-shaped configuration of brick buildings was easily viewed from the street. The most striking feature of those buildings was the way each of its window panes was cut in the form of a cross.

Behind such differences in style and wealth, the Cambridge colleges were alike in their essentially monastic structure. Around the lush inner courts of the larger colleges and behind the crosses in the windows of Catharine's Hall lived tight communities of single men, the "fellows" of the college. Twice a day they worshipped together in the college chapel; once a day they dined together in the college hall. Every year the colleges welcomed groups of students who were watched over by fellows designated as tutors, whose job it was to act in loco parentis for their charges. In Blackburne's day, St John's and Trinity boasted undergraduate classes of as many as forty-five students, whereas Catharine's Hall had only two or three.[3] In all cases, the tutors were responsible for pre-paring their students for the university degree exercises, which they faced after about three years of college life. Those who did well were eligible to become fellows, and so the cycle reproduced itself.

On this institutional level, Cambridge colleges were equal, but their stu-dents were not. Most were like Blackburne, fee-paying "pensioners" who were the sons of clergymen, the younger sons of country gentlemen, or as the eigh-teenth century progressed, the sons of an emerging business class. Whatever their origins, the pensioners' prospects were not glittery; the most successful among them might rise to high positions in the university or the church, but the majority were looking forward to quiet lives in country parishes.[4] There were social layers both above and below the pensioners. On the low end were the charity students or "sizars," who had considerable incentives to study and often were the most academically accomplished of Cambridge students. Isaac Newton began his career as a sizar. This and other illustrious examples did little to ameliorate the day-to-day difficulties of their lives, however. Eighteenth-century sizars wore special identifying robes and earned their keep by serving their social betters. As the twentieth-century historian Winstanley put it, "Poverty was the badge of their race, and ... they were despised for it."[5]

The same world that despised the sizars revered the aristocrats and fellow commoners, who constituted the social levels above the pensioners. The aris-tocrats, blood members of the peerage, wore special identifying hats and on state occasions sported elaborately embroidered gowns. Fellow commoners were distinguished by their wealth and paid fees high enough to earn them special privileges; the name "fellow commoner" refers to their practice of din-ing at the fellows' table. Aristocrats were often entitled to university degrees simply by virtue of their birth, and fellow commoners were similarly excused from tutors' lectures and participation in university exercises. Fully a third of the students in Blackburne's Cambridge were aristocrats and fellow commoners who lived in a world of social privilege that was essentially inaccessible to the son of a Yorkshire stocking merchant.

Blackburne's status was reinforced by the sociopolitical atmosphere of his university, which was dominated by Tories. Staunch believers in the divine right of kings supported by the hierarchically organized Anglican Church, their power could make it hard to distinguish Blackburne's Cambridge from the sixteenth-century institution that Elizabeth had supported. For them, the solid Elizabethan towers of Trinity and St John's testified that after a complex and violent seventeenth century of religious and political dissent, the English had returned to a hereditary monarch—and a neo-Catholic Anglican religion.

The more simple architecture of Catharine's Hall may be taken as a reminder that in the midst of the political and religious strife of the seventeenth century, England saw the rise of a new approach to the world that emphasized the power of human reason. In the 1660s a group of men formed the Royal Society of London and pursued a series of investigations on the assumption that events in the natural world followed regular laws that were accessible to human understanding. In 1687, the Trinity College fellow Isaac Newton laid out a view of a mathematically ordered cosmos in his *Philosophiae Naturalis Principia Mathematica* or *Mathematical Principles of Natural Philosophy.* For England's most forward-looking thinkers and churchmen, Newton's success in understanding the essential workings of the cosmos stood as a staggering testament to the power of human reason. In the decades after the *Principia* was published, these men began spinning out the implications of his triumph for their country, their church, and their world. There was no real difference between religion and politics in the English eighteenth century, and by the time Blackburne arrived at Cambridge in 1722, the connection between the new science and the Whig political party had become so close as to be called a "Holy Alliance."[6] John Locke was the most powerful spokesman for the post-Newtonian Whig position. In his work, the power of reason formed the cornerstone for an anti-Tory, democratic vision.

One of Blackburne's clearest memories was that as he was entering Catharine's Hall he heeded the advice of "a worthy old lay gentleman, who said, 'young man, let the first book thou readest at Cambridge, be Locke upon government.'"[7] The book thus recommended was *Two Treatises on Government,* in which Locke constructed a view of government based on the power of reason. Locke began with a model of human beings, who in their original state of nature were constrained only by natural law. States arose from this original chaos when people recognized that total individualism was not adequate to enforce the justice that reason required. This meant that the fundamental role of the state was to protect the basic rights of life, liberty, and property. In Locke's construction, political power rested on the consent of the governed, as opposed to divine right.

Blackburne was a fiercely independent young man who was quickly convinced that humans were essentially defined by the power of their reason. Politically, this meant that by the time he was ready to be elected to a fellowship in Catharine's Hall in 1728, he had become a strong proponent for "principles of ecclesiastical and civil liberty." Unfortunately, however, his liberal Whig views were completely unacceptable in his college, which was dominated by royalists, who were fierce defenders of hereditary rights.[8] It took some finagling for the Tory fellows of Catharine's Hall to deny the young Yorkshireman a position that he was preeminently qualified for, but they found a way. Their refusal to elect him to a fellowship was Blackburne's first taste of the consequences that could attend his commitment to a Lockean view of human reason.

It was a bitter blow. Blackburne had already left the university for a year to fight a depression brought on by the death of his younger brother, Thomas. Now, once again, he returned all but debilitated to his relatives in the north, where he found an antidote to his black moods in fox hunting and the partying that accompanied it.[9] He later claimed that it was "to these relaxations he owed a knowledge of men and manners, highly serviceable on many occurrences of life, and such as it is impossible to learn from books,"[10] but that is to put a positive spin on a raucous period that in the privacy of old age he saw himself as lucky to have survived. It was not until 1739, by which time he had caroused through eleven years waiting for a church position to come open, that he finally succeeded one of his uncles as the rector of Richmond.[11]

When Blackburne took over the parish of Richmond, the lovely town below the towering castle was becoming a magnet for England's well-to-do. The surrounding wild lands provided ample space for fox hunting, and summer horse races—the Richmond Races—were attracting the fashionable set.[12] Over the course of Blackburne's forty-year tenure as rector, the cultured classes in the area grew large enough to support a theater that is still known as an eighteenth-century jewel. Such high-society activities were of little interest to the man whom Tory Cambridge had rejected, however. He was a sobering thirty-four years old when he took over his uncle's position and quite ready to turn his attention to the work of his parish and his church. In 1744 he married a widow, Hannah Elsworth, who brought into their marriage a little girl of the same name. In the years to come, Francis and Hannah were blessed with five additional "pledges of my heavenly Father's love" in the form of Jane, Francis, Sarah, Thomas, and William.[13] Under the watchful eye of the wife he lovingly referred to as "my Dear girl,"[14] the Blackburnes thrived in the vicarage in Richmond.

As Blackburne's family grew, so did the demands on his income. Within the Anglican Church, the accepted solution was to take on more appointments, so in 1750 Blackburne accepted the position of archdeacon of Cleveland. In addition to increasing his income, becoming archdeacon was a significant promotion, but Blackburne did not particularly like his new status. He was viscerally egalitarian and convinced that the essential insight of Protestant Christianity was the recognition that all men should be free to reason their own way to God. It was precisely this point—that religion must be free from any human authority—that Blackburne thought distinguished truly reasoning Protestant Anglicans from Catholics, and it made his lofty position as archdeacon of Cleveland highly uncomfortable. Although he could not wholly escape the administrative demands that left him fuming against the "the oppressions of my heart under the load of trash, which my soul abhors," he did what he could to minimize them.[15] He fulfilled whatever duties his higher position entailed from his Richmond home or on his annual trips to official church councils. The rest of the time, he lived as a "Hermit on the Swale,"[16] dividing his time among his family, the people of his parish, and the theological studies of his library.

Blackburne was a rather private person who did not confide easily, but as his explosion about the "oppressions of his heart" suggests, in the years after he became archdeacon he was sharing his thinking with at least one other person. His confidant was Theophilus Lindsey, a young parish minister who came into Blackburne's orbit just after the older man had accepted his new position. On the face of it, the two could not have been more different—Lindsey was a very sociable young man, fast rising within the church on his high-society connections—but they were alike in intellectual intensity. In the years after they first met, that intensity fueled a passionate discussion of the meaning and implications of Locke's ideas for their lives in the Anglican Church. Slowly, over the course of several years, Blackburne's and Lindsey's common embrace of reason joined them in ways that proved more fundamental than any of their differences.

The engraving of Lindsey that forms the frontispiece to the *Memoirs of the Late Reverend Theophilus Lindsey* provides a striking contrast to the frontispiece of Blackburne's collected works (figure 2). Instead of a plain backdrop, Lindsey is standing in front of a curtain partially drawn back to reveal a distant landscape. The rest of the image is equally indirect; he is facing left but looking to the right in a way that almost avoids the viewer's eye. His frock coat and white cravat are not unambiguously clerical, and his oblique gaze imparts an almost flirtatious quality—at least as much courtier as priest. Little besides his powdered wig would seem to connect the shifting messages of Lindsey's image to the sober directness of Blackburne's.

The ambiguities of the image reflect important realities in Lindsey's early life. Born June 20, 1723, in Middlewich, Cheshire, he was the son of Robert Lindsey, a textile merchant and financier of some local salt works. More to the point, his mother was a Spencer, a member of the family that two hundred years later would include Lady Diana. Before Theophilus was born, his father had made some poor business decisions, which meant that the family relied heavily on his mother's connections to get by. Thus, when their third and final child was born, the Lindseys named him in honor of the Earl of Huntingdon, who served as the infant's godfather. Young Theophilus thrived as the godson of an earl. His "amiable manners, cheerful disposition and unaffected humility"[17] reinforced all impulses to take care of him, and throughout his childhood and youth he was supported by two Huntingdons, Lady Betty and Lady Ann Hastings. At the age of seventeen, he entered St John's College in Cambridge, where he was so virtuously studious that his classmates dubbed him "the old man."[18] The "piety and exemplary conduct"[19] by which he earned this mocking moniker marked him as a treasure for those in power, and he was easily elected fellow of St John's in 1747. The contrast with Blackburne's Cambridge experience could not have been more striking.

The differences continued thereafter. From an early age Lindsey wanted to be a minister, and he was well situated to achieve the goal.[20] Ministerial positions, or "livings," in the Anglican Church were often called "preferments" because they were dispensed at will by the people who controlled the estates on which they were located. Blackburne's unemployed years in the wilderness reflect the kinds of difficulties a dour Yorkshireman could face in the effort to secure such a living; that the ultimate solution was to place him in a position previously held by a member of his family was also typical. Lindsey's situation was different. The combination of his easy affability and aristocratic connec-

REV? THEOPHILUS LINDSEY, M.A.

Figure 2: Theophilus Lindsey. Stiple engraving by Giovanni
Vendramini. D14260 © National Portrait Gallery, London.

tions made him a perfect candidate for preferments. Lindsey was completely
trusted by his noble patrons because of his "prudent and exemplary conduct"
and the "suavity of his manners"; regarded "not with the distance and coldness of a dependent, but with the liberality and affection of a friend."[21] The
prospects of such a man were essentially without limit, and as soon as he was
ordained by the bishop of London, aristocratic women began all but throwing positions at him.

Lady Ann Hastings was the first to court Lindsey's favor with the living
of a chapel in Spital Square. Within a year, however, the Duchess of Somerset plucked him out from under that Huntingdon wing in order to bring him
home as the domestic chaplain to her husband, Algernon, Duke of Somerset.
While he was there, one of the duchess's friends was so taken with the young

man that she wanted him in her parish, but, sadly for her, her vicarage in
Chew Magna was already filled. Undaunted, she instead bequeathed Lindsey
"her right of alternate presentation,"[22] that is, the power to appoint the next
vicar. In the meantime though, the Duchess of Somerset remained in control.
Lindsey stayed in her house until after her husband died, at which point she
sent him to chaperone her nine-year-old grandson on a two-year European
tour. By the time the pair returned, the valuable living of Kirkby Whiske (now
Kirby Wiske) in the North Riding of Yorkshire had come open, so the duchess
could bestow it upon him.

Lindsey was thirty years old when he accepted the living of Kirkby Whiske
in 1753. The rectangular church of St John the Baptist was a solid fourteenth-
century structure on an eleventh-century foundation, but that was not enough
to tempt the London socialite to spend much time there. However, he did
come often enough that he met his archdeacon, Francis Blackburne, who
was in Richmond, just ten miles to the north. There the young man's charm
proved to be as powerful as it had been in London, and soon "the hermit on
the Swale" was pouring his frustrations with their church into the young man's
sympathetic ear. Lindsey was well equipped to understand the archdeacon's
concerns as their common Cambridge education made the crusty Yorkshireman and suave socialite intellectual equals.

The bedrock of Blackburne's and Lindsey's early conversations lay in
Locke's *Reasonableness of Christianity*. In this work, Locke followed his rational human beings beyond the worlds of politics and science into a Christian
world that rested on the relationship between a loving God and the reasoning creatures God had made. Locke's analysis began with the Old Testament,
which he saw as the history of an essentially reasonable agreement between
God and the Jews. "God out of the infiniteness of his Mercy, has dealt with
Man as a compassionate and tender Father. He gave him Reason, and with
it a Law: That could not be otherwise than what Reason should dictate," he
concluded.[23] The New Testament posed a somewhat different challenge. The
story of Jesus was the story of a person whom God had sent as "a Deliverer" to
free his erring people from the strictures of Jewish law. The condition of this
Christian dispensation was that "whoever would believe him [Jesus] to be the
Saviour promised, and take him now raised from the dead, and constituted
the Lord and Judge of all Men, to be their King and Ruler, should be saved."[24]
On the face of it, pledging allegiance to a person might not appear to be reasoned, but for Locke, that the Christian message could be framed as a "plain,
intelligible proposition" to which one could assent rendered it reasonable.

Having formulated the issue in this way, the fundamental challenge Locke faced in *Reasonableness* was to explain what it meant to believe that Jesus was "the Saviour promised," and he turned to the scriptures for his answer. Jesus's message, he avowed, was to be found in "what our Saviour and his Apostles proposed to, and required in those whom they converted to the Faith,"[25] which material was contained in the four Gospels and Acts. This sharply focused approach led Locke to the conclusion that being a Christian entailed "*believing that Jesus was the Messiah;* giving credit to the Miracles he did, and the Profession he made of himself."[26] For Locke, that the New Testament God worked miracles or raised Jesus from the dead was not a challenge to reason; God was all-powerful and could do whatever was required to make a point about the special nature of Jesus. The essential task for a reasoning Christian was to determine and embrace "the Profession he made of himself" in the Bible. The only way to achieve clarity about the religion Jesus preached was to focus determinedly on the words he used to speak it. The Old Testament and the Epistles might be useful for clarification, but they were no substitute for Jesus's basic message, which was confined to the four Gospels and Acts.

At the time Lindsey began visiting Yorkshire, Blackburne was beginning to bring the insights of Locke's *Reasonableness* to bear on his position in the Anglican Church. The focus of the older man's concern was the Thirty-Nine Articles of the Anglican Creed. These Articles were the intellectual counterpart of the soaring towers of Elizabethan Cambridge: a determined show of English power. Originally set out by the Convocation of 1571, these Articles were formulated to meet Queen Elizabeth's goal of carving out a via media, or middle way, between Catholics and Protestants. Anti-Catholic articles rejected transubstantiation, denied purgatory, and disavowed clerical celibacy; anti-Protestant ones pushed back against claims for predestination, affirmed infant baptism, and defended the legitimacy of swearing oaths. Despite various challenges, in Blackburne's England subscribing to the Thirty-Nine Articles remained essential for entrance into the Anglican fold.[27] Students at Oxford had to subscribe before they entered the university; those at Cambridge subscribed before they took their degrees; ministers subscribed each time they accepted a position or promotion. Blackburne had himself subscribed upon being ordained in 1727, upon taking his position as rector of Richmond in 1739, and upon being named archdeacon of Cleveland in 1750. Lindsey had subscribed upon being ordained in 1747, upon accepting the living of Spital Square in 1748, and upon coming to Kirkby Whiske in 1753. However, by the time Lindsey and Blackburne first met, Blackburne was beginning to see that

subscribing to the Thirty-Nine Articles was an essential violation of the scripturally based Christianity that Locke had so clearly laid out in his *Reasonableness.* The more Blackburne thought about it, the more he was convinced that
a true Christian was essentially bound to "*stand fast in the liberty wherewith*
Christ *hath made him free,*"[28] and that doing so required him to reject all
merely human doctrines. This meant that it was simply wrong "to allow any
creedal statement," including the Thirty-Nine Articles, "to shape and delin-
eate the meaning of the scripture."[29] He insisted that true Christians should
be free to read and interpret Jesus's message for themselves.

Blackburne was not alone in his discomfort with the Thirty-Nine Articles;
many of his liberal-minded compatriots had for decades been resisting the restrictions thus placed on membership in the Anglican Church. In the years
after Locke wrote his *Reasonableness,* some had found ways to embrace the
minimalism of his scripturally grounded vision even as they continued to subscribe. Members of this group, quite a few of whom Blackburne had greatly
admired as a student, were often referred to as "latitudinarians" or "broad-
churchmen." Their way around the constraints imposed by the Articles was
to interpret them loosely. Thus, in his introduction to *Scripture Doctrine of the
Trinity,* Samuel Clarke insisted "the church of *England* did not mean more by
subscription, nor require more of subscribers, than that they should conform
their opinions to the true sense of scripture."[30] Others pointed out that since
in some places the Articles seemed to contradict themselves, this kind of loose
reading was essential.[31] Taken to its limit, as it was in a 1719 pamphlet by *Phi-
leleutherus Cantabrigiensis,* the latitudinarian position meant that all that was
required of the subscriber was to believe the Thirty-Nine Articles "in *any* sense
the words will admit of."[32]

Blackburne and Lindsey agreed with the latitudinarians' goal of inclusivity but disagreed with their approach to that goal. The impetus for their
disagreement came from Locke's word-focused interpretation of reason. That
truth was to be found in maintaining the close connection between words and
their meanings was essential to Locke's program of finding Jesus's religion in
the words he had spoken. But accepting the essential validity of this linguistic form of reason carried implications beyond reading the Bible. Recognizing that truth was to be found in respecting the close ties between word and
meaning applied to the Thirty-Nine Articles as well. From this perspective,
the latitudinarians' loose readings of the Articles were willfully dishonest. As
Blackburne put it "to believe one thing, and to profess another, the Christian
religion calls *hypocrisy,* and under that name severely censures and condemns

it."[33] Both Blackburne and Lindsey thought it was wrong to allow the linguistic foundations of reason to be undercut by latitudinarian obfuscation.

This demand presented the two churchmen with the challenge of envisioning a church whose inclusiveness was supported by linguistic clarity rather than willful obscurity. They agreed with Locke that the way to salvation lay through strict readings of Jesus's words in the scriptures and those words alone. They also agreed that a wonderful freedom was to be found in the apparent rigidity of this requirement. Accepting it meant that "faith and conscience, having no dependence upon man's laws, are not to be compelled by man's authority."[34] In such a world, everything Jesus did not explicitly prescribe, which included virtually everything in the Thirty-Nine Articles, would be legitimately open to an enormous range of Christian opinion. The result would be a broad church but with a difference. Whereas the latitudinarians' loose readings expanded the boundaries imposed by the Thirty-Nine Articles until the church canopy they supported covered a huge area, a reasoned church of close readings took the form of an umbrella, which balanced an equally large sheltering canopy on a tightly defined scriptural handle.

Blackburne and Lindsey rejoiced together in this vision of a church of reason that was broad enough to include all Christians. But even as they were coalescing around their shared vision of a truly inclusive church in which Jesus's message was wholly unsullied by human intervention, very human aspects of the actual church to which they had each pledged their lives intervened. Lindsey's patrons in the Huntingdon family had long been irritated that their young favorite owed his position in Kirkby Whiske to the Somersets. So in 1755, only two years after he accepted the living in Yorkshire, the Huntingdons claimed their rightful "honour of providing for Mr. Lindsey" by transferring him to the living of Piddletown (now Puddletown) in Dorset.[35] The Huntingdons' insistence on moving Lindsey to the south of England raised a serious issue. Just as Lindsey was agreeing with Blackburne that it was wrong to subscribe, accepting his new position required that he do so. This created a major crisis. Lindsey's new bishop, John Conybeare, agreed with Blackburne that subscription required a strict reading of the Articles. The difference was that whereas Blackburne had concluded that this meant it was simply wrong to subscribe, Conybeare found subscription essential. Conybeare's God was essentially mysterious. He saw wholehearted subscription to the Thirty-Nine Articles as the qualifying mark of someone who would willingly give up his reason and his will to the church's authority.

Lindsey was miserable as he traversed the hundreds of miles between his reasoning friend and his new Tory bishop. The move was another step up the ladder of church preferment he had always aspired to climb, but his discussions with Blackburne had created an enormous problem. Their conversations had opened to him a wonderfully empowering view of himself as a reasoning being who was loved by a reasonable God. But inextricably bound to his conviction that true Christianity was to be found in Jesus's words was the corollary that a true Christian would not subscribe to a humanly constructed creed. A panicked Lindsey could see no way out but to duck. As he knelt to subscribe before his new bishop, he took the latitudinarian approach by silently adding "of my own record, so far only *as they are agreeable to the Word of God.*"[36] Neither Conybeare nor the reason Lindsey embraced recognized this kind of prevarication to be a truly honorable option, however, and the new vicar of Piddletown knew it.

Lindsey's distress at what he had done is immediately evident from the letters—scores of them—that flowed in his direction from Richmond. Lindsey's half of this correspondence has not survived, but the near desperation of Blackburne's responses suggests the intensity of the younger man's struggle. In the first of the letters, written on January 2, 1756, Blackburne spent pages trying to calm his conscience-stricken friend, who was in the process of deciding that the only honorable response to his situation was to leave the Anglican Church entirely. Blackburne devoted the first of his pages to answering the direct question of how Lindsey could best explain his decision to leave the church to his new congregation. Blackburne devoted the rest of the letter to persuading the young man not to follow through on that decision. Lindsey's immediate duty, Blackburne explained, was not to leave his congregation but rather to support them in their efforts to live as Christians. "There is your Post," he insisted, and there you should remain "till the guard is relieved."[37] And, he went on, Lindsey must remember that "if our Heart condemn us, God is greater than our Heart" and may know when "we condemn ourselves where we don't deserve it."[38] Whatever might be the opinions of the churchmen around him, Blackburne assured his young friend that God would understand and forgive his prevarications.

Lindsey was not so easily comforted or persuaded, however. Just four days after he had sent his first letter, Blackburne wrote another lamenting, "I see you will certainly go out."[39] This time the Yorkshireman was alarmed because he was beginning to recognize that the viability of his own situation was inextricably tied up with Lindsey's decision. He questioned how he could possibly stay in the church when it was his arguments that were leading Lindsey out. Blackburne had persuaded himself that having subscribed and realized it was a mistake, one could legitimately stay, but he also believed he had to take responsibility for his actions. If Lindsey quit the church as a result of embracing Blackburne's ideas, Blackburne felt he had to do so as well. Through more than a year of letters, the archdeacon scrambled to find an acceptable response.

Blackburne's first idea was the rather extravagant fantasy of emigrating to the colonies. Over the course of several letters, he explained to Lindsey how he had managed to get his wife to agree to give up everything they had built up over the fifteen years of their marriage in exchange for a new life on the other side of the Atlantic. Despite the joys of teaching the gospel free of Anglican restriction, it is rather difficult to read all of Blackburne's plans as more than a way to avoid the enormity of what he was facing. Pulling up stakes in the town that had been his home for his entire life required considerably more than his wife's permission, and three months after first entertaining this solution the Blackburne family was still firmly rooted in Richmond.

Blackburne explained his immobility to Lindsey as a financial issue. Providing for a family was expensive, and even with two church positions the Blackburnes found themselves with "five helpless Babes—and ourselves in a Situation where we can lay up no thing."[40] The archdeacon was churchman enough to know that his financial concerns were worldly and that he should somehow rise above them. Still, he moaned, it was one thing for a single man like Lindsey to contemplate leaving the security of a church position, but he had a family to support. As he thought about leaving the church while his children were happily playing in the next room, Blackburne found the "strings" of his heart to be "nevertheless some way bound about them," in ways he could not just ignore.[41]

The financial troubles Blackburne described to Lindsey in the spring of 1756 were more than a convenient explanation for his rootedness. They were real enough that various of his friends began looking for church promotions that would offer him "a better Subsistence."[42] However, their efforts to be helpful served instead to make the situation worse. In his first letter to Lindsey, he had boldly proclaimed he "would not put down my name any more to the articles for any thing the world has to give me."[43] When, only three months later, he received a tentative offer of an additional position from Lord Rockingham, who had heard of his troubles and was trying to help, he was brought face to face with his vow. "It was no Trial," he boasted. "I had not the least difficulty to say 'some particular Reasons determined me not to accept any more, or any other Preferment.'"[44] These were brave words, but no one could understand why he would flatly turn down a promotion that would solve his obvious financial problems. Lindsey was the only person with whom he had shared his understanding of the demands of reason, and the younger man's response was all the evidence he needed to show their dangerous power. All of a sudden, Blackburne's entire world was threatened. He could remain silent with his friends, but if his bishop were to ask, he would have to come clean. His home, his family, his career—all seemed to be hanging in the balance.

The case of John Jones added to Blackburne's misery. Just seven years earlier, in 1749, Jones had published a pamphlet arguing that the Anglican Church should stop requiring its members to subscribe to the Thirty-Nine Articles. Developments had moved slowly from there, but they were coming to head as Lindsey and Blackburne were discussing their options. Jones "hath received notice to resign after a few months and will be utterly destitute, being determined to subscribe no more,"[45] Blackburne fretted in March, and for the next several months he watched in helpless horror as his gloomy prognostications came true. A year after he had first brought up Jones's case, the poor man's increasing desperation remained as a powerful demonstration of the destitution that awaited anyone who rejected the Thirty-Nine Articles.

As he contemplated the possible consequences of being forced to confess his thinking, Blackburne found himself struggling "at the Borders of despondency," where "all my aspirations to do the will of my Heavenly Father are attended with a Deadness." He clung to Lindsey's support as he struggled to contain the "impure spirit lurking in the unswept mansions" of his mind.[46] His despair only increased when his oldest son Frank fell ill. Fearing his own prayers would not be answered, he wrote to beg Lindsey for his, and three days later, after his son recovered, he set out to be "sufficiently and properly thankfull for the multitude of [God's] mercies."[47] Slowly, he regained his equilibrium enough to feel the love of the God who refrained from condemning those who condemned themselves.[48] But the whole experience showed him that there were limits to how far he could accede to the demands of reason.

Lindsey stood faithfully by through all of Blackburne's darkest hours, but it was another six months before he too stopped exploring ways to leave the church. "Good Angels guard yr Pillow,"[49] Blackburne soothed, as he argued it was better to work within the church than to suffer without. "How I care and grieve for you I cannot say upon Paper," the archdeacon wrote comfortingly as Lindsey finally began to accept this discouraging reality. "Heaven keep you in Peace, and send you more reasonable creatures to deal with,"[50] Blackburne prayed.

When Lindsey finally stopped struggling against the Anglican conditions of his ministry in Piddletown, he faced the question how best to work within it. Here, Blackburne was ever ready with advice. From the very beginning of their correspondence, the older man had been clear that, whatever the confusions introduced by the linguistic demands of reason, when he subscribed he had given himself "to the Ministry, by a personal Stipulation with God." That he would devote his life to bringing the Christian message to the people of his parish was the condition he "must fulfill with all his heart."[51] A major justification for Blackburne's remaining within the church was his recognition that teaching the gospel was "best fulfilled from the pulpits of the church,"[52] but he also saw the fundamental importance of taking Jesus's message beyond the well-dressed people in his pews.

Blackburne's efforts at outreach were supported both by the universal reach of reason and the example of John Wesley, now known as the founder of the Methodist Church. From Locke's biblical focus came the essential insight that Jesus's message was given to the poor and uneducated. Locke's reasonable Christianity was "a Religion suited to vulgar Capacities" that even "the labouring and illiterate Man may comprehend."[53] Its grounding in reason meant that eighteenth-century Christianity was the same as it was when Jesus first preached to the multitudes, that is, straightforward and accessible to all. John Wesley was a model of what it meant to act on this insight. As Blackburne was fretting, Wesley was responding to his own struggles with the church by taking the message of Christ's love and redemption directly to the people. And while Blackburne and Lindsey were corresponding, Wesley was traveling around England, preaching three or four sermons a day, wherever and whenever people gathered. Blackburne was both fascinated and inspired by Wesley. "As long as there is a barn or a common, and a people who will come to hear,"[54] he vowed, he too would spread the gospel message to all he could find.

The essentially human nature of reason inspired Blackburne's teaching. In catechism classes he treasured the "awkward Simplicity in children's Answers" and encouraged them to employ "their little thoughts" in order to catch and reinforce any "opening in their own understanding of their own making."[55] He tried to do the same for adults in a series of Bible studies, which, he bragged to Lindsey, drew an audience as large as Wesley's.[56] When two months later the group had grown in size and diversity, Blackburne rejoiced in it as a sign of divine support "when I had given over all hopes of any such thing."[57] Bringing Jesus's message to as many people as possible was the best antidote for Blackburne's black moods. Teaching the people in his parish to reason their ways to God gave him great joy.

All of Blackburne's talk of "barn" and "common" was at first somewhat foreign to Lindsey's life in Piddletown. His church was the lovely fourteenth-century St Mary's, but the young man was spending most of his time in what he called "the Colony"[58] of high society that gathered in the elegant drawing rooms of his patron's Islington House. From the beginning of their correspondence, Blackburne's understanding of himself as a parish priest pulled against the "chattering"[59] of Lindsey's aristocratic connections. The Yorkshireman was careful to respect the ties of friendship that bound Lindsey to his patrons, but he could also be quite clear: "I once more do not like yr Patronesses of the Colony."[60] In letter after letter, by word and example, he suggested the kinds of things that might be "part of God's purpose" in sending Lindsey to Piddletown. Over time Blackburne's urgings began to have an effect, and by his third year in Dorset, Lindsey too began searching for ways to move his religion out of the drawing room and into his parish. Blackburne was thrilled. "I hope, wish and pray you will now find satisfaction and success," he wrote encouragingly in 1759.[61]

For Lindsey, taking Jesus's message into the parish was one step toward adopting Blackburne's way of life; getting married would be another. In July 1759, Blackburne raised the issue. "But seriously do think of it," he urged. "Get a wife who fears God—You would not have so much leisure to macerate on things, which your solicitude will not mend."[62] Here, as in so many other areas, Blackburne had a palpable effect on his young friend, and within six months he was rejoicing at having persuaded Lindsey of "the necessity of a help-meet" for his life and work.[63] Nine months after that, on September 29, 1760, Blackburne presided over the ceremony in which Theophilus Lindsey married Hannah Elsworth, the stepdaughter who had come into Blackburne's family at the age of four. It was a blessed and happy day because it guaranteed that the reason Blackburne had taught his stepdaughter and shared with Lindsey would flourish into the next generation.

The Lindseys may be seen as the first couple of the family of reason, but their wedding day was not that family's founding moment. The ceremony in St Mary's Church in Richmond marked the Lindsey's entrance into Blackburne's family, and the older man did all that he could to bind the couple close. But over the course of the next fifteen years the Lindseys began to feel constrained, until they finally broke away from Blackburne's increasingly confining arms to form a family of their own. Reason was a constant goad and companion in all of their journeys to this momentous, and in many ways tragic decision, but new friends and new circumstances also played critical roles. The road to the break between the Lindseys and Blackburne deserves a chapter of its own.

# CHAPTER 2

# Breaking Away

Lindsey did not keep any of the letters that accompanied his courtship. Before it began, Hannah appeared in Blackburne's letters only as one of his children; when it ended, she was "your wife," whose mother was complaining because she wasn't writing home.[1] Lindsey's decision to destroy the three years of correspondence in between may be read as an attempt to erase the undoubtedly complicated negotiations that surrounded his decision to marry Blackburne's stepdaughter. The gap may thus stand as a testament to the strength of the personal bond he and Hannah forged in their marriage.

Whatever preceded it, the Lindsey marriage was one of two complementary personalities. Hannah Elsworth was a clear-sighted woman who brought to their union all of the cragged independence of her Yorkshire roots. She was twenty-one years old when she married Lindsey, and for seventeen of those years her stepfather had been educating her, both at home and in catechism classes. One of the results can be seen in her firm determination to minister to the whole range of people; in Piddletown she was long remembered for her schemes "for the temporal and spiritual benefit of parishioners, and especially of the poor and ignorant."[2] Another was a restless intellect, powerful enough that at the time of her marriage her stepfather valued her thinking as much as he did Lindsey's. As he said when sending some of his writings to Lindsey for comments, "I expect yr criticism (hers especially)" in the next post.[3] It is not clear how the Lindseys responded to this rather peremptory request. Blackburne's success in raising Hannah to think for herself meant that she was not a person to be bossed around. Many of Lindsey's friends found her abrasive and opinionated, but her crisp decisiveness made her truly a "help-meet" to her husband. Over the course of almost forty years of marriage, the combination of Hannah's strong-mindedness and Lindsey's charismatic warmth spawned a movement for reasoned religion whose effects are still felt today.

The Lindseys' recognition of the mission that reason required of them built up slowly over the first decade of their lives together. From virtually the moment they met as adults, Hannah was made aware of the role that her husband's understanding of reason was playing in his life. Her stepfather may have succeeded in persuading Lindsey to accept the latitudinarians' pragmatic peace in the case of the Thirty-Nine Articles, but that peace did not extend to the more fundamental project of finding his true religion. In private, he continued to explore the implications of reason, and by the time Hannah moved to Piddletown as his wife, his tight focus on Jesus's words was leading him to a deeply heretical position.

At issue was the nature of the Christian God, who since the fourth century had been recognized to be a Trinity, in which God the Father, Jesus the Son, and the Holy Spirit were somehow mysteriously combined into one. However, as Lindsey followed his reason into the Bible, he found nothing Trinitarian in the God that Jesus worshiped. Nor did he find any Trinitarian claims among Jesus's immediate followers. Instead, the Christian embrace of a tripartite God was the outcome of an enormous conflict between Trinitarian Athanasians and anti-Trinitarian Arians that first burst into the public eye with the Council of Nicea in 325 CE.

The battle over whether the focus of the Christian religion should be the mystery of a God who was at the same time singular and composed of three coequal beings was bitter and prolonged. Central among the many issues at stake were questions about whether Jesus was essentially human or divine and, relatedly, whether a priesthood was necessary to interpret his central message to the laity. Slowly over the course of several decades the Trinitarians consolidated their power and pushed their opponents to the peripheries. By the seventh century, the anti-Trinitarian Arians were effectively suppressed throughout the Roman Empire.

The Nicean Creed, the Apostles' Creed, and the Athanasian Creed—all of which are included in the Thirty-Nine Articles of the Anglican Church—originated as increasingly belligerent statements of allegiance to Trinitarian Christianity. That the Trinity they extolled was not an essential part of Christian doctrine until two or three hundred years after the death of Jesus was well known in Lindsey's world. His Anglican colleagues interpreted this time lapse as necessary in order for the raw religion Jesus preached to develop to maturity. Lindsey was not impressed by this argument. From his Lockean perspective, all people are essentially alike in their capacity to reason, which meant that there was no place for history to affect religious insight. When God sent Jesus to the earth, the goal was to deliver an immutable message that could be understood by all people. This meant that the words Jesus spoke were intended as much for eighteenth-century Englishmen as they were for first-century Galileans. Centuries of history may have separated the two populations, but they were all human beings. What could be truly understood by one, could be equally understood by all.

Subsequent religious history further supported Lindsey's conviction that the Trinity was a travesty forcefully imposed on Jesus's original religion. When the straightforward, monotheistic position reemerged during the Reformation it was violently repressed: Michael Servetus was burned at the stake in Geneva in 1553, Francis David died in a Transylvanian prison in 1579, Faustos Socinus was stoned out of Krakow in 1598.[4] The anti-Trinitarian positions these men defended were nonetheless powerful enough to live on among small groups protected by the isolation of eastern Europe, but if their ideas ever flared into more public view, they were again forcefully attacked. For Lindsey, one of the evidences that the Trinitarian God who thus prevailed could not be Jesus's God was that the Trinitarian God depended on violence. In England, the question of whether God was one or three had come to a head in the so-called Trinitarian controversy of 1687. The date marks the publication of an anonymous pamphlet, a *Brief History of the Unitarians or Socinians,* which led to a fierce ten years of pamphlet wars within the Anglican Church. The upshot was twofold. On the one hand, the Blasphemy Act of 1697 made it an offense for anyone educated in the Christian religion to deny the Trinity. On the other hand, a latitudinarian truce within the church allowed enough freedom in reading the creeds, which were part of the Thirty-Nine Articles, that anti-Trinitarians could be absorbed into the body of the church.

Fifty years later the truce still held, and in Piddletown, Lindsey tried to take the advice of those who recommended accepting it: "If I could in *any way of interpretation* reconcile the prescribed forms with the scripture in my own mind, and make myself easy, I was not only to be justified, but to be commended."[5] But he nonetheless, found it very difficult to do that. The more he searched his Bible, the more convinced he became that the Christian concept

of the Trinity was a fundamental perversion of Jesus's core message. As he
put it, "In the end, I became fully persuaded, to use St. Paul's express words,
1 Cor. 8:6. 'That there is but one God, the Father, and he alone to be worshiped.'"[6] Jesus remained the son of God, whom God had sent to save suffering humanity, but being the son of God was not the same as being God.
For Lindsey, dissolving the mystery that lay at the center of traditional Christianity returned Jesus to his full humanity and reaffirmed the power of the
reason that was essential to the human beings God had created and loved.

Blackburne was interested in Lindsey's arguments, but the latitudinarian
truce that had saved him from his demons was far too important to allow the
discussion to be more than intellectual. Lindsey had also allowed himself to be
pulled back from the brink, but he was considerably younger than Blackburne
and his views were less settled. By the time he married Hannah, Lindsey was
well on his way to becoming an anti-Trinitarian, and she quickly recognized
the power of his position. From the earliest years of their marriage, the Lindseys were joined in their embrace of a heresy so radical that it placed them
totally beyond the pale of Christianity. The Lindseys at first maintained a public silence about their views. Theophilus supported Hannah's commitment to
work for all of the people in their parish, while at the same time maintaining
good relations with his Huntingdon patrons.

In 1763 the Duke of Northumberland asked the couple to come with him
to Ireland, where he had just been named lord lieutenant. It was not yet clear
what church position the duke could provide—for the moment the Lindseys
would just live in the palace with him and his wife—but it was no leap to expect that a bishopric was in the offing.[7] Blackburne was appalled. For years
he had been planning to bring his brilliant stepdaughter and beloved friend
closer, but now he faced the possibility of losing them both forever. In more
than twenty letters he scrambled to find a way to bring the Lindseys back to
Yorkshire. His focus was on Catterick, a town of about five hundred people
just five miles to the southwest of Richmond. Catterick revealed traces of its
ancient history as the Roman station of Cataractenum in the fading foundations of an ancient castle, the crumbled ruins of a Roman amphitheater, and
its position as a major thoroughfare with a bridge over the River Swale. There
was enough traffic to support a large inn, which housed not only a stream of
travelers but also vacationing noblemen who "with their grooms and footmen
used sometimes to spend whole weeks, for the purpose of fox hunting."[8] It
was an appealing place for a young couple to live.

In addition, Catterick was the center of a sprawling parish that included three chapels as well as St Anne's, a solid fifteenth-century edifice flanked by the vicarage that stood on a hill in the middle of the town. A seventeenth-century vicar had endowed a grammar school for the poor children in the town and a hospital for the care of six poor widows. The position was not a bishopric; it could not match the luxurious possibilities that the duke was offering. But the new vicar of Catterick would be wonderfully positioned to follow Blackburne in his determination to bring the Christian message to people of every walk of life. At least equally to the point, Jeremiah Harrison, the vicar of Catterick, was seriously ill in the early months of 1763, and Blackburne's letters took on a somewhat ghoulish quality as he waited for the man's demise.[9] When Harrison died in midsummer, everything was finally in place for Lindsey to exchange the vicarage of Piddletown for the vicarage of Catterick. Blackburne thrilled at his success in bringing the Lindseys home to his Yorkshire fold.

The couple's move to Catterick effectively ended the correspondence between Blackburne and Lindsey—there was no need for letters between people who lived only five miles apart—while at the same time generating a different source for information about the lives of the Lindseys. The daughter of the deceased Jeremiah Harrison, Catharine Harrison (later Mrs. Catharine Cappe), was nineteen years old when she found she could no longer live in the vicarage where she had grown up. The death of her father, which Blackburne had awaited with such anticipation, reduced her to sharing a rented house with her mother that was so tiny the two women could barely fit. Forced, therefore, to visit various friends, the young woman quickly learned how hard it could be to depend on the kindness of others.

Catharine had just escaped from a particularly unpleasant visit when Hannah, who had known her from childhood, invited her to Catterick. It was as if she had been "suddenly transported into a new world."[10] Lindsey's sympathy to her plight left her hardly able to "persuade myself, he was a being of the same order with those, in whose family I had lately been an inmate."[11] She attributed the power that elevated Lindsey to "a distinct class in the moral and intellectual scale of human being"[12] to his conviction that God truly cared for his reasoning people. That Jesus was a person, who was like all people in being a child of God, made him a model to which Christians were bound to aspire; it turned his life into an invitation to all people to "look up to God as their Father and Friend."[13] Encountering the Lindseys transformed the young woman into an absolutely committed anti-Trinitarian. Her response to the couple who had forced her out of her childhood home is both our only source for the nature of their lives in Catterick and a vivid testament to the magnetic quality of their religious vision.

The God Catharine found in Lindsey's home was loving, like Theophilus, but also clear and firm in expectation, like Hannah. The Lindseys' life in the Catterick vicarage was a minute-by-minute fulfillment of their obligations to their God. They eschewed "every species of unnecessary indulgence, whether of food, of outward appearance, or of mere amusement,"[14] to focus on the people of their parish. Theophilus spent mornings in his study, "except when the duty of visiting the sick, comforting the afflicted, or exhorting the impenitent claimed a portion of his time."[15] Hannah devoted hers to the flowers and herbs she had planted along the roads, around the vicarage, and in the churchyard. She also maintained a considerable library which supported her work as a medical practitioner who mixed and dispensed "the best drugs from Apothecary's hall" with notable success. Under her care, the plain little vicarage of Catterick came to wear "an appearance of great comfort"[16] that reflected the spirit of the couple within.

Sundays were the Lindseys' busy days. In addition to the morning and afternoon services in St Mary's, Lindsey held catechism classes for his parishioners and Bible classes for the more than one hundred boys of the grammar school, and for young men and women on alternate Sundays.[17] He lived and breathed his work, and as he spoke to his parishioners about the joys of following Jesus's path, "his eyes would sparkle with delight."[18] Hannah was equally present, participating in the services and holding her own Sunday school classes—for girls one week, for boys the next—in which she taught young children how to read. The work could be exhausting, but "to the Christian, who knows that in the eye of God all his rational offspring are equal,"[19] opening people to the power of reason was manifestly worth the effort.

It required considerable resolve for the Lindseys to live the way they did, separate from the foxhunters and country gentlemen who surrounded them. Continually saying no to invitations was a relatively straightforward process for Hannah, who was remarkably self-directed. It was much harder for her husband, who was always loath "to give cause of offence or pain to any human being."[20] Hannah's was the more powerful voice in this instance, so instead of socializing after their morning work, the Lindseys took long walks together through the fields or along the banks of the Swale. The only thing possibly missing in the harmony of their life together in Catterick was that year after year the couple remained childless. There was, however, plenty of family nearby, and the Lindseys spent one or two afternoons a week with the Blackburnes in Richmond.[21]

The understanding of reason that supported all aspects of the Lindseys' life in Catterick also acted as a powerful snake in their rural Eden, however. Theirs was an Anglican vicarage, and Lindsey had had to subscribe yet again when he accepted his new position. Fraught though it may have been, that was just a onetime event, but as the vicar of Catterick he was being forced to prevaricate constantly. Hannah could, and most probably did, say precisely what she thought, but her husband was not so free. Being an Anglican minister entailed presiding over services that were permeated with Trinitarian messages. Every time the prescribed order of service entailed speaking of the Trinity, leading the congregation through one of the creeds or praying "through Jesus Christ our Lord," Lindsey squirmed, and no amount of latitudinarian justification could salve his conscience. He had committed himself fully to reason's linguistic approach to the scriptures, and now every instance of loose reading felt like a sin. Before he could honestly live within the church, that church would have to change.

Lindsey's first efforts to accomplish change in the Anglican Church were indirect. Blackburne had long argued that the best way forward was from within. For all of the time he and Lindsey had been struggling with the problem of Lindsey's subscription, Blackburne had been formulating his arguments against the practice. By the time the Lindseys arrived in Catterick, Blackburne's *Confessional, or A Full and Free Inquiry into the Right, Utility, Edification, and Success, of Establishing Confessions of Faith and Doctrine in Protestant Churches* was wholly written, but lying silent among his papers. The example of the unfortunate John Jones was powerful enough that Blackburne had no plans to take his ideas any further.

Hannah, however, was always ready to take a stand, and her husband was uncomfortable enough that he too thought it was time for the ever-wary archdeacon to make his position known. So Lindsey leaked word of the manuscript's existence to Blackburne's friend Thomas Hollis. The latter responded as Lindsey had hoped, by discussing it with a publisher, who in turn traveled from London to Yorkshire to ask Blackburne about it. In response to a clear call for publication from all of these people, Blackburne finally agreed to allow the *Confessional* to appear—anonymously—in London in the spring of 1766. After more than a decade of struggle, the archdeacon of Cleveland finally broke cover.

Blackburne's arguments against the legitimacy of creedal statements in Protestant churches were powerful, and his *Confessional* generated a blizzard of pamphlet response.[22] The range of opinions that swirled around it was certainly broad enough to include Lindsey within the church, and for a moment he found comfort in knowing that he was not alone. But being in good company was not enough to make his latitudinarian obfuscation feel right. In 1768 or '69, a serious illness drove the point home. As he tossed in his bed of pain, he saw ever more clearly that his latitudinarian position entailed "a blameable duplicity, that whilst I was praying to the one God the Father, the people that heard me were led, by the language I used, to address themselves to two other persons of distinct intelligible agents."[23] For Lindsey, this duplicity was of central importance: "As the one great design of our Saviour's mission was to promote the knowledge and worship of the Father, the *only true God*,"[24] it was surely wrong for him to lead services that he believed were in direct contradiction to Christ's teachings. Lindsey arose from his sickbed deeply chastened and once more determined to find some way to change his church.

At this point, an unexpected development began to show Lindsey a new way forward. In June 1769, he met Joseph Priestley.[25] Priestley later described the friendship that ensued as "the source of more real satisfaction to me than any other circumstance in my whole life."[26] For Lindsey it was, if possible, even more significant. Priestley was a dissenting minister in Leeds when he and Lindsey first met. That he was a dissenter meant he was one of the rather small minority of England's Christian population who would not subscribe to the Thirty-Nine Articles. Dissenters were legally tolerated in eighteenth-century England, but they were not embraced; they could not hold public office, for example, nor could they take degrees from either of England's two universities. Otherwise, however, they were free to pursue their lives in whatever ways they saw fit. In the depths of Lindsey's despair, Priestley modeled a possible life outside of the Anglican Church in which he had been chafing for more than a decade.

Priestley's life reflected the positive side of the dissenting experience. Born in 1733 in a small town outside of Leeds, he was a brilliant child whose family supported him as he rose through a variety of dissenting schools, tutors, and academies. He emerged from this education as well-versed in Greek, Latin, and mathematics as any Anglican. In theology his education was arguably better; because the dissenting community was not bound by a single orthodoxy, he learned a wide variety of perspectives. This range of choice supported the young man as he tried to find his own way to God.

Priestley began his journey as a member of a Calvinist family for whom the essential mark of a true Christian was a conversion experience. As a young man, however, a serious illness forced him to confront the reality that he had not had such a transforming experience. In this defining moment, something rose up to defend him against the unfairness of having his essential worth judged on something so completely beyond his control. He rejected his parents' Calvinism in favor of Locke's position that becoming a Christian required assent to the "plain[,] intelligible Proposition" that Jesus was "the Saviour promised."[27] From this beginning, reason took Priestley down the same path through the Bible that had led Lindsey to his anti-Trinitarian position.

Following the dictates of reason transformed Priestley as much as it did Lindsey, but it had very different institutional implications. In the dissenting community, there was no equivalent to the Thirty-Nine Articles, nor to the fixed orders of service of the Anglican communion. Priestley was free to speak the truth as he saw it as long as a congregation was willing to hear him. His earliest efforts to preach his religion were hampered by a speech defect, but here again the freedom of dissent saved him. After six years stuttering from a pulpit in Needham, he found himself a more comfortable position as the master of a dissenting academy in Warrington. From 1761 to 1767, the institution Priestley directed thrived so dramatically that many Anglicans chose to send their children there. Yet despite his resounding success as a teacher, the ministry was always Priestley's calling, and in 1767 he was happy again to mount the pulpit as the minister of a dissenting chapel in Leeds.

Priestley's and Blackburne's paths first crossed when Blackburne sent his oldest son, the Francis Blackburne who would later become Sarah Frend's father, to Priestley's Warrington Academy. Soon afterward, when the new minister arrived in Leeds, he wrote to renew their acquaintance. The impulse came from the *Confessional,* which Priestley had read with passionate interest. From his dissenting point of view, the archdeacon's "full and free inquiry into the right, utility, and success of establishing confessions of faith and doctrine in Protestant churches" promised liberation from all of the slights, large and small, that England's dissenters faced every day. Vanishingly little divided Priestley's religious views from those of the many Anglicans who were hiding their objections behind latitudinarian readings. The only real difference was that the latter had been raised within—and were often supported by—the established church, whereas he had not and was not. Priestley had no desire to enter a church that required him to subscribe to things he did not believe in.

But if Blackburne succeeded in lifting the subscription requirement, Priestley and his fellow dissenters could practice their religion without discrimination. It would be a significant liberation.

When he realized that Blackburne had written the pamphlet, Priestley and his friend William Turner, a dissenting minister from Wakefield, traveled to Richmond to talk over their ideas with Blackburne. Lindsey and Hannah joined them and for an extraordinary few days in the early summer of 1769, the three Anglicans and two dissenters followed reason's lead up and through the heights of theological discussion. None of them ever forgot the experience. Blackburne regretted that he couldn't "talk and even dispute on horseback"[28] the way the young people did, but nonetheless emerged bubbling with new ideas and energy. Hannah agreed with her stepfather about Priestley's and Turner's "various and uncommon mental endowments," but she was even more struck by "the innocent, and even playful cheerfulness of their conversation." When she offered this observation to Lindsey, he just looked at her sadly and replied, "Your observation is just [true] but they are at ease!"[29] Hannah's theologically tortured husband was most certainly not at ease, but his long weekend with Priestley and Turner gave him a glimpse of the freedom to be found outside of the established church.

In response, Lindsey began prodding his father-in-law to use his considerable institutional power to effect real change within the Anglican Church. Once again movement was slow, but by July 1771 he had persuaded Blackburne to travel to London in order to meet with a group of supporters. Blackburne served as the senior leader of the group that met in the Feather's Tavern. Notable among the younger men were Lindsey; John Disney, the vicar of Swinderby in Lincolnshire; and John Jebb, a fellow at Peterhouse in Cambridge. By the end of the day, the group had drafted a petition—always thereafter known as the Feather's Tavern Petition—asking that Parliament lift the subscription requirement for membership in the Anglican Church.

Getting Parliament to pass the Feather's Tavern Petition was Lindsey's last best hope to find ease within his church. In the months after it was drafted, the petitioners set to work gathering signatures. Jebb canvassed Cambridge, while Lindsey covered two thousand miles on horseback to get the support of his far-flung network. On February 6, 1772, they presented more than two hundred signatures in support of the Feather's Tavern Petition to Parliament; that afternoon the House of Commons voted it down by a vote of 217 to 71.[30] Blackburne was not defeated. He had already in the *Confessional* recognized "how well disciplined the forces that are brought into the field against *Re-*

*formers;* how able the generals that head them, and how determined the whole body not to yield an inch, even to the united powers of piety, truth, and common sense."[31] The battle against such foes could only be won by "persisters" who are "like *Abraham*" in their ability *"against hope, to believe in hope."*[32] So Blackburne settled back down in Richmond to wait for the next best opportunity to try again.

Lindsey turned his tired horse toward Catterick. That Hannah had been ever "willing to run all hazards, *me currentem incitans* [keeps me going],"[33] her proudly loving husband had declared as he rode around England in pursuit of signatures. Now he returned home with his pockets full of pennies to reward the little parishioners who had been brave enough to submit to the smallpox inoculations she had been pushing on her flock. He tried to follow Blackburne's example, proclaiming, "Though defeated we sing a victory; as truth and reason were all for us, and over-powered [merely] by power; and we are not disheartened, but in high spirits."[34] As weeks turned into months, however, the reality of failure began to set in. A month after Parliament rejected the petition, Priestley asked Lindsey why he would let anyone have that kind of power over him. "As the disciple of a Master whose kingdom is not of this world, I should be ashamed to ask anything of temporal powers, except mere peace and quietness,"[35] he declared. When a stung Lindsey lashed out at his dissenting impertinence, Priestley instantly backed off: "I am truly sorry that I made the observation in my last on your application to Parliament,"[36] he wrote a week later. Priestley's apology was undoubtedly sincere, but its necessity demonstrated that his words had found their mark. Lindsey struggled on in England's established church for another year, but by the spring of 1773 he had finally and irrevocably decided to leave.

The complications that attended Lindsey's decision went far beyond leaving a job. To be an Anglican minister in England was a sacred trust, and willingly to reject such a position was a blamable violation of that trust. Jones's desperate plight was a stark example of the very real consequences that attended ministers leaving the Anglican Church. Lindsey was nonetheless convinced that it was his only possible option. He had made a good faith effort to change the Anglican Church from within; now he had no choice but to try to change it from without. Priestley convinced him that people would gather if he formed an independent chapel in London, and Hannah agreed. So the Lindseys determined to make the effort.

The first step was for Lindsey to resign his position in Catterick. This process began at home. Hannah stood firm against the tearful supplications

of her mother, while Lindsey faced her father. Blackburne was clear that his
subscriptions had been in error but equally clear that they were in a past that
need not be repeated in order to serve his parish, and he saw no reason for
Lindsey to take a different position. His-son-in-law's decision at first irritated
him. "I have yours of the 7th," he wrote in May 1773, "which brings me nothing but the mortifying conviction that you have taken a resolution which will
be proof against anything I might further have to offer by way of controvert-
ing the principles upon which it is formed."[37] Week after week thereafter, as
it became clear he was not going to change Lindsey's mind, Blackburne's irri-
tation ripened into a fury that covered a hurt that made him feel "he had lost
his right arm."[38] He judged Lindsey's decision to leave the church as "the utter
ruin"[39] of the Feather's Tavern petitioners' plan to effect gradual reform within
the church. He never forgave Priestley for taking the Lindseys away and final-
ized the break by destroying all of the letters he had received from Lindsey.

With Hannah and Priestley at his back, Lindsey had the support he needed
to stand his ground. "Such things are to be expected," he sighed. Facing Blackburne's rage "may be of service to prepare for coldness, neglect, misrepresen-
tation, and unkindness from the world, and to lead us to depend only on him
who never faileth those who in well doing put their trust in him."[40] Hannah
was even less moved. She was completely convinced of Lindsey's position, and
for her, "the determination in itself was right, and she regarded very little of
what others might feel or think."[41] More than ten years after their marriage
had drawn them together into Blackburne's embrace, it was time for the Lindseys to strike out on their own and establish a community in which they could
follow a religious path of reason without apology or dissimulation.

Breaking away from the family was not the Lindseys' only wrench; leaving
the people of their parish was almost as difficult. In the very first of his Piddletown epistles, Blackburne had laid out for Lindsey what would be necessary
if he left a congregation: "If I should go out of the church," Blackburne had
written almost twenty years earlier, "I should think myself indispensably
bound both in justice to myself and for the Satisfaction of all Christian people
to declare my reasons in the most publick manner, and to give an honest and
ingenuous account of my whole conduct."[42] Lindsey might have been leaving
Blackburne's family, but as he extricated himself from his ministerial position,
he tried to fulfill his archdeacon's charge by publishing *The Apology of Theophilus
Lindsey, M.A. on Resigning the Vicarage of Catterick, Yorkshire.* And, on November 28, 1773, he honored his more personal commitment to the people of
his parish in Catterick by explaining himself in a sermon.

The individualist component of reason stood at the forefront of the explanation Lindsey offered for his determination. He emphasized that his decision was his alone and that each of them was free to come to God in his/her own way. That they might not agree with his position did not change his situation, however. If a person thinks something is "contrary to God's word, and sinful," he asserted, it is dishonest for that person to comply with it. In his case, he went on to explain all prayers "addressed to the Trinity, to Christ, to the Holy Ghost, or to any other person, but God himself" appeared sinful to him.[43] The members of his congregation were completely within their rights to come to different conclusions, but when he had allowed reason to lead him to God, he had arrived at his own. However difficult it might be to leave, Lindsey simply could not stay in the Anglican Church.

In Catterick the Lindseys tried to comfort parishioners who were "stupefied with grief."[44] Hannah found her husband's presence made her "as happy as I can be among these sorrowing people,"[45] while Theophilus was deeply grateful for the support and comfort of a wife who was "worthy of a better fate in worldly things than we have a prospect of."[46] Still in the throes of these emotions, on December 6, 1773, the resigning vicar and his wife left their Yorkshire family, house, and garden behind and headed for London.

The city the Lindseys entered could not have been more different from their Catterick parish. On the order of 10 percent of the population of England—about 800,000 people—resided in or around the capital city. It was very difficult for Hannah, who had lived her whole life in Yorkshire with a brief stopover in Piddletown, to find herself in the center of its filthy crowded streets. When the ever-faithful Catharine visited them there a few months later, Hannah's temper showed the strain of trading her mother, her gardens, and her beloved parish children for a tiny unfurnished apartment in the middle of Europe's largest city. Yorkshire-raised Catharine also saw "nothing enviable" in their situation. Theophilus, on the other hand, was "cheerful, easy and contented." The couple had sold most of their possessions to finance the trip, but he had been able to bring his library with him and was quite willing to sit on one pile of books and write on another. The trip took Hannah away from home, but for Lindsey it was a move from the periphery to the center. At last secure in "the testimony of a good conscience," he was eager to establish the truth of reasoned religion in England's capital city.[47]

Lindsey's major supporter was Priestley, who was also newly arrived in London. While Lindsey was involved with the Feather's Tavern Petition, Priestley had been pursuing the study of natural philosophy, which was one of the features of his dissenting education that was all but completely lacking in Lindsey's Anglican one. Priestley's investigations of electricity and magnetism culminated in 1772 in a paper in the *Transactions of the Royal Society* that was so rich in new ideas and insights that the Society awarded him its highest honor, the Copley Medal, in 1773.[48] Many who read the paper were convinced that Priestley's work was far too important to be pursued only as a sidelight to a ministry. The naturalist Joseph Banks tried but failed to have him included as an astronomer on Captain Cook's second voyage to Tahiti, and Benjamin Franklin tried in vain to find a position for him in the colonies. The problem was solved when Lord Shelburne offered him a purposely open-ended position as companion, which provided Priestley both time and equipment to freely pursue his investigations. While working under Lord Shelburne's wing, Priestley succeeded in isolating oxygen and identifying the properties of what he always called "uncommonly good air," but from his point of view, the Lindseys' project was considerably more important. As soon as they arrived in London, Priestley threw himself into the effort to raise money to rent space for a chapel, while Lindsey revised the Anglican prayer book for services,[49] and both worked to bring as many people as possible to their cause of reasoned religion.

It was a difficult challenge. Lindsey's fondest hope was to attract dissatisfied Anglicans, but although many of his acquaintances had listened to and sympathized with him for decades, few were willing to take such a drastic step. Some never spoke to him again, and others went to great lengths to change his mind. When one of these asked "if he had the right to subject Mrs. L. to so many inconveniences and hardships,"[50] Lindsey was shattered but Hannah was not. Displaced though she was, she remained as steadfastly committed to reason's program as he was. Night after night she muted her impatience with the social niceties of dinner parties so that her husband could deploy his conviction, his determination, and his charm to generate support. He shone in those drawing rooms that would open their doors to him, and slowly a new group began to coalesce around him.

Lindsey performed his first non-Anglican service in a second-floor room on Essex Street off the Strand on April 17, 1774, less than five months after his departure from Catterick. Having failed to persuade many of his Anglican contacts to attend, he was rather apprehensive about the kinds of people who would actually show up and was relieved to preach to "a larger and much more respectable audience than I could have expected, who behaved with great decency."[51] He was a powerful speaker, and within months of his first service, the group of committed and curious who came to hear him preach was spilling into the hall outside of the small rented room. In 1775, John Jebb left his fellowship at Cambridge to join Lindsey in London; in 1782, John Disney left his positions as the rector of Panton and the vicar of Swinderby to join him as well. A number of dissenting ministers, including Priestley, Turner, and Richard Price of Newington Green coalesced around these fallen Anglicans to form a loose network of like-minded reasoned dissenters. People labeled the group Unitarian because they denied the Trinity, but those who gathered around the Essex Street Chapel saw themselves as pure Christians who were using the reason God had granted them to come to a true understanding of the religion Jesus preached.

By the time the Disneys arrived in London in 1783, ministering to the Essex Street congregation was a big enough job that Lindsey offered his friend a position as assistant minister. Disney had married Blackburne's daughter Jane in the same year that the Lindseys had left the church, so when he accepted Lindsey's offer, the Essex Street Chapel truly became a family affair. Blackburne saw the defections of the two men as creating a fundamental rift that effectively destroyed his family, but his children did not. When Blackburne's daughter Sarah's Anglican husband was looking for a position, Lindsey simply gave him the Anglican living of Chew Magna that his admiring patron had bequeathed to him more than twenty years earlier. Hannah's deep commitment to Unitarianism certainly did not stand in the way of her relationships to the Anglicans in her Yorkshire family. Her Blackburne siblings and their offspring were always welcome in her London apartment, and year after year she firmly brought Lindsey back to visit the home of her youth. Though in public Blackburne's family was divided between Anglican and Unitarian, in private the division seemed not to matter.

Bridging differences between denominations was relatively easy among women. That none of them could hold church positions meant that none of them were forced to proclaim or defend their religious positions publicly. As a result, their religious positions really were personal issues which could be raised or ignored at will. That women had these chameleon possibilities did not mean that those clustered around the Essex Street Chapel were any less committed to reasoned religion. Long before her husband left his Anglican positions, Jane Disney had "expressed upon all occasions her high approbation of the step which Mr. Lindsey had taken" and "defended the principles and the conduct" of her stepsister and brother-in-law.[52] When Jane arrived in London, she found herself in the midst of a powerful community of women that included Hannah; Jebb's wife, Ann; and also always present in spirit and by letter, Catharine Harrison, who became Catharine Cappe when she married a Unitarian minister in York in 1788.[53] In addition, one of Lindsey's earliest converts, Mrs. Elizabeth Rayner, quickly became a stalwart benefactor and paid his salary. These women left significantly fewer historical traces than did the men they sustained, but their support was essential to the movement's success.

Rayner was only the most visible of the people who provided the Lindseys with financial support. Although several among them were loath to stand with a Unitarian in public, in private they were generous enough that just four years after the Lindseys first arrived in London, they were able to buy the building on Essex Street and convert it into a chapel and a living space. Lindsey's charm and social connections were the source of their funds, but it was Hannah's practicality that kept the Essex Street Chapel organized and functioning. It was she who figured out "how to make the most of the small allotted space" in the Essex Street building, and she who daily supervised the workers who were doing the remodeling.[54] Many of the men who gathered around her warm and thoughtful husband found her forcefulness irritating, and she herself recognized she was "more of a useful than a loveable creature."[55] Lindsey's warmth and conviction drew people together, but it was Hannah's "prudence, activity and energy of mind"[56] that created and sustained the institution that turned those people into a movement.

Hannah saw her mission to be one of protecting her husband from "every secular care" so that he could be "at perfect liberty to devote all the powers of his mind" to "the great object in which his whole soul was engaged," — the creation of a Christian church of reason.[57] In his London years, two different — and sometimes contradictory — strands were entangled in the vision that Lindsey shared with his followers. The first was a universalist strand spun from the conviction that the power of reason constituted the essence of what it was to be human. In his first sermon, he based his call for inclusivity on Ephesians 4:3, in which Paul called upon Christian churches to endeavor "to keep the unity of the Spirit in the bond of peace."[58] Lindsey's vision of a church that was large enough to include all under its canopy rested as much on a Lockean ideal of reason as it did on the Christian ideal of Christ's love. "God never designed that Christians should be all of one sentiment," he explained. There would always be "different sects of Christians and different churches,"[59] but in Lindsey's Christian church of reason, all would be welcomed to stand together on the common ground of divine love and human reason.

Even as Lindsey called for an all-embracing church, the specifics of his interpretation of reason pulled in the opposite direction. His insistence that reason's truth required a literal reading of the Bible had already demonstrated its divisive power when it pulled him out of the Anglican Church. He dismissed the problem in that case as the result of civic meddling; problems caused by the subscription requirement could surely not be attributed to "the mild and gentle doctrine of the gospel."[60] But even as he preached inclusion, Lindsey was committed to following what he saw as the appropriate form of reason for Christians. His desire to include Anglicans meant that he kept the form of his services close to theirs, but there was always a difference. In the close of his first sermon, his anti-Trinitarian message peeked out when he explained that his purpose in "forming a separate congregation distinct from the national church is, that we may be at liberty to worship God *alone,* after the command and example of our Saviour Christ."[61] At first glance, this may seem an anodyne statement, but it was not. Lindsey's use of the phrase "God *alone*" and his nod to the "command and example of our Saviour" were not-so-coded references to his insistence that God was a single, as opposed to triune, being and that Jesus was a man, whom God had sent to save the human race. Lindsey was reaching toward a universal Christian church of reason in which all would be equally welcome, but the gospel he preached from the pulpit of the Essex Street Chapel was always strictly Unitarian.

Not everyone found Unitarian doctrine as liberating as Lindsey did; Blackburne considered the strictly literal form of reason that led Lindsey to his conclusions to be more confining than freeing. In 1773, his daughter Jane's marriage to the Anglican Disney had somewhat mitigated the older man's sorrow at Lindsey's defection, but nine years later, Disney's decision to resign reopened the wound. Blackburne retired to his study to explain himself in an essay entitled "Answer to the Question: Why Are You Not a Socinian?," that is, a Unitarian. There the old man wearily admitted that there were many problems in the Anglican Church. But, he went on, Lindsey's and Disney's moves into Unitarianism were based on readings of the scriptures as rigid and narrow as any to be found among Trinitarians. "What should I get," he asked, by "putting my faith into the hands of a Socinian Doctor [the reference is to Priestley], who is fully as peremptory in asserting his own interpretations as the old Archbishop, and would no more suffer me to theologise in my own way (perhaps not so much) than the Church of England does?"[62] By the time he was asking this question, the religious splits among his children had taken their toll on Blackburne's desire to be heard, and he left his pamphlet to lie unpublished among his papers.[63] But negotiating the tension between inclusivity and dogmatism that the elderly man felt so acutely would challenge Lindsey and his followers throughout their lives.

For the family of reason, the most important of Lindsey's followers was the fragile eighty-year-old man who was so pleased with his daughter's wedding in 1837. William Frend was not yet nine years old when Blackburne published his *Confessional* in 1766 and was still a schoolboy when Lindsey moved to London in 1773. But slowly, over the course of the next fifteen years, he found his way to Lindsey's church, and from then on, his life was defined by the power of reason he embraced as a sacred duty. Frend was the head of the middle generation of the family of reason, and the path he followed over the course of more than fifty eventful years connected the Lindseys' eighteenth-century world to the De Morgans' nineteenth-century one.

# CHAPTER 3

# Education of an Anglican

There are few hints of radical religion or politics to be found in the early life of William Frend. His grandfather, George Frend, was a younger son of a country gentleman who moved to Canterbury from the countryside to make a living selling clothes. By the time William was conscious, his grandfather was a distinguished man about town, and his grandmother was "a fine, stately, antient dame with a long ivory headed cane in a stiff silk gown which would stand upright when left to itself."[1] William's father, another George Frend, was a powerful and convivial man, who "loved his pipe & his bottle."[2] He served for years as a Canterbury city councilor and was twice elected mayor, in a process that involved ever more raucous festivities and a considerable amount of bribery.[3] These forbearers were pioneers in a middle-class world of commerce and enterprise that over the course of England's eighteenth century was slowly rising to become a recognized realm of social respectability.

William was born on November 27, 1757, in George's house on Burgate Street. Out the front door, the walled city bustled, as it had for at least a millennium, at the merging of two great Roman roads to London. At the back of the house sprawled the venerable cathedral. Through the middle of the house ran the wall that divided the city from the cathedral close. On a day-to-day basis the Frend family moved easily from one side to the other through a passageway. But when the boundary between the secular and the sacred was significant, George chose the secular side. This was noticeably the case when it

came to the birth of his children. Anyone born within the city walls of Canterbury had specific rights, so George carefully recorded that all of his eight
children—George, Elisabeth, Catharine, William, Sophia, Richard, Jane, and
Mary—arrived in the front room that overlooked Burgate Street. That they
were born on the Canterbury side of the wall meant that when they reached
their majority at the age of twenty-one, each of the Frend children could claim
his or her rights as a free-born citizen of the city.

The family thus constituted was, by all accounts, a warm and supportive
one. William remembered his father as at times "severe" but "when in good
humour no one [could have been] more affectionate and indulgent." William
and his older brother George delighted in their father's municipal honors.
They were equally impressed by the exploits of their Uncle Southby, who
was making a fortune as a wine merchant in the rapidly expanding world of
international trade. He had a much-appreciated habit of showering his nieces
and nephews with exotic gifts, "jars of raisins, boxes of Portugal Plummbo
and a vast variety of other good things of London," even as he expounded
on the value and rewards of hard work. William had no memory of his biological mother, née Catharine Fitch, who died two children after him, but he
vividly remembered the second Mrs. Frend, née Jane Kirby. She was a voracious reader who recognized her young stepson's brilliance and encouraged
him to study "by every indulgence"; he responded by loving her "as tenderly
as if she had been my mother by nature." The William who grew up on Burgate Street was thus a complex mixture of his father's sociability, his uncle's
independent spirit, and his stepmother's quiet bookishness.[4]

William's father credited his successes both in business and in politics to
the connections he had forged while a student at King's School, Canterbury,[5]
and he made sure that his boys had the same advantage. William found little
to say about his time at King's, though he did make two friends, James Tylden
and James Six, with whom he remained close for decades. He was equally engaged with Six's father, a natural philosopher, who happily shared his enthusiasms and experiments with his son's friends after school. The whole added
up to good preparation for "the severer studies of the university,"[6] and both
Six and Tylden went to Cambridge after they finished their studies at King's.

William's father had other plans for his middle son. He had already sent
brother George to learn the lessons of hard work with Uncle Southby in Lis-
bon. William was to travel even farther, to work in a mercantile house in
Quebec. The first step was a visit to the Catholic community of Saint-Omer
in northwestern France, where he could learn French. The next was to cross

the Atlantic to Quebec, where Mr. Hey, one of the friends his father had met at King's, was lieutenant governor and had agreed to find William a job. The year, however, was 1775, and tensions were building between England and the American colonies. The English were thin on the ground in Quebec, and no sooner did the eighteen-year-old boy stagger off the ship than Mr. Hey offered him a commission in the king's army. But William had no desire to be a soldier, and in an initial show of the stubborn resolve that was to characterize him throughout his life, he refused. By December, he was back in England, determined to follow Tylden and Six into a career in the Anglican Church.[7]

George responded to William's rebellion with more indulgence than severity. Supporting his son's desire to become a minister required George to turn from the bustling world of shop and town at the front of his house to the traditional world of the cathedral at its back. There, the archbishop, Dr. Moore, confirmed that the first step toward William's goal was a university education. Dr. Moore recommended that William begin his new career studying at Christ's College in Cambridge.

———•———

When William and his father entered the elaborate sixteenth-century gate of Christ's College on St Andrews Street in Cambridge, it was taking in from two to six students each year. It is not completely clear why Dr. Moore recommended that William go there, but perhaps it was because it boasted one of the most effective and beloved tutors of the eighteenth-century university, William Paley. Certainly Frend's first encounter with Paley on that cold December day in 1775 was memorable. In a letter to his daughter, written more than fifty years later, he was still grateful to the man who put him at his ease despite "the little agitation which a young man naturally feels at an examination."[8] After about an hour's conversation, Paley dined with father and son in the master's lodgings. He invited the Frends for supper there as well, but George had already made plans to dine with Richard Farmer, one of his hard-drinking Canterbury friends who had just been named master of Emmanuel College. By the time the Frends left Cambridge the following morning, William had been entered on the rolls as a pensioner at Christ's College.[9]

Socially, the Cambridge University that accepted William through this highly personal process was very like the one that Blackburne and Lindsey had known. Fully one-third of the students were aristocrats or fellow commoners who had money to spend and little reason to engage academically. Drunkenness was a perennial problem, prostitutes swarmed the nighttime streets, and

wandering bands of undergraduates were "scarcely less ferocious" than their
working-class counterparts.[10] Daytime activities were perhaps more benign
but equally energetic: "A very common practice, during the spring and summer months, was for a party to divide into two sets, one on a shooting scheme,
and the other on a boating and fishing expedition, both parties agreeing to
meet and dine" at the end of the day.[11] This kind of outing could shade into
more intrusive performances like galloping through the gates of one college
after another, "shouting most vociferously" racing around the court, and then
riding off "laughing heartily at the exploit."[12] Students who broke into private
rooms could be expelled, but few other limits were placed on their exuberance.

Life was generally quieter, but no less worldly, within the sheltering walls
of the colleges. All fellows and students attended two chapel services daily, but
the primary focus of each day was the college dinner. This communal event
became ever more elaborate over the course of the century; in Frend's day,
the white waistcoats and white silk stockings that students wore were looked
after by "persons who washed them for us as things too special for a common
laundress."[13] Hair styling was such a concern that Frend later commented, "it
seemed that as much pains as were employed by our tutor to furnish the inside, our frizeur should take to adorn the outside of our pates."[14] After what
was routinely two hours of preparation, the fellows, aristocrats, and fellow
commoners dined on their special food at their special tables. Pensioners like
Frend dined on their food at theirs. The lowly sizars served, only sitting down
to eat the leftovers when all others were done.

These elaborate rituals covered some major fault lines that ran between
the cocky young aristocrats, the wealthy fellow commoners, and the monkish
college fellows who were required to look after them. Fellows were known to
refer to their entitled students as "Empty bottles," "Useless members," or "The
licensed sons of ignorance,"[15] but not to their faces; their young charges were
far too powerful for that. In 1786, a grateful student left his tutor £20,000 —
a staggering sum in a world where a college tutor earned on the order of £100
a year.[16] What is more, after they left the university, the aristocrats and fellow commoners might well be dispensing the church positions, or "livings,"
that many fellows were anxiously hoping to attain. The pressures to land a
lucrative one were as much domestic as they were financial; college fellow-
ships required celibacy, but a parish minister could be married, as could the
masters of colleges and university professors. In 1780 all of the students of
Christ's College, including William Frend, watched with salacious fascination
as the engagement of their tutor, Thomas Parkinson, to a young woman "of

surpassing beauty" dissolved when he failed in his bid to become the master of Christ's.[17] Parkinson's plight was just a particularly visible example of the perennial struggles of Cambridge fellows to find livings outside of college. With the shape of their whole lives at stake, they took care to respect the prerogatives of their young charges.

In the intellectual realm, Lockean liberalism had made considerable inroads in Cambridge since Blackburne was refused a fellowship, but the process was slow. At the 1771 meeting that drew up the Feather's Tavern Petition, John Jebb was charged with gathering signatures in Cambridge, but he found it a difficult task. He had no trouble finding signatures at Peterhouse, where he was tutor to Blackburne's second son, Thomas, and where Blackburne's friend and confidant, William Law, was master, nor in the equally liberal Queens College. By contrast, Jebb collected none in Emmanuel College where the high Tory master promised punitive action against any fellow who signed. In the rest of the colleges the masters were less directive, and their indecisiveness left room for considerable difference of opinion and action. Frend's tutor, Paley, for example, would not accept the petitioners' insistence that the Thirty-Nine Articles be read literally. Surely, he argued, the Elizabethans who brought the Articles together did not expect "the actual belief of each and every separate proposition contained in them"; they were just trying to exclude "the members of such leading sects or foreign establishments as threatened to overthrow our own."[18] He "could not afford to keep a conscience,"[19] was Jebb's bitter assessment, but there is a more charitable way to understand Paley's position. He was supporting diversity within the church by following the broad church approach of loose reading.

In the end, some form of Paley's latitudinarian liberalism prevailed in Cambridge, and the Feather's Tavern Petition did not carry the day; nonetheless, Jebb continued to battle for the cause of reason at the university. He fought on both a negative and a positive front: *against* the specifically Cambridge requirement that students had to subscribe to the Thirty-Nine Articles in order to take a degree and *for* a meritocratic examination system that could cut through the class stratifications that structured the university. Jebb had some success in these efforts. Although the subscription requirement remained, he managed to water it down,[20] and his proposal to require annual examinations for all students, including fellow commoners and aristocrats, came within a hair's breadth of being accepted in 1772. When Jebb refused to leave it at that, his relentless bull-headedness began to take its toll, until "every plan or proposal, however good in itself, provided it comes from him, is sure to be rejected."[21] Only then did Jebb leave Cambridge to join Lindsey in London in 1775.

None of this was of particular interest to the William Frend who arrived at Christ's College in the following year. His decision had been to enter the Anglican Church, not to join those who were attacking it. He later remembered that "a fondness for society led me to pass many hours in company," and a portrait from his Cambridge days captures him in full curls and powder.[22] He flirted with young women whose mothers squired them to Cambridge, in winter he "scated" on the river Cam, and in spring and fall he joined minor hunting expeditions. Ultimately, though, Frend was still the earnest young man whose bookishness his mother Jane had encouraged. He was at Cambridge because he wanted to be, and with neither a powerful name nor fabulous wealth, his future depended on his doing well. He studied willingly and hard.

Studying with Paley was an important part of Frend's initial impetus. "You should have seen him, Sir, as we did, when he stept out of his little study into the lecture room, rolled from the door into his arm chair, turned his old scratch [a short informal wig] over his left ear, and his left leg over his right, buttoned up his waistcoat, pulled up a stocking, and fixed a dirty, cover torn, ragged Locke upon his left knee, moistened his thumb with his lip, and then turned over the ragged leaves of his book, dogs eared and scrawled about, with the utmost rapidity."[23] When Paley left the celibacy of his fellowship for marriage and a parish, Frend was left to learn moral philosophy from John Barlow Seale and mathematics from Thomas Parkinson. Frend never said a word about either Parkinson or Seale in later life, but his experience of Paley stayed with him.

The "dirty, cover torn, ragged" book that Paley carried into every class was Locke's *Essay Concerning Human Understanding*. In this bible of eighteenth-century reason, Locke explained that human beings encounter the world as a tangled mass of sensations—sights and sounds, smells and tastes—which are transformed into ideas when they enter our minds. The faculty of reason is the power that enables us both to analyze our complex ideas into their simple components and to synthesize those components into new ideas. Knowledge gained only through authority, is "borrowed wealth, like fairy-money," which will "be but leaves and dust when it comes to use,"[24] Locke explained, but understanding gained through the combination of experience and reason promised all the "advantages of ease and health, and thereby increased our stock of conveniences for this life."[25] It came too late for Blackburne, but by the time Frend was at Cambridge, Locke's anti-authoritarian philosophy of reason was moving toward the heart of the university's curriculum.

Paley reinforced Locke's message about the essential value of reasoning for oneself with his teaching style as well as with the substance of what he taught. More than two decades later, Frend vividly remembered the way his tutor would first pose a question and then wait quietly while the students squirmed toward a response. Personal reflection was just as important as class participation for the Lockean learning experience of Frend's Cambridge, since adjusting to new ideas required considerable time beyond what the classroom could afford. "Read one hour and think two," he was told.[26] Frend spent hours at his desk, filling large copybooks with notes on Locke, natural religion, and moral philosophy, and then, when he had exhausted his patience for scribbling, "indulged much in solitary walks."[27] Paley encouraged all of these activities as part of a concerted effort to help students learn what it meant to think for themselves.

Frend was at least as likely to have been contemplating mathematics as moral philosophy on his treks through the Cambridge countryside. In his first year, he was required to master the first six books of Euclid and Colin MacLaurin's *Treatise of Algebra* in preparation for more advanced work in mechanics, which culminated in at least the first mathematical book of Newton's *Principia*.[28] This was an ambitious curriculum that moved Cambridge students from the most elementary foundations of mathematics to its most powerful cutting edge.

That studying the *Principia* lay at the heart of a curriculum designed to educate Anglican ministers reflected an English religious tradition that dated to Newton himself. When, within five years of the publication of the *Principia* in 1687, Richard Bentley, the bishop of Westminster, asked Newton's advice for a series of lectures on natural theology, the cosmologist was happy to help. "When I wrote my treatise about our system, I had an eye upon such principles as might work with considering men for the belief of a Deity," he explained.[29] The view of the world Newton shared after this opening was so powerful that it undergirded Anglican natural theology for the next two centuries.

Locke was not mathematician enough to understand the specifics of Newtonian reasoning, but he recognized that reasoning mathematically was a significantly stronger and more powerful exercise than was reasoning on the natural world. The combination of sensation and reason may teach us much that is useful, but as a basis for true understanding, its claims are weak. We can never be certain about anything that we learn through finite experience because that kind of knowledge is essentially limited by that experience. Mathematical knowledge, on the other hand, transcended the limitations of experience. For Locke, mathematics was rooted in intuitive ideas, which are so immediately clear "the mind is able to perceive that they agree or disagree as clearly as that it has them."[30] Our grasp of intuitive ideas is "certain, beyond all doubt, and needs no probation [proof], nor can have any; this being the highest of all human certainty."[31] Locke illustrated the dynamic with the example of a "young lad" who intuitively knows that his whole body is larger than his little finger. When this Lockean lad studied geometry, he would learn to express this piece of intuitive knowledge as "the whole is larger than a part."[32] Once thus expressed in language, what had before been an isolated bit of intuitive knowledge became embedded in a coherently reasoned system that encompassed an enormous range of understanding. A whole new level of certainty was available to those who experienced the truths of geometry as immediately as they did their intuitive knowledge of themselves.

In the *Principia,* Newton tried to appropriate the certainty of Euclidean geometry for cosmology by formulating his argument in the Euclidean terms of axioms, theorems, and proofs. The result, he claimed and Locke agreed, was a knowledge of the workings of the universe so certain that to understand it was to understand the mind of the God who had created it. Some of Frend's contemporaries experienced this insight directly: "Language sinks beneath contemplations so exalted, and so well calculated to inspire the most awful sentiments of the Great Artificer," one cried as he reached the conclusion of a mathematical exercise.[33] For others, studying mathematics was just a way to lay the reasoned groundwork for an experience of the divine. However they experienced it, Frend and his classmates were studying Newtonian cosmology as a way to understand the God they were being educated to serve.

For Locke, theology was the only area besides geometry in which this kind of certain knowledge was attainable. God's existence was to him "the most obvious truth that reason discovers";[34] his problem was to transpose this certainty from the sermons and music that had long proclaimed it in order to deduce it "from some part of our intuitive knowledge."[35] Faced with this challenge, Locke began with his immediate, intuitive knowledge of his own existence. When combined with the equally intuitive recognition that nothing comes of nothing, this insight meant that some real being created him. Locke's reason then led him forward to further conclusions. First, this being must have existed before anything was created; second, this being must be more powerful than any created being; and third, since it is "repugnant to the idea of senseless matter, that it should put into itself sense, perception, and knowledge," this being must have those qualities as well.[36] In this way, Locke reasoned himself to the conclusion that *there is an eternal, most powerful, and most knowing Being.*[37] Even as he presented his argument as a proof of God's existence, he recognized that only those who followed their own faculties of reason through the steps would see its power. He found those who refused to recognize an eternal, all-powerful thinking being greater than themselves to be "sillily arrogant and misbecoming,"[38] but he also knew that recognizing the presence of the divine was a profoundly and essentially private experience. His proof represented a reasoned effort to point to an existence that was immediate to him but which others might or might not experience.

There is no reason to count Frend among those who would deny Locke's conclusion, but he found God's presence to be more effectively communicated in Handel's aria "I Know That My Redeemer Liveth" than it was in the philosopher's proof. The young man from Canterbury learned mathematics as a necessary step in following the career path he had chosen. His dutiful approach was supported by the examination system by which he would be evaluated at the end of his student career. In the university as defined by the Elizabethan statutes, the major process of evaluation and promotion was through a series of logical disputations, followed by the "accustomed examination."[39] Through the seventeenth and into the eighteenth century, it is not completely clear what was included in the "accustomed examination," but by the time Frend graduated, the logical disputations were withering under the shadow of a mathematically focused Senate House examination, commonly known as the Tripos. Students were still required to "keep an Act," that is, to debate academic questions under the watchful eye of a moderator, in the spring of their third year, and again in the fall of their fourth year.[40] But what mattered for their futures was their performance on the mathematical Tripos.

The growing importance of the Tripos is yet another sign of the inroads Locke was making into Frend's Cambridge. From Locke's point of view, logical disputations were noxious because they encouraged people to argue from words rather than from the ideas those words referred to. Often the best way to win a logical argument was to "perplex the signification of words" until they became even "more obscure, uncertain, and undetermined in their meaning than they are in ordinary conversation." This dynamic depended on the kind of slippage among our experiences, our ideas, and the words we use to represent those ideas exemplified by the use of the words "red," "crimson," or "magenta" to describe a single experience of color. Locke was willing to grant the name of knowledge to the ideas we might draw from such experiences, but the ambiguity of the language with which we describe them is a troubling sign of the essential uncertainty that characterizes our understandings of them. There are no such ambiguities in geometry, however. There the connection between immediately known intuitive ideas and the symbols that represent those ideas is absolute; there are no alternative words for "triangle," or "line," or "right angle." For Locke, this linguistic exactitude was the crucial mark of epistemological certainty, and it goes a long way to explaining why Frend's performance on a mathematically focused final examination was the culmination of an education preparing him for a career in the Anglican ministry.[41]

Frend's Tripos was a primarily oral four-day examination. In the weeks before it began, the approximately seventy-five students who were to take it were divided into groups based on some combination of their tutors' assessments and their performance in their Acts. On the Monday and Tuesday of the examination week, university examiners and moderators asked mathematical questions of the students in each group; they also held special evening sessions with more difficult, sometimes written questions for those in the first and second groups. The focus shifted from mathematics to moral philosophy on Wednesday, only to return to mathematics on Thursday in order to finalize class rankings.[42] Locke would certainly have found the mathematical focus of this examination to be a vast improvement over the logical disputations that had preceded it.

That most of the process took place in conversation allowed examiners to ascertain the depth of the students' understanding based on a back-and-forth dynamic. It also means that we have no record of either the questions or of Frend's responses. What we do know is that he emerged as second wrangler, that is, second in his class. He also finished second on the Smith's Prize examination, which had been established a decade earlier to reward the best mathematical student at Cambridge.[43] These stellar achievements assured him a good position in the Anglican world he was so determined to enter. In the following year, he turned down an extravagant offer to be tutor to Archduke Alexander of Russia in order to accept a position as fellow and mathematics tutor at Jesus College.[44] He was ordained as a deacon of the Anglican Church in 1780 and named the vicar of the small neighboring parish of Long Stanton on December 22, 1783. The eager young man's stubborn refusal to follow the path his father had laid out had taken him where he wanted to be, comfortably situated in the quiet studious life of his college and his church.

——•——

At Cambridge, he was always "Frend," but to the family in Canterbury that applauded his accomplishments he was still "William." "Success attend you my lad!" brother George enthused; "we are all upon the same scent, independence is what we aim at!"[45] But holed up in his Cambridge rooms after a Christmas at home, William found himself homesick enough to admit that, "a letter or two from my sister puts me in mind of other scenes & recalls me to more pleasing imaginations."[46] This note of appreciation marks the beginning of a correspondence between twenty-seven-year-old William and seventeen-year-old Mary, which is our only direct source for what were arguably the most critical five years in his life. At its beginning, he was a fast-rising minister in the established Anglican Church; at its end, he had broken from the church to become a ragingly radical Unitarian. It is difficult to construct the issues that propelled these changes from within a correspondence undertaken as a diversion, but some of its broad outlines can be gleaned from the siblings' letters.

The correspondence begins in earnest in spring 1784, when Mary set out to visit London with her eighteen-year-old sister Jane, and William wrote to guide them through the vibrant metropolis. In England's capital city, the gentry lived and reigned around the palace in Whitehall, the destitute poor huddled in the squalid east of Whitechapel, and between them England's entrepreneurial Uncle Southbys plied their wares. Boats from all over the world brought goods to the wharves that lined the Thames, and England's merchants sold their wares from elaborate showrooms and shops on Fleet Street and the Strand. Darwin's grandfather, Josiah Wedgwood, was selling his china from a showroom on Great Newport Street. This middle London was positively vibrating with energy in the 1780s.

In their 1784 letters, both William and Mary were deeply engaged in establishing their proper place within the carefully circumscribed social world of a nascent middle class that was exploding into power in England.[47] Even as he excitedly recommended that his sisters go to hear Handel in "the grandest concert that ever was perform'd," he cautioned them to save their money for good seats where they would be in "genteel" company.[48] Even among their own relatives, it was important that Mary and Jane consider carefully "in what manner you mean to treat them or upon what footing you are to be with them."[49] He was particularly concerned that they might be staying with their cousins the Salts, who were undoubtedly "very good and kind people" but lived in such a "dreadful" and "dreary" lane that he feared the drag of their "situation and manner of life."⁵⁰ With all of these social quicksands to be navigated, Mary's brother breathed a sigh of relief when his sisters moved into the safety of brother George's ample house on Greville Street, in a decidedly more reputable neighborhood.

Once satisfied with his sisters' living arrangements, Frend stepped up to guide them through the bustling city. He eagerly awaited Mary's descriptions of the bridges, the Monument, the splendors of Somerset House, and the activity in the Pantheon.⁵¹ "Fashionable people" might dismiss such sights as "vulgar," but Mary and Jane should take in Westminister Abbey, St Paul's, and the Tower.⁵² Notably lacking in William's list of places to visit is any mention of the Essex Street that ran to the river just west of where Fleet Street met the Strand. Every Sunday of the four months that Mary and Jane spent in London, Lindsey was there preaching about the "one true God" to a congregation large enough that he was actively searching for an assistant. But the newly minted Anglican minister in Cambridge was certainly not going to suggest that his sisters visit the dissenting group gathered around Lindsey's Essex Street Chapel.

As the weeks went by, William received Mary's reports of an art exhibit, the Crown Jewels, and the charming "figures of Queen Elizabeth, the page, and the horse" in the small armory of the Tower.⁵³ He responded with tales of rowing on the river Cam and of dancing "till four o'clock in the morning, every person in good spirits, everyone pleas'd & everyone wishing to see his neighbour in good humour."⁵⁴ Nothing in these letters suggests that the Frend siblings were anything but satisfied with their respectable lives.

By July, Mary was luxuriating in the peace and quiet of Canterbury after four months of the London swirl, and William was engaged with commencement, "the gayest season in the University."⁵⁵ After months of quiet, he reported a plethora of concerts, musical parties, a ball, private dinner parties, and "promenades every evening upon a Common where there are temporary Booths for the fair."⁵⁶ As soon as these festivities were over, he set off with a friend, their horses, and the all-important hair dresser on a "tour of considerable length" to visit the "North Britons."⁵⁷ At summer's end, Frend settled comfortably again into his life both embedded in and molded by his Anglican college, university, and church.

Meeting eligible people was an important subtext of Mary's and Jane's trip to London, and in the months after her return home, Mary was full of romantic ideas about marriage. William tried to moderate his young sister's visions with a more reasonable description of his ideal wife: a woman with a "tolerable share of good sense that could endure the little mortifications of life without peevishness, that could consider home as the place of her happiness & that could love her husband."[58] The whole conversation was highly theoretical as neither sibling had any real marriage prospects. Nonetheless, the ideal that William then painted is a rather good description of the woman he eventually married. Little else in his life matched twenty-seven-year-old William's image of himself, however. In the more than twenty years that transpired before he and Sarah Blackburne were married in the Essex Street Chapel, his pursuit of reason led him to be expelled from the Anglican Church, "rusticated" from Jesus College, and banished from Cambridge. This future, however, is nowhere to be found in the contented letters of Mary's older brother as he moved into a new year among the books and students at Jesus College in fall 1784.

# CHAPTER 4

# *The Road to Unitarianism*

For the next year, letters may have continued to flow between William and his sister, but none of them have survived. It is only in November 1785, a full year after William outlined his marital hopes, that a new set begins. In the interim, the siblings had begun to move into their roles as adults, earnestly and actively engaged in the world around them. This set of letters begins with Mary asking her brother to recommend books that she can read to educate herself.[1] But after William enthusiastically responded, she confronted a host of problems—from Jane's absence in search of a husband, to the discomfort of studying before the fire was lit, to "making mourning for my Grandmother Kirkby, who died last Friday fortnight"—that prevented her from following through.[2] She treated these interruptions as somehow her fault, and in the spring warmth of April again proclaimed her determination to read; sadly, this time her good intentions again collapsed when Jane became engaged.[3] Mary's experience powerfully reflects the difficulties women faced pursuing an intellectual life without institutional support, but her brother was struggling as well. Over the course of the next two years, the internal pressures generated by his efforts to fit himself into the established Anglican world of his church and his university would catapult him out of that world entirely.

Frend's conversion began in his local community. In response to the Industrial Revolution that was both literally and metaphorically gathering steam around them, a growing number of England's middle-class citizens were feel-

WILLIAM FREND, M.A.
Fellow of Jesus College, Cambridge.

from an Original Drawing by S. Harding taken
Nov. 1789 in the Possession of the Rev.ᵈ D.ʳ Disney.
Pub. Aug. 1ˢ 1793 by S. Harding Publisher.

Figure 3: William Frend as a fellow at Jesus College. Stiple
engraving by Andrew Birrell, after Silvester (Sylvester) Harding.
D7887 © National Portrait Gallery, London.

ing the impulse to help the poor. Bolstered by Locke's optimistic view of children as open books, innocent tabulae rasae ready to take in all that was written upon them, many parishes created Sunday schools to teach their youngest
members.[4] In 1785, a well-meaning group of Cambridge residents decided to
set up such a school in the small adjoining parish of Maddingly. That fall,
Frend pulled himself out of his study to accept the responsibilities of director
and teacher (figure 3).[5]

Frend found himself quickly consumed by the joys of teaching children to read. At the height of his involvement in the Maddingly Sunday School, he seems to have left the preaching in Long Stanton to someone else so that he could spend all of his Sundays in his school. Week after week from 9:00 to 5:00, he and his co-teacher, Miss Cotton, devoted themselves to their little flock. Most of their students were entirely illiterate, and the more advanced could barely read a sentence. But a year of Sundays spent learning the alphabet, reading short phrases and then longer sentences rendered the best among them able to move fluidly through a whole chapter in the Bible.[6] What the children took away in literacy they gave back in music. One of their first reading challenges was Watts's hymns for children, which they learned to sing in a chorus led by a blacksmith music director and accompanied by Frend's "miserable flute." Within a short time, they were singing well enough that not only their parents but people from adjoining parishes came to hear them.[7] By Christmas they were accomplished enough to carol through Cambridge, whose charmed inhabitants rewarded them with gifts.[8] Frend was profoundly affected by the whole experience. More than thirty years later, a simple question from one of his many correspondents triggered a flood of memories of the little school he had directed in Maddingly.[9]

Mary was also deeply involved in the Maddingly Sunday School. In the summer of 1785, even before the school had opened its doors, family friend Dr. Farmer was on summer break in Canterbury. As master of Emmanuel College, he was an avid supporter of the project and took Mary through her father's shop to buy materials for clothes for the children. In December she was again hard at work, making kits so that the children could do the sewing on their own. "Tho' I am not passionately fond of cutting out, yet I ashure you I took great pleasure in doing these," she valiantly declared as she packed up individual parcels of linen, thread papers, needles, and one partially completed shift to act as a pattern.[10] Mary put considerable effort into this project, but her brother was so totally immersed in teaching that he barely noticed. Having received the package of sewing kits, he simply asked whether she could write lesson books for the children.[11] When, in a rare moment of decisiveness, his harried sister refused,[12] Frend was left to write his own "little books of the history of the Bible with questions at the end of each lesson."[13] Neither Frend's texts nor Mary's clothes have survived, but their efforts were certainly of value to the children of Maddingly. They were also at least as important for Frend, laying the groundwork for what was later to blossom into a vocation writing texts for children and their mothers.

Mary kept hoping that her brother would pay a visit to Canterbury, but by the end of a year divided between his duties as mathematics tutor at Jesus College and Sunday school director in Maddingly, he was eager for freedom and an adventure. So in mid-June 1786, he passed up all graduation festivities and left Cambridge for a trip to the continent, accompanied by his Canterbury school friends, Six and Tylden, and a dog named Dash. He had no fixed address in Europe, which means he could not receive letters. From then on, Mary appears in the historical record only as the silent recipient of her brother's travelogues.

Frend's continental trip began with three days in Paris, which he and his companions approached with touristic enthusiasm. They went to the King's Gardens and to the Cabinet of Natural History, attended lectures on botany and "chymistry," listened to disputations in theology, took in the Grand Opera, visited two theaters, and viewed the art collection of the Duke of Orléans.[14] All of this cultural exploration was just the hurried prelude to a much less conventional adventure. There was no friseur on this journey; there were no horses either. Instead, the three men sent their luggage ahead and left the city on foot. With their essential possessions slung over their shoulders on a stick and Dash frisking about them, "the road was open to us as the whole world to Adam when he left Paradise — not that I mean to compare Paris with Paradise for perhaps it more resembles Pandemonium," Frend wrote to Mary. "Unmolested by the cares of horse & chaise, postilions or servants or baggage, we were now free to go even to Constantinople if we pleased."[15] In the event, the little troupe got only as far as Germany and Switzerland, but the effects of the trip on Frend were as significant as if they had traversed the world.

It was not unusual for Englishmen to travel in Europe in the summer of 1786, but their peripatetic approach set Frend's group apart. In France their odd appearance elicited no particular response, "for the French are more used to sights than the English, and do not, like the latter, think themselves entitled to laugh at the costume of any country."[16] Farther off the beaten track, they created more of a stir. In the Rhine valley, they were entertained by a woman "who, presuming we were princes in disguise, gave us clean feather beds for counterpanes, sheets edged with lace & worked pillowcases."[17] Others took them for "officers or thieves"[18] and were considerably less welcoming. Later, deep in the Alps, their "clean," "neat," and "civil" hosts found it so peculiar that strangers would visit their country out of curiosity that they "supposed that we had lost our way."[19] Throughout the journey, Frend's letters reveal that he was watching all of these people as much as they were watching him.

That Switzerland was the group's destination was also unusual; most late eighteenth-century Englishmen traveled to Italy in an effort to complete their classical education. Although the group did bring a copy of Caesar's commentaries along with them, Frend's letters to Mary focus on the present: "The smell of the new made hay, the contented & happy looks of the country people to whom this luxuriant harvest seem'd to add new spirits, the wide extended fields with the promise of a golden harvest, the sides of the hills cover'd with vines, the delightful banks of the Marne."[20] He was equally struck by the towns: the cathedral at Rheims, the remarkably regular streets of Nancy and Luneville, the cathedral spire in Strassbourg, whose "lightness & elegant origlants" were beyond description, the just-discovered Roman baths in Basel where they "were very comfortably parboiled."[21] Frend's effusive descriptions reflect his pleasure in the European world he was encountering for the first time.

The natural world the little group encountered was at least equally impressive and at times rather more insistent than the cultural ones. Until they reached Zurich, the weather was good, but after that the group encountered rain. In and between towns they handled the problem by riding in carriages, but as they headed toward Glarus they passed beyond villages and cows into the forests, where "the majesty of nature appeared in full grandeur" with a crashing thunderstorm. The intrepid band pressed on "sometimes on our legs, sometimes on our backs" through torrential rain and hail until finally the sun broke through and "lighted up the lake" of Wesen. The scene, "with the abrupt rocks on each side falling perpendicularly into the water" flanked by snow-covered peaks, opened Frend up to "true ideas of the sublime and terrible."[22] His description of the trip to Wesen is as romantic as he was ever going to get, but it was in the mountains above the Glarus valley that the hardy three encountered nature in all its naked power.

The Glarus valley expedition began with a scramble over rocks to the snowline and the hut of a single herdsman, where Frend and his companions spent the night in a ramshackle construction of stones piled up in such a way as to give "free vent to both wind and smoke." At four the next morning, after a quick breakfast of milk and meal, the group sleepily followed their herdsman guide into the snow-covered mountains. After breaking a path for several miles, Frend noticed himself suddenly sinking and threw himself forward onto the snow. Having recovered his footing, he looked around for the others and saw Six in front of him, the guide to his left and "a small aperture thro' which Tylden had perhaps slidden into an infinite abyss." The diminished group, which was itself standing "on the brink of ruin," had neither the ropes nor the expertise to rescue Tylden. They crawled away from the treacherous crevasse, and then walked for miles in search of others who could help. When they finally returned with some better-prepared mountaineers, they were vastly relieved to see Tylden limping toward them. Apparently after his initial shock wore off, he managed to chisel his way up the sheet of ice until he was "again restored to the light." Gratefully reunited, the shaken group descended the slope, Frend and Six placed their very stiff companion on a horse, and they all left "the horrid mountains behind." Almost losing Tylden struck far too close to home to indulge in romantic thoughts of the sublimity of nature. Instead, what Frend took from this experience was the memory of a "moment the most terrible I ever experienced."[23]

There is as little reverence toward God as toward nature in Frend's response to Tylden's accident. Nowhere in his narrative does God or providence either smile or frown on Tylden or his friends. In fact, there is a striking absence of either struggle with or reliance on the divine in all of the young minister's letters to his sister. But the public religious expressions he witnessed attracted considerable attention. Even before he got to Paris, he had been suspiciously noting signs of French totalitarianism and Catholic superstition. While he proclaimed himself pleased with the tomb of a commoner in St Dennis, because "we naturally [want] to pay the due tribute of our respect to real merit," he was pointedly unimpressed by a monument to a Catholic priest who devoted himself to a statue of Mary for fifty years. "A long attachment this," he admitted, but he was not at all tempted to admire the devotion of someone "addicted" to superstition.[24] Although he had not previously expressed such a view, Frend was apparently already determined to hold religious expressions to some basic standard of reason.

With religion as with nature, it was in the Swiss Alps that Frend found the most food for thought. As he struggled through the rain, the neat and clean appearance of the Protestant canton of St Gall, Switzerland, led him to extoll the values of democracy. When, only a day later, he contemplated the small and dirty houses of equally democratic Appenzel, however, he was forced to recognize that other things besides government were involved in the happiness of a community. He attributed the difference to their responses to industry: "In the one, every person seems employed & their cotton manufactury brings them a great supply of wealth, in the other they say Masses too often & lose half their days' work in the churches."[25] Even as he railed against Catholic "superstition," Frend marveled at the sight of "two small churches, Protestant and Catholic, within a stone's throw of each other," in a little village lit by the setting sun.[26] Although his England prided itself on it religious toleration, when it came to genuine acceptance of difference, Frend's European experience suggested that he and his country still had a long way to go.

Ironically, considering his consistently anti-Catholic comments, the only personal religious experience Frend seems to have had occurred in a Catholic monastery. At the very end of his trip, Six and Tylden took off for a more traditional tour of Italy, leaving their companion to make his own way back to England for the fall term. Before racing off to Cambridge, he decided to pay a visit to the Catholic community in Chartreuse. At the end of a six-mile walk past rivers and through rocks and forests as lovely as any he had seen in Switzerland, he was greeted by a group of "venerable fathers, totally abstracted from worldly cares & yet full of civility & attention to strangers." He was overwhelmed by their warmth and welcome, telling Mary that it was impossible to "rest unmoved by their manners and devotion."[27] His sister did not have any trouble understanding the transcendent holiness her brother found in Chartreuse. She had been similarly struck just six months before by a visit to the equally Catholic monastery in Saint-Omer. "I was perfectly enchanted" she had then exclaimed. "How good they were!"[28] Despite all of Frend's condemnations of Catholic superstition, neither he nor Mary would find the same kind of spiritual uplift within the respectably decorous confines of England's established church.

---

When Frend returned to Cambridge, he found that his months wandering as he pleased had left him "as unwilling as a headstrong freshman to be brought under the discipline of a College." His little school in Maddingly still felt "of more consequence to me than Versailles," but when it came to the rest of Cambridge life he was glad to have "paid a visit to the Chartreuse & contracted a more than ordinary desire for solitude."[29] To Mary, he attributed his restlessness to "a redundancy of health"[30] but there were other forces at work as well. He had been unsettled by the vastness of the world outside of Cambridge, deeply impressed by the viability of foreign social structures, and strongly affected by the overwhelming power of natural forces. When he set out, he had been completely self-assured, his judgments crisp and confident, but by journey's end, a day trip to Chartreuse was enough to shake him to his very core. He was never again comfortable in the smug respectability of his Anglican world.

In retrospect, Frend attributed the changes that overtook him to the influence of reason. He never referred in his published writing either to Tylden's accident or to his visit to Chartreuse; rather, he recorded that before leaving Cambridge "I looked round our college library for a small pocket volume to muse me on the road, and the *Racovian Catechism* falling into my way, seemed well adapted to the purpose."[31] The nonchalance of this statement is disingenuous because from the point of view of the Anglican Church in which he was a minister, the *Racovian Catechism* was a highly heretical text. First published in Poland in 1605, its defining feature was the anti-Trinitarian conclusion that Faustus Socinius had culled from reading the Bible directly. The *Catechism* found its way to England in the midst of the country's highly unstable religious and political seventeenth century, when its arguments were so inflammatory that in 1652, the English Parliament declared it to be "blasphemous, erroneous and scandalous" and resolved that all "printed copies of the book" should be burnt.[32] Against this background, the most plausible part of Frend's explanation for his choice concerns the size of the book he picked up; because the *Catechism* was technically illegal, many editions were designed to be small enough to be hidden in a sleeve or pocket.

Whatever led Frend to select the *Catechism* from Cambridge's all-inclusive libraries, perusing it as he walked through Europe affected him deeply. As he struggled to make sense of one unfamiliar experience after another, he found that "the plainness and clearness" of its doctrines "made an impression on my mind,"[33] so in the winter after his return from Europe, he plunged into a systematic rereading of the Bible.[34] This program of rigorous biblical study did nothing to restore the Anglican minister to his previous complacency. On the contrary, it left him increasingly disturbed by the differences between the religion Jesus preached and the Trinitarianism of the Anglican Church. In Mark 12:29, Jesus told his disciples that the first requirement of his religion was to "love the Lord thy God with all thy heart, and with all thy soul, and with all thy mind, and with all thy strength." As he did so, however, he never said that *he* was God; he never asked them to worship *him*. As Frend contemplated these discrepancies in the dark winter after his European tour, his Trinitarian religion began to fall apart.

Frend always described his journey away from Anglican conformity as a deeply personal one, but he was not alone in making it. Tylden had also been radicalized in Europe, and when the two of them searched out Jebb's friend Robert Tyrwhitt in Jesus College, the latter brought them up to date regarding the Essex Street Unitarians. By the spring, their discussions began leaking out into the larger University community. Unitarianisim "has gained some ground here," a fellow of Dr. Farmer's Emmanuel College commented, "three of the fellows of Jesus are avowedly of the persuasion and some others are thought to have a tendency towards it."[35] Lindsey's insistence that the way to a truly reasoned Christianity lay in a strictly literal reading of Jesus's words in the Bible was beginning to make inroads in Cambridge.

Exploring the implications of reason in college conversation was one thing; acting upon its dictates quite another. When Frend confided his heresy to Miss Cotton, she pleaded with him not to take action, and he agreed to lie low until the end of the Sunday School year.[36] Almost fifteen years later, he still vividly remembered "the pain, the uneasiness, the disquiet of mind" of that spring, "all the struggles between hope and fear, preferment and poverty, honour and disgrace" that dominated his thinking.[37] But, with each week it became more difficult for him to conduct Anglican services that required him to direct prayers to Jesus, much less lead a congregation through the Athanasian Creed, which proclaimed that "except a man believe truly and firmly [in the Trinity], he cannot be saved."[38] Frend's experience in Chartreuse had profoundly impressed him with the power of complete religious conviction, making it as clear to him as it was to Lindsey that prevarication was a sin. In June of 1787, Frend resigned his position as vicar of Long Stanton and director of the Sunday school in Maddingly. His bishop, James of Ely, knew better than to try to change the stubbornly determined young man's mind. Following the demands of "an honest tho erroneous conscience will at last be rewarded," he wrote gently, and in the meantime promised to "recommend you in my prayers to the protection of an all-knowing and gracious providence."[39] The words were mild, but their import stark: with them, Frend's career as an Anglican minister was ended.

———•———

Frend's resignation from the ministry was a relatively quiet decision that did not disrupt his situation as a fellow and a tutor at Jesus College. Neither position required that he lead Anglican services, but in his newly awakened state he was uncomfortably aware that both rested on his subscription to the Church of England. He recognized that latitudinarians like Paley would have let this situation rest, but like Lindsey he could not; there was no question that his status at Cambridge depended on a subscription that for him had become a lie. In the fall of 1787, Frend decided that the solution to this problem lay in removing the subscription requirement from Cambridge. He found a companion for this endeavor in yet another fellow of Jesus College, Thomas Edwards, who had just received his degree in law. Edwards discovered that the subscription requirement was not part of the Elizabethan statutes but rather an added requirement established by a vote of the university senate. That it had been adopted by a vote rather than by royal decree meant that it could also be removed by a vote. So Edwards moved to have the subscription requirement repealed by the university senate, and Frend published a pamphlet in support of the action. In *Considerations on the Oaths Required by the University at the Time of Taking Degrees, and on Other Subjects Which Relate to the Discipline of That Seminary,* Frend laid out his vision for his university.

Frend's first foray into print addressed the policy aspect of the proposal to lift the subscription requirement. He began by emphasizing that the mission of Cambridge was confined to the education of youth and the maintenance of a quiet scholarly life, which meant that reforms of the university need not be treated as if they were of national importance.[40] Having thus emphasized the modesty of his goals, he proposed the kinds of reforms Jebb had advocated so forcefully a decade before: raising admission standards so the university did not have to accept "any blockhead, who has neither knowledge nor abilities";[41] erasing the social orders within the university; increasing the number of tutors to combat large class sizes; adjusting the Tripos to support a more flexible curriculum; and, finally, abolishing the subscription requirement.

In support of his final point, Frend claimed that the subscription requirement forced many of England's finest minds to pursue their educations abroad. The result was a group of young men who returned to their native land with "foreign manners" and "foreign vices," chattering with "indifference" and "contempt" about the "most boasted bulwarks of British freedom." In addition, their exposure to superstitious Catholic rituals—Frend's word was "mummery"—rendered them ready to disparage all religion.[42] Dropping the subscription requirement and thereby opening the Oxbridge Universities to everyone was, he insisted, the best way to ensure that all of England's youth was educated safely, at home.

In the event, Frend's arguments did not succeed in rallying the university to the cause. Instead, in December 1787, Edwards's petition to abolish subscription was allowed to die in committee. This brush-off marked a major turning point for Frend; when all of the careful arguments of *Considerations* were simply ignored, he exploded in righteous wrath. In May 1788, he poured out his pain in a frontal attack on England's established church. In *Thoughts on Subscription to Religious Tests* he described the Anglican Church as an institution that refused "access to the Philosophick enquirer" and delighted in "pouring down damnation upon every unbeliever."[43] In the past, its cardinals, bishops, and monks had burned those who disagreed with them, he fumed; now, "instead of whips, and tortures, and fire, and faggot, they are content to strip a man of his coat and waistcoat, and then send him to graze on the Common without Money or Friends."[44] Clearly the cases of John Jones and others like him weighed heavily on Frend's mind, as he doggedly followed the directives of reason out of the Church of England. He pointed to two essential issues that fundamentally separated him from his Anglican past. The first was the intolerance, so clear both in modern church policy and in the Athanasian Creed, that consigned anyone who did not fully accept the doctrine of the Trinity "to everlasting perdition." The second was the irrationality required of anyone who embraced that doctrine. Having laid out these objections, he asked his readers to accept his current efforts as "atonement for the errors of his conduct," committed before he had come to see the light of reason.[45] By writing *Thoughts,* he had finally cleared his conscience.

Having thus broken cover, Frend wanted others to join the cause. And so, on September 10, 1788, he followed *Thoughts* with *An Address to the Members of the Church of England, and to Protestant Trinitarians in General, Exhorting Them to Turn from the False Worship of Three Persons to the Worship of One True God.* The language of Frend's *Address* is, if anything, more extreme than that of *Thoughts,* and it is not clear how many converts he won with his zeal. Nonetheless, the *Address* provides a window into the kinds of issues that drove him out of the Anglican Church.

Frend's major position remained biblical monotheism. He began with Mark 12:29, where Jesus answered an inquiry about his God with "Hear, O Israel; the Lord our God is one Lord." From this starting point Frend moved to reconsider the place of Jesus in that monotheistic world. The problem with allowing Jesus to somehow be absorbed into that one God was that it erased the "difference between the great Being, who sent and the man who was sent." Frend insisted on the distinction "between him, who is the God and Father of Jesus Christ, and that same Jesus, who declared that his Father was greater than he [John 14:28]." For Frend, Jesus remained the man whom God had sent to save his people, but that Jesus was the son of God did not mean he *was* God.

This position constituted an assault on the fundamental mystery of the Trinity that lay at the heart of Christian doctrine. "Mystery" Frend spluttered "means something hid; revelation means the discovery of that which was hidden and unknown."[46] Jesus himself had said: "For whatever is hidden is meant to be disclosed, and whatever is concealed is meant to be brought out into the open."[47] And so, Frend concluded: "There are no mysteries in the religion of Jesus Christ. His gospel is plain, simple and clear." This description of Jesus's message resonates with the clear and distinct ideas that essentially mark the truth of the axioms of Euclid's geometry. That it arose in the middle of scriptural argument reflects Frend's conviction that reason's rightful place lay at the heart of Christian doctrine. There was nothing "plain, simple and clear" in Trinitarian doctrine, Frend declared: "The notions annexed to it in your creed, of three persons, each of whom is God, making but one God, is rank nonsense."[48] And for him, there was no room for nonsense in the reasoned religion that Jesus preached in the Bible.

An immediate result of the essential straightforwardness of Jesus's understanding of God was the accessibility of the Christian message. Frend's experience teaching in the Maddingly Sunday School had reinforced his conviction that anyone could come to the true Christian religion, and the Jesus he embraced as a model was a man who could speak to everyone. "Brethren," he asked in his *Address,* "to whom did Christ preach? Was it not to the poor? Was it not to the unlearned?"[49] Frend railed against highly educated churchmen who were blocking access to the truly accessible God that Jesus had served. "Trinity is a Latin word, not to be found in the [Greek] scriptures,"[50] he pointed out. The moment at which this intruder word was allowed into Christian discussion was the moment that the corruptions of human historical change first sullied the pure simplicity of Jesus's religion.

Frend closed his *Address* with a call for action: "If anyone who reads this address, believes with me, that there is one God only, the God and Father of our Lord Jesus Christ, I call on him to forsake the temples, where they have set up other objects of worship."[51] This support for apostasy was too much for Dr. Beadon, the master of Jesus College. Frend's fellowship was a lifetime appointment, but Beadon would not allow students in his Anglican college to be taught by someone who declared those who "offer up prayers to Jesus Christ, to the Holy Ghost, or to the Trinity" to be "highly criminal."[52] And so, within two weeks of Frend's clarion call, signs appeared in Jesus College announcing that henceforth Thomas Newton would be the mathematical tutor there. Frend was incensed: cannot "the same lessons of prudence, temperance, frugality, sobriety, industry, submission to their superiors, conformity with the Statutes of the College, of piety to God and benevolence towards men, be inculcated by a Unitarian as well as by a Trinitarian Christian?" he asked.[53] Seething, he penned *A Second Address* in which he further elaborated his position "that the whole doctrine of the Trinity is a libel on the scriptures and an

insult to the understanding of mankind."[54] Thus backed by reason, Frend
again appealed the decision to the college visitor, who was the same James,
bishop of Ely, who had been so gentle eighteen months before. This time the
bishop was less supportive as he firmly upheld the position of Jesus College. A
furious Frend published the entire proceedings as an example of "ecclesiastical
tyranny," proclaiming his refusal to allow anyone to take away his right as an
Englishman to speak his mind as to "whether Jesus be or be not God, or even
whether he be or be not the Messiah."[55] This explosion availed him nothing,
however; he was no longer the mathematics tutor at Jesus College.

Losing his tutorship as well as his ministry meant that Frend's total income
had been reduced by two-thirds, and "the sensation of comparative poverty
was by no means agreeable."[56] Still the consequences of his anti-Anglican dia-
tribes were mild; he may have been stripped "of his coat and waistcoat," but
he had actually not been sent "to graze on the Common without Money or
Friends."[57] On the contrary, he still had room and board, a fellow's stipend,
and free access to both college and university libraries. And, having been
pushed to the sidelines of the Anglican world, Frend was now free to follow
his reason in whatever direction it might lead.

# CHAPTER 5

## *Exercising Reason*

The Essex Street Chapel was Frend's first port of call as he was breaking out of his Anglican world. In December 1787, as his and Edwards's petition against the Cambridge subscription requirement was being allowed to die in committee, he traveled to London to attend one of Lindsey's services. When the young man first introduced himself, Lindsey was impressed by his "great ardour of mind" but was still unsure "how far his scruples may go."[1] By June, however, he and Priestley were reading *Thoughts* as proof of the writer's "virtue and integrity and sound mind and judgment, such as must make the friends of truth rejoice and its enemies tremble."[2] As a Cambridge-educated fallen Anglican, Frend was precisely the kind of person Lindsey had hoped would join him when he set out to reform England's established church. "The greatest good must accrue from your conduct and example,"[3] he gushed as he welcomed Frend into the fold.

The Essex Street community that Frend joined in 1788 had grown in both size and sophistication in the fifteen years since the Lindseys first arrived in London. Their touchstone remained the insistence that the essence of Jesus's religion was to be found by attending closely to the words of the Bible, but time revealed how difficult it could be to agree upon the precise meaning of those words. By 1782, this challenge had become significant enough that Priestley and Jebb founded a Scriptural Society to establish Jesus's message.

They spent years pushing back into ever-earlier manuscripts to fix the true meaning of Jesus's words.

The scriptures were not the only place that the members of the Essex Street Chapel were pursuing their literalist view of reason. It was truly impossible to separate the religious from the political in eighteenth-century England, and for the Unitarians gathered around Lindsey's Essex Street Chapel, reason defined the proper approach to both. Jebb served as the primary link between the religious world of the Essex Street Unitarians and the political world of William Pitt the Younger. Pitt was first elected to Parliament in 1781 and became prime minister in 1783 at the remarkably young age of twenty-four. In the early years of his political career, American attempts to create effective representative government were challenging the English to reconsider the situation in their own country. Within a year of his election to Parliament, Pitt began trying to reform the English electoral system in which more than half of the members of Parliament (MPs) represented districts with fewer than one hundred members while burgeoning new cities like Manchester essentially had no representation at all. Jebb had already joined forces with Richard Price, the radical dissenting minister of Gravel Pit in Hackney, to form a Society for Constitutional Information, whose mission was "to diffuse throughout the kingdom, as universally as possible, a knowledge of the great principles of constitutional freedom."[4] Fair representation was a central concern for Jebb's and Price's Constitutional Society from its inception, and they eagerly backed Pitt's efforts at reform.

The conviction of the Essex Street Unitarians that reasoning humans could create and sustain representative democracy was fundamental to these reform efforts. Pitt and his supporters saw themselves as revolutionary in an eighteenth-century understanding of the word, in which historical change was cyclical rather than linear or progressive. Revolutionaries in such a world were not engaged in innovation but rather in attempts to return to a previous Edenic state. As Pitt put it in a 1782 speech, "There was a defect in the frame of representation, and it was not innovation, but recovery of constitution to repair it."[5] "Purity" was the watchword for the past to which Pitt and the Constitutional Society were trying to return. According to Pitt, "that beautiful frame of government which had made us the envy and admiration of mankind," had by his day "so far departed from its original purity, as that the representatives ceased, in a great degree, to be connected with the people."[6] The goal of his efforts at reform was to return the country to the state of pristine clarity that had become obscured by the vagaries of historical development.

The "purity" Price and Jebb were reaching for in the Constitutional Society was the same that Jebb and Priestley were seeking in their biblical studies, the purity of an uncorrupted translucent language that captured the realities of the world immediately. Thus, at roughly the same time that Frend in Cambridge was reasoning his way to an understanding of the divine by studying Hebrew and ancient history, Jebb in London was reasoning his way through politics by studying "the saxon language, the anglo-saxon laws, English history and antiquities, with a view to examine our criminal code, and particular points of liberty."[7] For the Unitarians in the Constitutional Society, the ultimate legitimacy of the reform measures Pitt introduced in 1782, 1783, and 1785 could be traced to the true meanings of England's founding documents.

Lindsey was supportive of all of these efforts, but his major focus remained on spreading the work of Jesus's reasoned religion. Frend's fall from Anglicanism meant that he was no longer welcome in the home of Sir John Cotton, but Lindsey introduced him to two other country squires, John Hammond of Fenstanton and Richard Reynolds of Little Paxton, who supported Lindsey's efforts.[8] Frend did what he could to spread the word by establishing "a theological lecture at a private house in the town," delivering occasional sermons at a non-Trinitarian church in Fenstanton, and distributing pamphlets from his college rooms.[9]

In the years after he declared his apostasy, Frend acted as a colonizer for Lindsey's religion, but in many ways the most powerful model for his rebellion against the Anglican Church was not a Unitarian at all but rather the minister of the Stoneyard Baptist Chapel, Robert Robinson.[10] The Baptist's background was dramatically different from Frend's, Lindsey's, or even Priestley's, but the religion he preached and practiced exemplified the inclusivity they found in reason. Born in Norfolk to a penurious Baptist minister who disappeared from his life before he was ten, Robinson had only a few years of grammar school before he was apprenticed to a London hairdresser at the age of fourteen. His subsequent education entailed listening to as many dissenting sermons as he could, which confirmed him in the faith of his father.[11] When he was released from his apprenticeship, an impoverished church in Cambridge invited the twenty-four-old to their pulpit. The community that invited him into their midst was as shaky and experimental in its way as was Lindsey's Essex Street Chapel. Its congregation was so tiny and so poor that Robinson collected less than four pounds for the first six months he was their minister, but he was a powerful preacher. By 1775, he was drawing hundreds of people from all walks of life to the newly built Stoneyard Baptist Chapel just two blocks from Jesus College.

Robinson rivaled Lindsey in his ability to draw people together with a powerful religious vision. He also shared Lindsey's political positions. During the time that he was pulling himself up from extreme poverty, he remained acutely conscious of and frustrated by the legal restrictions that denied him political standing. Because he was a dissenter, he could not sign the Feather's Tavern Petition, but he had avidly advocated for it, and when that national initiative failed, he had stood behind Jebb's efforts to eliminate the subscription requirement at Cambridge. By the time Frend met him, Robinson had been a firm supporter of Unitarian positions for years.

Robinson differed from Lindsey in one essential way, however; in his world, reason did not define what it was to be human. He was a wholly empathetic person, who was wont to proclaim that "nothing so much humanizes the heart, as bearing with the infirmities of others."[12] In practice, he was known for his efforts to offer some combination of shelter and employment to homeless vagrants. Reason had little to do with the way Robinson served the guileless children of God who wandered across his path.

Robinson's radical acceptance stretched beyond the needy. Frend was one of a varied group of intellectual seekers who relished the freedoms to be found in Robinson's dissenting chapel. For his part, Robinson enjoyed their theological discussions, but he experienced them more as a pastime than as a way to understand the divine. He was liable to shut down doctrinal disagreement with comments like, "Brother, I have delivered my present sentiments; but I am going to feed the swans at the bottom of my garden: on my return, I perhaps shall think differently."[13] As a dissenter, Robinson applauded Lindsey's break from the established church, but he was a Baptist, both by upbringing and by personal conversion. Lindsey's anti-Trinitarian doctrines left Robinson cold, and in the decade before Frend's conversion, the Baptist and Unitarian had engaged in a bitter pamphlet war about the divinity of Christ.[14] Nonetheless, they remained joined by their opposition to the Anglican establishment, and by the time Frend was leaving the Anglican Church in 1788, Lindsey and Robinson were committed to mending their fences.

Frend was at once an agent and beneficiary of this thaw. As he was falling away from the Anglican Church, he slipped easily into a lively and shifting group of intellectuals who gathered for discussion after services in Robinson's church. From as far away as Germany and as close as Frend's own college, the goal of this group's discussions was less to fix opinions than to open up all of God's complexities. In their company, Frend worked to be flexible and to adjust his thinking without "the prejudice which is supposed to possess the mind of one, who has publickly maintained them."[15] This openness did not alter his anti-Trinitarian stance, but it did show him that some combination of reason's rational individualism and Christ's love required him to respect the common humanity that joined him to those with whom he disagreed. It took years for Frend to grow into this kind of acceptance, but by the end of his life, one of his greatest joys was finding reason's common ground in all the people he encountered.

Frend's first and most productive experience of religious difference was with the Jewish community of London. The restless young man had been making sporadic attempts to learn Hebrew at least since 1784, when he had first declared to Mary that he was studying divinity, but the level of discussions at Robinson's church showed him the limits of what he could learn through self-study. Therefore, at some time in the late 1780s, he spent several summer months studying Hebrew in London's Jewish community.[16]

Frend's decision to study among the Jews was truly radical. In eighteenth-century England, Gentiles routinely turned to Jews for financial support and advice; otherwise, they were pointedly avoided. In the 1780s, this general distrust was exacerbated by Lord George Gordon, a radical MP whose heated rhetoric had, in 1780, set off a series of anti-Catholic riots that cost more lives in London than did the 1789 uprisings in France. Then, as this episode was fading from memory, Gordon flamboyantly converted to Judaism in 1785. For months, London was transfixed by the sight of his dramatically growing beard and titillated by the thought of his circumcision, but vanishingly few were interested in following the example of his conversion. Most preferred to have as little to do with the Jews as they could.[17]

Nevertheless, it would be hard to conceive of an environment more clearly designed to support Frend's Unitarian insistence on close biblical reading than the building where he studied Hebrew. The Bevis Marks Synagogue is still today a single room whose basic simplicity is reminiscent of the Quakerism of Joseph Avis, its major architect. The soaring clear-glass windows letting in light from all four sides are much like those to be found in the churches Wren was erecting all over London during the same period. However, there is neither a Quaker nor a Wren counterpart for the brass chandeliers that hang low over the dark oak pews, ready to shine their light on the words of the sacred texts.[18] The focus on reading embodied in the architecture of the Bevis Marks Synagogue stands as a testament to the importance of words that undergirded both Jewish and Unitarian religious understanding.

Frend was not the only person who recognized the deep connections between Judaism and the reasoned religion being practiced in the Essex Street Chapel. The Unitarians' insistence on worshipping Jesus's non-Trinitarian God placed them in the same legal category as the Jews, who were explicitly excluded from the protections of the 1688 Toleration Act. Lindsey and his followers chafed against the idea that such exclusions could be imposed upon them (although actually none ever were), but over time they came to welcome the association between the two groups. In the decade after Lindsey's first call for a church of reason large enough to include all Christians, the Essex Street Unitarians' vision of a truly open church expanded to include the Jews. For several years Priestley and Lindsey unsuccessfully tried to persuade the Jewish community to join them in a single church of reason.[19]

Frend's direct approach proved a much more effective way to create connections. The Jewish community welcomed with open arms the young man who wanted to learn their language. They shared Shabbat with him in their homes and invited him to the services in their synagogue, until he could claim to have "been present I believe I may say at every feast & festival."[20] Almost no other English Gentile could have made such a claim, and very few would have wanted to, but for the rest of his life, Frend was at least as comfortable among the Jews as he was with any other group of people. "I have frequently been in company with Spanish, Italian, Eastern & Morocco Jews & a Jew once spent the day with me whose residence was on the shore of the lake of Tiberius [the Sea of Galilee]," he proudly proclaimed.[21] They certainly never joined him for worship in the Essex Street Chapel, but London's Jews opened for Frend an enormous world that stretched far beyond England's shores.

Given Lindsey's encouragement, Robinson's warmth, and Jewish acceptance, it might be tempting to think that Frend's move out of the church was somehow easy. It was not. There are no letters to or from Mary during this tumultuous period, but he undoubtedly knew he could count on his sister's understanding. He had less cause to be sanguine about his father, who he realized felt "severely the disappointment of his expectations, when my prospects of success in life were cut off by the freedom of my opinions," but George Frend once again proved himself more affectionate than severe and continued to support his son.[22] Nonetheless, Frend was insecure. After a year trying to make his way as a Unitarian in Anglican England, he "resolved to divert his thoughts for a time by a tour on the continent."[23]

With hindsight, the summer of 1789 was an extraordinary moment to pick for a soothing visit to Europe. As Frend was setting out, financial desperation was leading the king of France to call a meeting of the Estates General for the first time in more than a hundred and fifty years. By the end of June, that group was fracturing, and its middle-class members were protesting the dominance of the aristocracy and clergy at a Paris tennis court. On July 14, an angry mob in search of weapons succeeded in overcoming the guards at the Bastille. By August, a National Assembly had supplanted the Estates General as the governing body of France. None of this had any effect on Frend as he traveled through Belgium to northern Germany. Emerging from years of religious struggle, he remained primarily focused on ways to be open to the wide variety of religious points of view he was encountering.

The first of these appeared even before Frend's journey had properly begun. In the coach that took him to Calais, the man who had just spent a year exhorting his colleagues to examine the true meaning of their words found himself in a conversation with a woman who was completely unconcerned about them. Far from being distressed or dismissive, in a letter to Mary Frend admitted that "it always gives me pleasure to find among Catholicks & Protestants persons who tho' they rank themselves under one of these denominations scarce know the meanings of the terms & are really better people on that account."[24] Frend's acceptance of the unsophisticated religion of the woman in the carriage was reinforced by the openness he encountered when he stopped in Bruges to visit a friend of a friend, Mr. William Edwards. He was very impressed to find that Edwards had "three preceptors for his children—one an English Clergyman, the second a sound Papist, the third a Philosopher," while he himself adopted "the philosophy of Priestly & the liturgy of Lindsey." Frend reported happily that the company "chatted on Philosophy, on religion, on other subjects with the utmost freedom & did not decide to cut each other's throats when we happened to differ in opinion."[25] In Edwards's home, Frend was heartened to find an expression of reasoned religion that was as large as Robinson's empathetic one.

Even as he admired liberality in others, Frend could find it hard to adopt it in himself. Outside of the shelter of Mr. Edwards's household, Catholicism continued to gall him, as did authoritarian political structures. He attributed his experiences with bad roads to the hierarchies of Europe in which "it is not the interest of any person on the road to provide anything good except the Inns."[26] Even as he applied republican standards to his travels, he seems to have been all but oblivious to the political drama such ideas were fueling all around him. At one point he did comment that "rebellion" is "in fashion at present on the Continent," but the closest he came to encountering it directly was "a little debauch," where a few people "got drunk, broke each others heads

& called out 'Long live the States,'" in a small town close to Louvain. By September he did allow that he would return home through Paris only if he was confident that the roads were safe. Otherwise, he wrote simply, he would return through Holland.[27]

The view from England was more alarmist. "Heaven be praised," Lindsey wrote after Frend's return, "you will hardly believe how many known and unknown to you have been concerned lest you should have suffered for want of health, or amidst the perils and dangers now on our continent."[28] Subsequent history was to show that Lindsey was quite right to be concerned about the dark clouds gathering in Europe, but his immediate fears proved to be unfounded. For the next couple of years Frend and his Unitarian community remained focused on furthering the cause of their Christian church of reason.

Upon his return from Europe, however, Frend found himself faced with a different set of problems. His beloved sister Mary had traveled to brother George's house in Oporto "on the forlorn hope of recovery from a decline" and died soon thereafter.[29] Frend's grief was compounded by the death of his father in January of the following year. Frend faced sorrow in Cambridge as well, when Robinson died in the summer of 1790. With these deaths, Frend's personal center of gravity shifted to Jesus College. Although he was no longer the official mathematics tutor, he remained an important draw for a number of students, including, most notably, Robert Malthus and Samuel Taylor Coleridge. Although he turned his college rooms into a veritable distribution center for radical literature, he remained nonetheless integrated enough in his college to hold the elected office of steward from 1791 to 1792. All in all, Frend's experience in Jesus College stands as a testament to the liberality of Anglican Cambridge in the early 1790s.

It is hard to believe that Frend would have been so accepted had the members of Jesus College known about another project that he was secretly pursuing under their roof. Over the course of the 1780s, one of the recognitions that emerged from Priestley's and Jebb's Scriptural Society was that the King James Version of the Bible was a determinedly Trinitarian work, so in May 1789, Priestley began gathering a group to retranslate it. Lindsey was slow to endorse such a radically subversive undertaking, but by May 1790 he was wrestling with the first chapter of John, which opened with "In the beginning was the Word and the Word was with God and the Word was God [John 1:1]" and proceeded to "The Word was made flesh and dwelt among us [John 1:14]."[30] These lines were crucially important because they had long led both "learned and unlearned to believe Jesus Christ to be a great pre-existent being,

God or one next to him,"[31] but it was hard to know how to proceed. Frend sympathized with Lindsey's challenge from Anglican Cambridge where he was applying his Hebrew to the Pentateuch, while in Birmingham, Priestley was picking his way through the Psalms, Proverbs, Ecclesiastes, Ezekiel, and Daniel.[32] All agreed that bringing the Bible to its true Unitarian state was "a very laborious business,"[33] but nothing could be more essential to their project than a truly accurate translation of Jesus's message.

As a select group translated in secret, the Unitarians continued to pursue political goals in public. Jebb had died in 1786, but his wife, Ann, continued to uphold her husband's views in writing and in conversation. After Pitt's final effort at legislative reform failed in 1785, a new group of dissenting ministers, reforming MPs, and London businessmen came together around the more modest goal of repealing the Test and Corporation Acts. Made law in the religiously fraught years after the Catholic James II had been removed from the throne, these acts constituted the political version of the subscription requirement. They mandated that those who held civil, military, or other public positions had first to prove their loyalty to the state by taking the sacraments in the Anglican Church. It was because of the Test and Corporation Acts that England's dissenters could not vote, hold public office, or benefit from a wide variety of other lucrative and powerful positions.

Repealing the Test and Corporation Acts was an obvious goal for a group of dissenting ministers and businessmen who were being disenfranchised and disadvantaged by their restrictions. Self-consciously less radical than the Constitutional Society, the Repeal Committee nonetheless held the same cyclical view of history as did Jebb and Price. The difference was that they did not see the need to revert to Anglo-Saxon times in order to put their country back on track. From their point of view, the Test and Corporation Acts were a misguided piece of legislation perniciously spliced onto the otherwise sound political system that had emerged from the Glorious Revolution of 1688. The members of the Repeal Committee set to work to support their cause using pamphlets, persuasion, and perseverance, and at first it seemed that they might succeed. In March 1787 their proposal was defeated 176 to 98, but in May 1789 it was voted down by just 122 to 102. The change in these margins supported "a flattering prospect of future success" their minutes crowed as they set to work preparing for the next time.[34]

Lindsey saw the dangers surrounding the biblical translation, but he did not see the ones that attended the Repeal Committee's continued efforts to repeal the Test and Corporation Acts. In summer 1789, as these Englishmen were patiently preparing to submit their legislation to Parliament yet again, their disenfranchised counterparts in France were taking more decisive action. In June, the French Third Estate broke away from the Estates General to form a National Assembly, which on August 26, approved a "Declaration of the Rights of Man" that specified, "No one shall be disquieted on account of his opinions, including his religious views, provided their manifestation does not disturb the public order established by law."[35] England's religious radicals were thrilled. "The revolution in France is a wonderful work of providence in our days, and we trust it will prosper and go on, and be the speedy means of putting an end to tyranny every where," Lindsey exulted as he watched the established Catholic Church in France being stripped of its lands and privilege.[36] In a speech on November 4, Jebb's friend Price chimed in: "After sharing in the benefits of one Revolution, I have been spared to be a witness to two other Revolutions, both glorious." Price called on his hearers to celebrate their own version of the Glorious Revolution, which had rescued England's Protestant dissenters from persecution. Repealing the Test and Corporation Acts was, he concluded, the proper way to restore the country's triumph to its original form.[37]

Price's call to action found listening ears among the Essex Street Unitarians, and in the opening weeks of 1790, Lindsey's letters to Frend included running updates on a third attempt to repeal the Test and Corporation Acts. This effort came to a vote in March 1790, but this time it was defeated with a much larger margin of 294 to 105. Lindsey reported to Frend that the dissenters were not disheartened,[38] and Priestley proclaimed himself "not in the least discouraged,"[39] but they were whistling into a reactionary wind that was building to hurricane strength. None of them saw it coming, but Lindsey, Priestley, and Frend were standing on the brink of a new reality. All of their lives would be turned upside down by events powerful enough to destroy the cyclical view of history that supported their religious and political searches, while assigning a different and violent meaning to the word "revolution."

The winds of change could be felt in the strongest voice raised against the 1790 Reform Bill, that of Edmund Burke, who was the MP from Malton. Burke's opposition came as a bit of a surprise to Lindsey's Unitarians, who had long counted on him not to oppose their basic positions. In 1773, soon after the failure of the Feather's Tavern Petition, Burke had voted in favor of repealing the Test and Corporation Acts because he saw no reason to protect an institution as firmly established as the Anglican Church. But as the Essex Street Chapel Unitarians grew in power over the course of the 1780s, he became more ambivalent. By 1790, events in France had confirmed all of his fears about the dangers of rational dissent, so when the bill was again brought forward in March of that year, Burke firmly voted against it.[40]

Burke's vote itself was arguably less important than the justification he provided for it in *Reflections on the Revolution in France and on the Proceedings in Certain Societies in London Relative to That Event.* The proximate cause that produced this work was the organizational power that he saw coalescing around the reform agenda in groups like the Repeal Committee and the Revolutionary Committee. Within this larger context, the irritant that produced the pearl was Price's November sermon extolling revolution.[41] In his book, Burke rejected Price's argument that the Glorious Revolution was part of an ongoing series of developments toward a more representational government and fiercely denied the relevance of the American and French Revolutions to the politics of England. England was, he insisted, a completely separate and special case:

> Our political system is placed in a just correspondence and symmetry with the order of the world and with the mode of existence decreed to a permanent body composed of transitory parts, wherein, by the disposition of a stupendous wisdom, molding together the great mysterious incorporation of the human race, the whole, at one time, is never old or middle-aged or young, but, in a condition of unchangeable constancy, moves on through the varied tenor of perpetual decay, fall, renovation, and progression.[42]

Burke saw little room for reform in a political system whose essence was an "unchangeable constancy"; even tinkering with the "transitory parts" of a system mysteriously molded by the "stupendous wisdom" of the whole "human race" was a dangerous mistake. There was no place in Burke's England for the talk of reason, of rights, and of liberty that was powering new political approaches in America and in France.

Burke's *Reflections* was as much an argument about words as it was about political systems.[43] Himself a powerful orator who had long constructed his speeches in a linguistic space beyond reason's calls for transparency, he found Locke's insistence that each word must correspond to a clear and simple idea for a statement to be meaningful was simply not true. Words like "virtue" or "honour" did not correspond to Lockean clear and distinct ideas but were nonetheless very effective. The emotive power of Burke's soaring speeches stands as clear evidence of his ability to use such words to move an audience.

In his 1790 *Reflections,* he turned his defense of a rhetorical style into a frontal attack on Lockean reason.

The impossibility of arguing against a text in which there was "no fixed principle to refute" left Price, Priestley, and their fellow reformers spluttering with indignation. They found it all but impossible to respond to a man who by his own admission had allowed himself "to throw out my thoughts and express my feelings, just as they arise in my mind."[44] Price took issue with some of the specific ways Burke had misrepresented him but otherwise resolved to "submit myself in silence to the judgement of the Public";[45] Priestley responded more effusively in a series of letters arguing against Burke's position on *civil establishments of religion.*[46] Neither Price's nor Priestley's efforts had much of an impact because the two men were representatives of a world that was fast disappearing. Events in France were creating a dramatically new political situation in England, and it took an outsider to argue the cause of reason in the face of Burke's *Reflections.*

That outsider was Thomas Paine, who had already proven himself a master of political prose. The arguments in his pamphlet *Common Sense* were central to Americans' decision to stop negotiating with the British, and instead move toward immediate and complete independence. Paine's writing style was an essential component of his remarkable success. It was as simple and direct as Burke's was complex and flowery. This clarity of language was essential to reaching the enormous reading and non-reading public that soaked up his revolutionary arguments. The Declaration of Independence that the Continental Congress adopted on July 4, 1776, may be seen as a product of Paine's *Common Sense.*

When Burke published his *Reflections,* Paine was becoming ever more deeply enmeshed in the revolutionary events that were unfolding in France. In the spring of 1791, he published another pamphlet, *The Rights of Man,* in which he insisted on being clear about the English political system that Burke had spoken of so poetically. In response to Burke's flowing prose Payne briskly asked: "Can Mr. Burke produce the English Constitution? If he cannot, we may fairly conclude that . . . no such thing as a constitution exists or ever did exist."[47] In Paine's view, no government that did not have a clearly defined, written constitution could claim legitimacy beyond its present. The only way to set the English government to rights, he maintained, would be to follow the example of the Americans and the French, by electing a group to cut through all historical confusion by writing a clear English Constitution.[48]

Lindsey's Unitarians were thrilled by the way Paine took the wind out of

Burke's rhetorical sails, but his challenge to the validity of the English Constitution was as far beyond the Unitarian pale of reason as it was beyond Burke's flowery language. Both Jebb's determined effort to clear away the linguistic accretions that were obscuring the pure roots of the British political system and Price's more modest effort to celebrate and promote the Glorious Revolution had assumed the fundamental validity of an admittedly defuse, but nonetheless textually established, English Constitution. Paine's refusal to grant any credibility to that historically based entity undercut Unitarian efforts to use reason to reform the English political system as much as it did Burke's efforts to defend its "stupendous wisdom." Paine's work caught Lindsey's Essex Street community by surprise and placed them on the defensive.

Priestley was among the first whose life was completely upended by reactionary rage. In 1791 he proposed that a local branch of the Constitutional Society should meet in a local tavern on July 14 to celebrate the second anniversary of the storming of the Bastille. As the date approached, however, anonymous literature calling on people to prevent the gathering began to circulate through the town. In response, the dinner was rescheduled for earlier in the evening, and Priestley did not attend, but his absence did not placate the mob. The unruly crowd proceeded to break the windows of the hotel, moved on to destroy Priestley's meetinghouse, and then burned both Priestley's house and his laboratory. The mood was murderous enough that it was fortunate for Priestley they did not find him. After several days moving furtively from place to place under the cover of night, he found sanctuary with Lindsey in London.

Insofar as their goal was to quell dissent, the reactionary mob was not immediately successful. Although Priestley was deeply saddened that all of the manuscripts for the Bible translation were lost in the flames, he was not a man to be easily discouraged. "I doubt not, all will be for good in the end,"[49] he declared firmly, as he returned to preaching to large congregations from the pulpit in Hackney that had been vacated by Price's death in 1790.

But that was just the beginning. In February 1792 Paine printed a second part of the *Rights of Man,* which was even more inflammatory than the first. His attack on Burke in the first part had been constructed within the confines of a gentlemanly world that read and discussed arguments about language and constitutional structures. Having spent another year in revolutionary France, Paine broke out of that space to speak directly to the lower classes. The result was electric. Soon after the second part of the *Rights of Man* appeared, the essentially middle-class Constitutional and Revolutionary Societies began to be joined in their calls for reform by a network of Corresponding Societies.

At the first Corresponding Society meeting in a London pub in 1792, a shoe-maker, Thomas Hardy, joined eight other men in calling for a world in which "every adult person, in possession of his reason, and not incapacitated by crimes, should have a vote for a Member of Parliament."[50] With this de-mand, the working-class men of the Corresponding Society began reaching beyond the vision of the original Constitutional and Revolutionary Societies and toward Paine's world, in which all men, even "Tradesmen, Shopkeepers, and Mechanics,"[51] would have a voice in their government. Within a year membership in the London Corresponding Society had grown into the thou-sands, and there were branch groups all over the country.

This was a dramatically new development. Pitt had been willing to work with Jebb and Price in support of a more representative government in the 1780s, but none of them had been thinking in terms of a government of shoe-makers. It was not hard to see that the ideas being promoted in England's Cor-responding Societies were the same as those which had completely destabi-lized France. By fall 1792 the English government was beginning to prosecute people for advocating "Jacobin" ideas. All over the country reactionary groups burned Paine's books and hung him in effigy. A radical attorney, John Frost, was sentenced to eighteen months in prison for saying in a coffeehouse, "I am for equality.... Why, no kings!"; a printer was given four years for publishing a work of the Constitutional Society; a bookseller received eighteen months for selling the *Rights of Man*.[52] Pitt's government was convinced that the country was in danger and that controlling radical ideas was necessary to keep violent revolution at bay. It was to be almost forty years before the next bill to repeal the Test and Corporation Acts was introduced into Parliament. In the mean-time, the cause of reason in England had a long and winding road to travel.

# CHAPTER 6

## Trials in Cambridge

Frend was at first protected from the growing reaction by the isolation of Cambridge, but by the winter of 1792, conflict was seeping in there as well. An unfortunate grocer in the town, who was said to have spoken against the government, was flamboyantly hung in effigy. "Church and King" mobs gathered on dark winter evenings to break the windows of known dissenters as well as those of the Stoneyard Baptist Chapel where Robinson had preached. Some college fellows worked with the magistrates to disperse the rioters, but others approved their zeal. Groups formed within the university to oppose "Republicans" and "Levelers"; those against Pitt's government were stigmatized as "enemies to the constitution"; and on the last night of 1792, Paine was burnt in effigy on Market Hill.[1] Frend's Cambridge, like the rest of England, was becoming divided against itself.

In January 1793 Frend decided to introduce a note of moderation into the discussions that were convulsing his community and his nation with a pamphlet entitled *Union Recommended to the Associated Bodies of Republicans and Anti-Republicans.* His stated goal in this work was to calm the troubled waters by bringing the various contending parties nearer to one another. If the republicans would be moderate in their demands and the anti-republicans willing to accept some reform, he counseled, the result would be a government that could stand "as a center of union," a model of unity working to improve the country and increase "public happiness."[2] As he made these suggestions,

Frend presented himself as a reformer "in the true sense of the word," which meant that "the things to be reformed had been previously in a better state; and that the intention of the reformer is to bring them to their original destination."[3] He devoted the bulk of his pamphlet to considering ways to rectify various inequities and confusions that had crept in to obscure the basic validity of the system established in the Revolution of 1688. In all of these ways, Frend clearly identified himself with the moderate republicanism of the Repeal Committee.

Bringing the clarity of reason to the language of the English legal system was of central importance to Frend's program. He declared himself happy to loudly celebrate the virtues of the English Constitution, but he found the legal language of his country to be a disgraceful example of purposeful obfuscation. Misguided efforts to avoid disputes through the use of "infinite circumlocutions" had led to the situation in which "scarce an act of our legislature is intelligible to a man of tolerable capacity and the jargon of a profession, which ought to use the clearest and best terms, is now become proverbial."[4] The issue of clarity was particularly pressing when it came to laws that affected members of the lower classes. These laws, he insisted, "should be equal, clear, and decisive, such that a school-boy might read them, and be brought up with a sense of their propriety, and fear of offending them."[5] Frend was not critical of the content of English law. His criticism was directed against the pernicious unclarity of the way it was written, and he called on his countrymen to unite in reforming the language of the law in such a way that everyone could understand its workings.

Having thus laid out his vision of legal reform, Frend turned his attention to the relations of church and state. As he did so, he self-consciously avoided doctrinal issues in order to focus on the Anglican Church "only as a political institution."[6] From this perspective, the "believers or pretended believers of certain doctrines, and the dissients [*sic*]" became simply "political factions,"[7] while the Test and Corporation Acts became ways to control those factions. Frend argued that there was really no reason to believe that dissenters posed any danger to the powers of the established church, which meant that discriminating against them was "disgraceful in the extreme."[8] That the oppressors were purportedly Christian made the situation every more reprehensible. "It is time to cast away the leaven of party spirit, and to act as christians," he scolded.[9] For him, repealing the Test and Corporation Acts was the way to bring all of England's people together.

None of these arguments was particularly striking. His father's friend,

Richard Farmer, who was still the staunchly Tory master of Emanuel College, pronounced Frend's pamphlet "a poor business" but was convinced that it "certainly would not have been noticed at any other time."[10] However, the first months of 1793 were not like "any other time." On January 21, the French king, Louis XVI, was executed by order of the National Assembly; on January 24, the French ambassador was expelled from monarchical England; on February 1, France declared war on England, and as Frend was putting the finishing touches on his pamphlet, he realized that his country was just about to respond in kind. Horrified by the prospect of impending war, Frend hastily composed two appendices and added "Peace" to the title, so his pamphlet became *Peace and Union Recommended to the Associated Bodies of Republicans and Anti-Republicans.*

The first of Frend's last-minute appendices expanded his argument beyond "Union" to include the "Peace" that appeared in the new title of his published pamphlet. In "On the execution of Louis Capet" he argued that the execution of Louis XVI did not justify war with France; after all, he pointed out, the English had themselves killed a king only a hundred and fifty years earlier. The French should be allowed to do the same, he asserted, and in the long run the outcome might well be the same comfortable one that the British now enjoyed. This was a relatively standard argument among the dissenters at the time, and Frend's rendering of it was no more noteworthy than the rest of his pamphlet.

The same could not be said of Frend's second appendix: "The Effect of the War on the Poor." In this pacifist lament, he took the side of the poor in ways that violated a number of tacit agreements about the respectable limits of gentlemanly discussion. The essay turns on two short sentences Frend overheard as he was walking to St Ives to check on the proofs of his pamphlet. He was at the time in conversation with two men who were deciding that, with the war coming on, they were going to reduce the wages they paid for spinning. As the three men were discussing this development "a groupe of poor women going to market" overheard what they were saying and exclaimed, "We are to be sconced [docked] a fourth part of our labour. What is all this for?"[11] It was a simple question, easily ignored by the middle-class men who were about to cut the pay of their workers. But Frend's years teaching in Maddingly and conversing with Robinson had opened his eyes to the realities of the world of these peasant spinners. He saw immediately the horror of what a 25 percent pay cut would mean to them.

Frend also saw the enormous gap that separated these market women from

the men who were reducing their wages. He understood that the issues pro-
pelling the English into war with France—"the beheading of a monarch,"
"the navigation of the Scheldt"[12]—meant nothing to women whose lives were
dominated by a desperate attempt to make ends meet. At the same time, he
knew that the men who were deciding to enter that war "know not what a
cottage is, they know not how the poor live, how they make up their scanty
meal."[13] Members of Parliament did not understand what it meant to lose
"three-pence in the shilling for spinning" any more than the market women
understood what it meant to disrupt English commerce on the continent. For
Frend the solution to this impasse was obvious: let those who are supporting
the war "be sconced one fourth of their annual income to defray the expense
of it."[14] But he remained realistically and miserably aware that those who
stood to benefit from the war would not be the ones to pay for it, and that
"years of calamity" would pass before "a single dish or glass of wine will be
withdrawn from the tables of opulence."[15] He returned from his trip to St Ives
in despair. "Let others talk of glory, let others celebrate the heroes, who are to
deluge the world with blood," he cried, before he closed in solidarity with the
market women by echoing their question: "We are sconced three-pence in the
shilling, one fourth of our labour. For what?"[16]

Up to this point, Frend's *Peace and Union* could be read as the work of
a republican reformer, but when he respected the voices of some illiterate
market women, he effectively crossed the lines that separated moderate reform from the radical world of Paine. In the body of his pamphlet, Frend
had called on his countrymen to recognize that "the poor are the instruments
of the ease, comfort, and luxury of the rich, and it would be contrary to the
temper of englishmen [*sic*] as well as the spirit of Christians to be ungrate-
ful to those, from whom we all derive our support," but these words constituted a call to a gentlemanly audience.[17] When Frend brought the voices of
the poor directly into the discussion he essentially challenged the assumptions
that underlay this kind of conversation. "The Effect of the War on the Poor"
echoed the works of Paine with a clarity and eloquence that burst beyond the
politesse of gentlemanly discussion. The result was explosive. Mr. Watson, a
fellow of Sidney College, immediately accused Frend of overlooking and mis-
representing the shifting prices that were always involved in setting wages, and
the two bickered to the brink of a duel over who was qualified to discuss the
economics of spinning. Their argument is stark evidence that giving voice to
spinners was subversive enough to threaten the fundamental rules of gentlemanly discourse.

In the fearful world of England in 1793, it was more as well. Farmer spoke for many when he read Frend's appendix as an effort "to call up the mob."[18] When understood in this way, Frend's cris de coeur became an example of "seditious libel," a legal term for cases in which the king or state was libeled in ways that threatened civil disruption on a grand scale.[19] Others agreed with Farmer that this was the proper way to understand Frend's appendix. On February 21, just ten days after the pamphlet first appeared, England officially declared war on France. Afraid of being held responsible for the material they were disseminating, booksellers and publishers began tearing the offending appendices out of *Peace and Union*.

The reaction in Cambridge was swift. On February 22, just one day after England declared war on France, five fellows of Jesus College called on the vice-chancellor of the university to take action against the author of a pamphlet "written with the evil intent of prejudicing the clergy in the eyes of the laity, of degrading in the public esteem the doctrines and rites of the established church, and of disturbing the harmony of society."[20] Within days, a group of twenty-seven fellows, which Frend insisted on calling "the Cubicks" because their number was $3^3$, began combing university statutes to build a case against him. Even as they did so, the five fellows who had sounded the initial alarm pursued an independent case in Jesus College. There was little to no precedent for either of these actions, but reason and precedence had become irrelevant. England had just gone to war with France, and the Cambridge establishment was anxious to demonstrate its loyalty to the crown.

In Jesus College, Frend's detractors found it difficult to consolidate support for their position; only ten of the thirty-two fellows came to the meeting the master, William Pearce, called on April 3 to discuss Frend's position. Many of those who did come were motivated by animus, however, and by the end of the evening, this group decided in a six-to-four vote "that Mr Frend be removed from the college, that is from the precincts of the college, and from residence in it, till he shall produce such proofs of good behaviour, as shall be satisfactory to the master and major part of the fellows."[21] The proceedings that led to this sentence remain hidden behind the closed doors of the meeting, but the outcome was clear.

Frend immediately appealed to the same James, Bishop of Ely, who had, two and a half years before, upheld the college's decision to remove him from his position as tutor. He objected on two grounds: first, that he was subject to a decision agreed to by less than a fifth of the college's thirty-two fellows, and second, that he was subject to a decision come to without giving any specifics about either the objectionable parts of his pamphlet or the college statutes he had broken.[22] Surely the bishop could see that the procedure offended the clear reason of law.

The master responded to Frend's argument in a very different tone. In his letter to the bishop, Pearce recognized that only six fellows had actually endorsed the decision, but he nonetheless insisted that it "was virtually passed by a majority of the fellows."[23] He then explained that Frend's request for specificity was an inappropriate attempt to introduce reason into a procedure that was properly based on community standards of conduct. It was unnecessary for them be specific about Frend's transgressions because the college position rested on "the whole body of the statutes," not on one, and was supported by "the general design and intention" of its formation.[24] The college had "an inherent right" to punish members who acted *contra bonos mores,*" that is, against the customs of the community, the master insisted.[25] Reasoned argument had no part to play in such a situation.

The bishop paused. He was not ready to decide between Frend's quest for reason and the master's defense of community standards until he had seen the results of the larger case that was building against Frend in the university. So, for the moment, he ruled that Frend could remain in his rooms in Jesus College.

Central to the university-level proceedings was Isaac Milner, the vice-chancellor of the university and master of Queens College. He was a formidable adversary. Just a few years older than Frend, Milner had been rising one step ahead of him throughout Frend's university career. Milner had entered Queen's College as a sizar in 1770 and finished his undergraduate studies as first wrangler and first Smith's Prizeman in 1774. Within a year of Frend's entry into Christ's in 1775, Milner became a fellow and tutor at Queens; by 1778, he was also the rector of St Botolphs in Cambridge. Milner's particular academic strengths were in mathematics and science, and here he accomplished more than Frend. He was elected a member of the Royal Society in 1780 on the strength of two papers on astronomy submitted in 1778 and 1779. By 1782, he was giving lectures on "chymistry" and in 1787 was named the Jacksonian Professor of Chemistry at Cambridge. In short, Milner was an academic star.

Even as he rose at Cambridge, Milner was in his way as far removed from polite Anglicanism as Frend was. He was an evangelical Christian who held his religious views every bit as passionately as Frend held his, but those beliefs were very different. Instead of Frend's fundamentally compassionate God, who responded lovingly to the efforts of his reasoning creatures, Milner's God was essentially inscrutable. It was only with the gift of God's "constantly superintending grace" that anyone can hope to remain on "the path of Christian holiness."[26] There was no place for reason in this relationship.

One result of the evangelical view of the relation between God and humans was that although Evangelicals disagreed with many of the Thirty-Nine Articles, they had no trouble remaining within the Anglican Church.[27] Because theirs was a religion of faith as opposed to reason, they were quintessential latitudinarians who could say virtually anything before their bishops without experiencing the crisis of conscience that had propelled Lindsey and Frend from their comfortable church positions. For Evangelicals, any merely verbal protestations of belief were essentially irrelevant.

Even as the distinction between reason and faith theologically divided late eighteenth-century Evangelicals from Unitarians, the two groups were joined by their refusal to let religion be merely a polite commitment. A notable example of evangelical activism may be found in Milner's student William Wilberforce, who devoted his life to the abolition of the slave trade. In this area, the convictions of Frend and his fellow Unitarians meshed completely with those of the Evangelicals; the only trace of public action that has survived from the year before Frend published *Peace and Union* is a proposal he brought to the Cambridge University Senate in support of Wilberforce's first, and unsuccessful, effort to pass an anti-slavery bill through Parliament in 1792.[28] But in subsequent years, political reaction in England accentuated the differences between the Unitarians and the Evangelicals. In their efforts to create a more democratic society, Frend and his fellow Unitarians relied on a God who supported the projects of the reasonable beings he had created. Milner, in contrast, defended an inaccessible, monarchical God whose kingdom could be entered only through obedience.

Milner's authoritarian God upheld his High Church Tory view of the world, but the anti-intellectual aspects of his evangelicalism differentiated him from other Tories. So, for example, Farmer was well known for his "deep-rooted dislike of dissenters," but at the same time he was respected as an "honest man" who always "behaved with fairness and impartiality" and welcomed the company of anyone whom "he believed to be sincere and disinterested in their opinions."[29] Milner did not value this kind of discussion with people he did not agree with, and as soon as he became the master of Queens College in 1788, he did all he could to pack the common room with Evangelicals. In

1793, responding to Frend's pamphlet offered him the perfect opportunity to take another step toward making Cambridge conform to his image of the orthodox.

Constructing a legal case from the tangle of tradition and statute that was Georgian Cambridge was a formidable challenge, however. Technically the university was governed by two different sets of rules: those created by royal decree, most of which were Elizabethan statutes, and those created by vote of the university senate, which were called "Graces." These two sets of regulations did not always agree with one another, and by the eighteenth century Cambridge approached them with all the precision they accorded the Thirty-Nine Articles. It took Milner's group of twenty-seven fellows months to settle on two specific rules that Frend had violated: an Elizabethan statute known as *De Concionibus,* which forbade speaking publicly "against religion or to anything pertaining to it,"[30] and a Grace of similar intent, which had been passed by the university senate in 1603. On May 3, William Frend was called to appear before Milner and a group of the heads of college in the university's Senate House to defend himself from the charge that in *Peace and Union* he had violated these regulations.

There followed a makeshift trial, the account of which is filed under "Rare Trials" in Harvard University's law library. It's not a bad classification. Milner was a formidable opponent, but he was acting as the judge, and Frend's prosecutor, or "promoter," was Dr. Kipling, who was totally unsuited to the task. Frend, for his part, was a defiant defendant, whose stated goal was "to hold [his accusers] up to the ridicule and contempt of the audience."[31] The students in the balconies cheered and hooted their support as he challenged everything, from seating arrangements to the propriety of his accusers' clothes.[32] In the evenings, roaming bands of students emblazoned college walls with cries of "Free Frend!" and lit gunpowder to burn "Frend and Liberty" into the Trinity College lawn.[33] Through it all, Milner, who was so massively overweight that it took a special chair to hold him, sat impassive. He could be formidable in debate, but for most of Frend's trial, his role was as all-powerful judge who stood above anything that transpired around him.

In the midst of the chaos Frend also pointed out repeatedly, but to no avail, that violations of the Elizabethan statutes were supposed to be heard by the masters of the university rather than the group gathered in the Senate House. This group was appropriate for hearing violations of a Grace, but in one of the more dramatic moments of the trial, he revealed that the 1603 Grace under which he was charged had never been formally added to the university's books. Again and again, Frend showed that Kipling and his accusers did not have a clearly reasoned legal leg to stand on. Milner, however, was unmoved, and the trial moved inexorably onward.

Kipling faced similar difficulties establishing that Frend's calls for reform in *Peace and Union* were "against religion." Everyone knew that the real issue lay in the question of whether "The Effect of the War on the Poor" constituted a dangerous work of seditious libel, but those issues were technically off topic. They crept into the debate anyway, and Frend's quarrel with Watson over the economics of spinning took up close to half a day. Still Kipling was hard-pressed to show how anything Frend had written in "The Effect of the War on the Poor" was relevant to the charge that he had written "against religion."

In fact, over the course of four days of examining witnesses, Kipling had considerable trouble finding anything that *was* relevant. He brought in a stream of witnesses to establish that William Frend, whose name appeared on the title page, was actually the author of *Peace and Union*. Then he tried to prove that the man sitting in the Senate House was actually the William Frend who was a fellow of Jesus. At one point, lacking the record book to prove that Frend had a master's degree, Kipling turned to the testimony of a student "who was examined for his bachelor's degree by Mr. Frend, and as none but masters can examine for that degree, it follows that Mr. Frend was a master of arts." At this point, Milner objected that the proof was inadequate and adjourned for the day to give Kipling time to produce the university register that included Frend's degree. Frend watched it all in amazement. "The English are famous for that species of humor called caricature," he commented, but Kipling "with his groupe of familiars, delineated to the life, would exceed the boundaries of art."[34] The stakes were nonetheless high, and he needed to defend himself. Finally, after sitting through days of Kipling's machinations, Frend and Milner went head-to-head.

Frend was up first. He opened his address with an impassioned defense of his religious integrity. Surely, he protested, a man who firmly believes "there is but one living and true god, everlasting" could not be called an atheist. And surely, a man who "grounds his hope of salvation solely on Jesus Christ" could not be called an infidel. A close reading of Frend's positions reveals his anti-Trinitarian slant: his "living and true God" is one, not three; his savior is "a person sent from heaven," not a part of God. But in speaking to an audience of Cambridge Anglicans Frend stressed that he subscribed to the "essential points of a christian's [*sic*] faith" that there was but one true God, who had sent his son, Jesus, to save mankind.[35]

Having thus argued that he was a true Christian, Frend turned to what he described as the third essential element of his creed: "Thou shalt love thy neighbor as thyself." For Frend this meant that every Christian was required to "love his fellow creatures of every sect, colour or description." Disciples of Jesus included people of all ages, languages, and colors, who lived all over the world, he pointed out. Their beliefs and practices varied widely, but one essential doctrine bound them all together: "Wherever they are, they cannot persecute for opinion, they cannot treat their neighbour injuriously for any religious persuasion, they are connected together solely by the ties of universal love." This inclusivity, Frend insisted, was what had been forgotten by his prosecutors in the Senate House. The trial in which he was engaged amply demonstrated that members of the Anglican Church would "persecute for opinion" and "treat their neighbors injuriously." Clearly, Frend asserted, Milner and his coterie were not displaying those "sentiments of universal benevolence" to which they, as Christians, were "bound."[36] As he fought for his right to remain among them, Frend called upon his prosecutors to act as loving Christians should.

Having thus defended diversity as a fundamental Christian mandate, Frend went on to argue that the same kind of inclusivity was fundamental to a university. In the religious context, the position was mandated by Christ's teachings of love; in a university, it was grounded in the critical, reasoned pursuit of truth. "I have been long, Sir, of opinion, that truth cannot suffer by the fullest discussion,"[37] he explained to Milner. On the contrary, the only way for someone to come to the truth was through a process of discussion, in which it was approached from all sides. Such openness could only happen in a community that applauded everyone's research, encouraged scholarly zeal, did not hold mistakes up to public censure, but rather encouraged its members to correct their ideas when they were shown to be wrong. This was the kind of university that had supported the innovative thinking of men like Newton and Locke. It was the kind of university that would not fear being "overset by a shilling pamphlet."[38] It was the kind of university that could stand firm against the forces of unthinking reaction that were washing over England.

When Milner responded several days later, he left Frend's religious charges to be adjudicated by an inscrutable God. He proclaimed himself unafraid of those who "openly attack" religious principles but concerned about the danger presented by "those who are perpetually talking of candour and liberality, of thinking for themselves, of examining things thoroughly." Milner could see that this kind of critical discussion was "exceedingly captivating to the unsuspecting minds of youth," but truth was a delicate sprig that had to be carefully cultivated in order to thrive. The rough and tumble of reasoned discussion would trample the truths that students were just glimpsing, crippling their delicate understandings before they had a chance to grow strong. Respecting the wisdom of their elders was the only way the next generation could hope to come to that wisdom themselves. Whatever their hopes for the future, he charged the students to study hard, obey their tutors, attend church services, and, most of all, "take it for granted, that our forefathers had some good reason for steadily adhering to, and supporting, these venerable institutions."[39] Milner's truth needed the protection of a strong university in order to flourish. There could be no place for market women, for reason, or for Frend within its sheltering walls.

Whatever the respective values of Frend's and Milner's positions about the nature of truth, reason, and faith, there was no question that Milner had all the power in the Senate House, and he was comfortable using it. Having spoken his piece, he directed Frend to publicly confess his "errour and temerity" by signing a retraction. Frend said he would not sign anything that did not clearly spell out his error.[40] But Milner was not about to listen to any more objections. He summarily adjourned the proceedings in order to give Frend time to consider. When the court reconvened three days later and Frend proclaimed he "would sooner cut off this hand than sign the paper,"[41] the students cheered, but Milner declared him "therefore banished from this university."[42] Frend immediately appealed the decision to the members of the Court of Delegates, but on June 29 they upheld it, and on July 13 the bishop of Ely fell in line. Frend determined to take his case to a yet higher court, and by mid-September he had presented his version of the story to the House of Commons in *An Account of the Proceedings in the University of Cambridge, against William Frend M.A.* In the event, the House of Commons was as unwilling to interfere with the university's decision as had been any of the other authorities to whom Frend had turned for help. Frend's banishment stood.

Frend composed his *Account* in the quiet of John Hammond's house in the neighboring village of Fenstanton, where he was recovering "from the fatigues of academical warfare."[43] There he could see that he had not been abandoned. Robert Tyrwhitt from Jesus College, James Lambert, the head tutor of Trinity College, and Thomas Jones, a fellow of St John's had all sat with him throughout the trial. Lindsey's friend Reynolds donned academic robes to stand beside him on the day of his defense; his cousin Herbert Marsh refused to testify against him; and behind closed doors, Farmer honored his family friendship by being "instrumental in the mildness of the sentence."[44] And, under the circumstances, Frend's punishment was in fact remarkably mild. Being "rusticated" from Jesus meant that he could not reside there; being banished from the university meant he could no longer use its libraries. Nonetheless, he remained a fellow of Jesus College until his marriage in 1808, and a proud member of the university senate for the rest of his life. More immediately and concretely relevant, his college fellowship continued to carry a stipend, which remained his primary income for the next fifteen years.[45] In short, Frend may have been banished, but he had not been cut off, and his many friends and supporters would do their best to stand by him in the years to come.

Nonetheless, they could not reinstate Frend in the Cambridge community where he desperately wanted to stay. Throughout the summer, he continued to make needling appearances in the university that had rejected him. When Kipling appeared as the featured speaker at commencement, Frend took the opportunity "to walk up and down the Senate House as usual and laugh at him and his folly."[46] When in August he found the gates of Jesus College closed against him, he established his "perfect contempt of the master's orders"[47] by ringing the bell and talking himself past the porter. Frend presented his forays as a way to discredit his accusers by exposing "their folly to the utmost,"[48] but others saw it differently. "This poor man still hovers about the *University,* and now and then attempts to break into the College. He cries out for persecution," Farmer commented.[49] Pearce agreed and placed an iron chain across the door of Jesus College to keep Frend out.[50] Sadly alone with the reality that he could not get in, Frend finally gave up. At the end of September 1793, he left his Cambridge home and moved to London.

# PART TWO

# Defining Reason

# CHAPTER 7

# Trials in London

When William Frend arrived in London in 1793, he was almost thirty-seven years old and had spent all of his adult life in the small university town of Cambridge. The contrast between the two places could not have been greater. Throughout the eighteenth century, England's capital city was so filthy and overcrowded that annual death rates consistently outnumbered birth rates, sometimes by as much as two to one. Nonetheless, in the ten years since Frend's letters had guided his sister through the city's neighborhoods, rivers of migrants from the countryside had increased its population by more than 10 percent, or 60,000 people. By 1800, London's more than a million inhabitants made it twice as big as Paris and eleven times as big as Liverpool, the next largest city in England.[1] Frend entered this world with George Dyer at his side. An entirely unworldly, shy, stuttering Cambridge graduate, whom Robinson took in as a tutor to his children, Dyer had transferred his allegiance to Frend upon Robinson's death and was Frend's constant companion as he tried to find his place. Lindsey, who welcomed Frend into this sea of humanity, was also going through a major adjustment. In the July that Frend spent resisting his exile, Lindsey had retired and passed the ministry of the church he had founded to his brother-in-law John Disney. In the years that followed, Frend became an active member of a London-based radical community that stretched beyond the Essex Street Unitarians to include William Godwin, Mary Wollstonecraft, Thomas Holcroft, and Mary Hays.

Frend's life in London was filled with a politics as alien to him as was his new abode. In July, as he was digesting his situation with the Hammonds outside of Cambridge, a dissenting minister in Plymouth, William Winter- botham, was sentenced to four years in Newgate Prison for "seditious" state- ments he made in two sermons. At just about the same time, Thomas Muir and Thomas Palmer, who were leaders of a Scottish reform group were tried and sentenced to fourteen years of exile in New South Wales. That fall Mau- rice Margarot and Joseph Gerrald suffered the same fate.[2] The Essex Street Unitarians did what they could to offer support to those caught up in these trials—Winterbotham named one of his children Rayner, another Lindsey in gratitude for their help—but they were essentially powerless, and the worst was yet to come. Even before the boats departed to carry the Scottish pris- oners away, the reactionary forces in Pitt's government were turning their attention to radicals closer to home. Priestley did not wait to see what would happen next, and in February 1794, he fled England for America. His fears seemed justified when, three months later, the government jailed the leaders of the London Corresponding Society (LCS). In the fall of that year, those leaders—Thomas Hardy, Horne Tooke, and John Thelwall—were all tried for high treason. The stakes couldn't have been higher. In theory, if convicted the defendants faced a gruesome combination of genital mutilation, hanging, disembowelment, beheading, and quartering; in fact, they would probably just have been hanged. If the prosecution succeeded in condemning their first set of prisoners to this fate, the rumor was that they had a list of at least two hundred more—including Frend—who would follow. Pitt's government was completely committed to ensuring that revolution was not going to gain any ground in England.

The Lockean directive that for reasoned discussion words must be tied to clear and distinct meanings lay at the heart of the London Treason Trials.[3] The first defendant, Thomas Hardy, was accused of violating a statute dating to 1351, in which Edward III had declared it high treason "*to compass or imagine the Death of the King, provided such Compassing and Imagination be mani- fested by some Act.*"[4] In October, the lord chief justice of His Majesty's Court of Common Plea, Sir James Eyre, explained his interpretation of these words in his charge to the grand jury considering Hardy's case. Eyre claimed that Hardy—and by extension, the rest of the defendants—were guilty of "*entering into Measures which, in the Nature of Things, or in the common Experience of Man- kind, do obviously tend to bring the Life of the King into danger, [which is] also com- passing and imagining the death of the king.*"[5] Eyre's specific, italicized words and his interpretations of their meanings were essential to his position. For the lord chief justice and the government he represented, the meaning of the "Nature of Things" or the "common Experience of Mankind" had been radically altered by the fact that the French had, just the year before, killed their king soon after forming a republican government. The prosecution claimed that by advocating a more representative government, Hardy and his cohort had engaged in acts which did "obviously tend to bring the Life of the King into danger." Under Eyre's interpretation, anybody who called for any kind of republican government after Louis XVI's execution could not help but be "also compassing or imagining the Death of the King" and was therefore a treasonous regicide.

Eyre's interpretation of what it meant *"to compass or imagine the death of the King"* was so broad and open-ended that it easily included not only all members of the LCS but those of the Constitutional Society, the Revolutionary Society, the Repeal Society, and Frend. In fall 1794, the radical writer William Godwin was forced to confront this pernicious extension when his close friend Thomas Holcroft was added to the initial group of three arrested for treason. Godwin had already established himself as a major voice in the radical community with *An Enquiry Concerning Political Justice and Its Influence on General Virtue and Happiness,* which was published in London the day after Frend's *Peace and Union* appeared. In his *Enquiry,* Godwin argued that reason was strong enough that governments would wither away to be replaced by a world in which reason alone would operate. Within the year he presented his thesis in action in his novel, *Caleb Williams.* Both publications were very successful and earned him a substantial following.

In real life, however, it was becoming ever more difficult for Godwin to maintain his optimistic views. Even as reason was triumphing in the world of *Caleb Williams,* its author spent the summer of 1794 doing what he could to support the Scottish defendants and their families. He was with friends in the countryside recovering from these efforts when he heard that Holcroft was to be tried for treason. Horrified, he rushed back to London where he locked himself in his house for two days to compose a rebuttal to Eyre's interpretation of what it meant "to compass and imagine the death of the king." He emerged with *Cursory Strictures on the Charge Delivered by Lord Chief Justice Eyre to the Grand Jury, October 2, 1794,* which was published the next day both in the *Morning Chronicle* and as a separate pamphlet. His was a powerful defense of the need for strict literal readings within English law.

Godwin's primary argument echoed a grievance Frend had leveled against his accusers less than a year before: "Surely the law ought to be definite and clear: tell us what propositions are criminal, and the authours will then know how far it may be prudent to speak the truth."[6] Godwin too objected to an approach that left the implications of law indistinct and unclear. "Are we to understand that henceforth the man most deeply read in the laws of his country, and most assiduously conforming his actions to them, shall be liable to be arraigned and capitally punished for a crime, that no law describes, that no precedent or adjudged case ascertains, at the arbitrary pleasure of the administration for the time being?" he asked. To the extent that the answer to this question was yes, the law was nothing more than a "mere trap," whose "apparent clearness and definition" simply provided a false sense of security in the midst of a vast field of legal landmines.[7] Godwin's insistence that the tight correspondence between words and their meanings be maintained in the law proved decisive in the trials of Hardy, Thelwall, and Tooke. When none of them were found guilty, all of the other defendants, including Holcraft, were freed. In the law at least, the principle of clear definition had triumphed.

Godwin and Holcroft may have prevailed in the trials, but both were deeply chastened by their experience and pulled back from further participation in radical activism. Frend did not. He joined Thelwall, Francis Place, and John Binns as leaders of a revived LCS which grew throughout the summer of 1795. In June, he spoke to the group when tens of thousands—the LCS claimed 100,000—gathered in St George's Fields in support of republican demands for annual parliaments and manhood suffrage. Provincial centers saw similar events—10,000 were reported at a Sheffield gathering—and pamphlets poured from innumerable provincial presses. It seemed the LCS had emerged from the treason trials stronger than ever.

There was little cause to celebrate the continuation of the LCS, however. All the reasons for its strength were dire. The fall of 1795 can be seen as the beginning of the "years of calamity" that Frend had foreseen in "The Effect of the War on the Poor." As the year wore on, poor harvests combined with the necessity of feeding far-flung troops meant that England began facing severe food shortages. By the end of July, Richard Reynolds was writing to Frend from Cambridge about mobs that were stopping wagon loads of wheat going to port towns in order to keep their contents at home.[8] Frend's market women and their families were beginning to starve.

Frend responded with a pamphlet, *Scarcity of Bread: A Plan for Reducing the High Price of This Article,* that elaborated further on the positions he had initially cried out in "The Effect of the War on the Poor." He challenged those who attributed the problems England's people were facing to natural forces beyond their control. England's ruling classes did have control, he insisted. It was the war that had "occasioned the scarcity by destroying the usual supplies of corn to this country, and by sending immense quantities out of the country."[9] Having thus assigned responsibility, Frend developed an elaborate scheme through which the government could atone for its mistakes by subsidizing food for the hungry. Frend's *Scarcity* had no real impact, but Reynolds responded to its publication with a warning that this kind of pamphlet might well "bring inconveniences" to its author.[10] In summer 1795, however, Frend seems to have judged correctly that the triumph of reason in the treason trials gave him enough leeway to speak his mind without retribution.

Nonetheless, Reynolds was right when he cautioned that the government was not going to put up with dissent much longer. By fall 1795, people were starving in the streets of London. A protest in Copenhagen Fields, Islington, on October 26, attracted on the order of 150,000 people, and three days later, a crowd of as many as 200,000 mobbed the monarch as he rode through the streets to open Parliament. At the end of a harrowing ride, the king claimed someone had shot at his carriage, and the response was immediate. On November 10, Pitt introduced the "Two Acts." The first made it a "treasonable offence to incite the people by speech or writing to hatred of King, Constitution or Government"; the second required that magistrates with broad powers of arrest and dispersal be present at any gathering of more than 50 people.[11] Frend remained obdurate, determined to resist as long as he could. On December 7, he stood beside Thelwall at the last gathering of the LCS to argue that the proposed acts were unconstitutional, that they "should be disregarded," and that "juries, before whom persons may be brought, under the provisions of the proposed Bills, were bound as Englishmen to shew their contempt of the bills, by acquitting the accused persons."[12] But after the king signed the "Two Acts" into law on December 18, the Corresponding Societies disbanded and even Frend said no more. The law was now clear, and reason would not support his breaking it.

These dramatic events dominated Frend's first two London years, but as he faced them, he was negotiating other issues as well. In Cambridge, he had lived in a society of highly educated men, but in London, he was faced with every conceivable kind of person. Difference was always uncomfortable for eighteenth-century radicals, who found it hard to reconcile the view that all people were essentially reasonable with the reality that they could disagree in essential ways. Frend had begun to confront some of these challenges while

working with Robinson and in London's Jewish community, but the city provided innumerably more. That people were all defined by their reason meant
in theory that they were all essentially alike, but in practice they were not. In
fact, even within the relatively tiny group of London's radicals, Frend's Anglican upbringing and Cambridge education set him apart. A man who observed
him standing with Thelwall at the final December 1795 meeting of the LCS
described him as "a gentlemanlike looking Man: of good stature and bulk,"
who was "dressed in blue with a white waistcoat; and seemed in appearance ill
suited to those about him."[13] Although Frend remained outside of the Anglican establishment for the rest of his life, he never shed the clothing and de-
meanor of the Cambridge education he had received within it.

Education was always Frend's first line of defense when faced with people
unlike himself. He was ever eager to erase differences by bringing everyone
onto a common learned platform. In the case of the Jews, Frend's leveling
educational impulse took the form of learning their language, but in London
in the 1790s it was far more usual for him to act as the teacher. Francis Place,
a fellow leader of the LCS who was a tailor by trade, remembered spending
hours in conversation with Frend in his shop: "As I knew a little of mathematics and something of astronomy, Mr. Frend took pains to teach me as
much as he could of these two sciences, he also put me forward in Algebra."[14]
Frend's efforts to share what he knew with Place were typical of his approach
to bridging the considerable social and educational differences that separated
him from many of the people he was working with in London.

In addition to discrepancies in education, in London Frend faced differences in gender. He had been living in an all-male community for almost two
and a half decades when he arrived in England's capital city, and day-to-day
interactions with women were a novelty for him. He had been enjoying a
flourishing social life when he was writing to his sister in 1784, and though
details are lacking, the Miss Cotton who worked with him in the Maddingly
Sunday School was clearly an able assistant and trusted friend. But his break
from Anglican orthodoxy dealt a heavy blow to these kinds of sociability. Not
only did it render him highly suspect to the daughters of well-established Anglican families, it disqualified him from the positions in parish ministry that
were the accepted alternative to the celibate life of a college fellow. Some combination of being a socially outcast Unitarian and being confined to a celibate
community meant that nine years after he had looked forward to marriage in
correspondence with his younger sister, Frend was still unattached when he
moved to London.

Frend's approach to women was the same as his approach to all other kinds of difference: education. That he had emerged from his childhood in Burgate Street with a clear recognition that women were his intellectual equals is clear in his letters to Mary. He was always ready to share what he knew with his sister and saw no reason that she could not learn what he had to teach. In his eagerness, he was at once generous and myopic, happy to pour out his knowledge while remaining blithely oblivious to the enormous challenges his sister faced in her attempts to engage with it. Mary couldn't acknowledge the problems either. She saw her difficulties as evidence of personal weakness rather than manifestations of a world that did not value her efforts. The universalizing view of reason offered no clear alternative to that interpretation. When Frend arrived in London, he was still ready to welcome everyone into his world by sharing his knowledge of mathematics, astronomy, and philosophy, yet he was totally unable to see that parts of that world were accessible to him only because of his position of masculine power. It took Mary Hays, a marginalized woman who grew up on the wrong side of the tracks (in this case, the Thames) to recognize the forces that guarded the boundary between the lives of women and the reason that defined Frend's world.

Hays was one of the first people Frend met when he arrived in London. The two of them had already come in contact a year earlier, in the context of one of the many little skirmishes about proper Christian practice that were constantly exercising England's radical religious community. In this case, the issue was the value of public worship, and Hays entered the lists under the pseudonym Eusebia, with a pamphlet entitled *Cursory Remarks on an Enquiry into the Expediency and Propriety of Public or Social Worship.*[15] Frend was thrilled by Eusebia's arguments, and on April 16, 1792, sent a long letter "in hopes of being introduced on my next journey to London to the acquaintance of a Lady who entertains the highest esteem for the writings of revelation and examines them with that freedom of candor described by Eusebia in the first page of her elegant pamphlet."[16] What excited him was the possibility that incorporating women's voices in such discussions could serve to cut through petty masculine wrangling. "So much candor of sound reasoning cloathed in the insinuating language excited in us the hopes that the aid of the fair sex may in future be often called in to soften the animosity and fervor of disputation," he enthused.[17] Little did he know how confusing "the aid of the fair sex" could become.

The Eusebia whom Frend thus welcomed into discussion was in real life, thirty-three-year-old Mary Hays, who had been born to a dissenting couple in Southwark in 1759.[18] The defining episode of her early life came in 1778, when an intense relationship was ended by the untimely death of her fiancé. Nineteen-year-old Hays was completely devastated. Her first glimmer of hope for a life beyond sorrow came in the form of Robert Robinson, whom she encountered in 1781 on one of his rare trips to London. Robinson was immediately sympathetic to the plight of the miserable young woman, and with characteristic warmth, he encouraged her to follow him into the worlds of philosophy and theology. At the end of what became a ten-year correspondence course with Robinson, Hays felt herself to be strong and well-educated enough to dip her toe directly into the world of radical discourse as Eusebia. By the time Frend arrived in London, Hays was becoming friends with Mary Wollstonecraft and a contributing voice in a small community of feminists that was coalescing in England's capital city.

When Hays and Frend first met, he immediately fell into his comfortable pedagogical role. The result was unexpected. Totally captivated by the polished expertise with which Frend engaged in discussions of everything from astronomy to the evils of the slave trade, Hays fell in love. Frend, however, did not. All of the differences of social status that made him so attractive to Hays operated in the other direction for him. Hays's intellectualism may have been in keeping with the view of intelligent womanhood he had expressed in his letters to his sister, but it is difficult to fit Hays's assertive feminism into his image of a wife who could be counted on to "endure the little mortifications of life without peevishness" and would "consider home as the place of her happiness."[19] In short, Frend had far less cause to be attracted to Hays than she had to be attracted to him.

This mismatch in affections placed the onus of building the relationship on Hays, and there were no clear guidelines for how to proceed. She and Wollstonecraft and London's other radical feminists could be scathing about the traditional structures of courtship and marriage, but other options were not clearly laid out. This left Hays with the challenge of finding her own way. Letters had long been her preferred mode of escape from the confines of Georgian womanhood, so when she fell in love with Frend, she turned to her pen. One contemporary observer thought that Frend was initially intrigued by Hays, but whatever he may have felt at the beginning of their relationship soon withered in the face of her epistolary barrage.[20] By early 1796, Frend was determined to break off all contact. A devastated Hays turned for help to William Godwin. Godwin was not an obvious choice for amatory advice; he had not yet experienced love and even among his friends had a reputation for

being emotionally obtuse. He scolded Hays for wasting her time trying to engage with Frend, who was, on this subject, as "impenetrable as a rock."[21] As a
way of moving beyond her misery at Frend's rejection, Godwin suggested that
she write through her feelings in a novel.

The result was *The Memoirs of Emma Courtney,* which Hays published in
1796. The book is fictional, but the situation it describes is a thinly veiled version of that between Hays and Frend. In the novel Emma Courtney falls in
love with a man named Augustus Harley. One of the issues that stands in the
way of Courtney's hopes for marriage is that Harley's income comes from a
bequest, which, like Frend's Cambridge fellowship, will be cut off if he mar-
ries. A more central problem is that though Courtney was in love with Harley,
he does not seem to be in love with her, and in response to his lack of interest,
Courtney writes him long and passionate letters. The parallels continue when
Harley finally refuses to engage any more, and Courtney turns in despair to
her friend, Mr. Francis, who advises her to give up her pursuit of Harley. The
fictional part of Hays's book embeds this dynamic in a larger romantic world
that involves a number of dark nights, pouring rains, heaving bosoms, light-
ning flashes, and a denouement in which a remorseful Harley dies in Courtney's loving arms.

A series of letters from Courtney to Harley and from Courtney to Mr.
Francis lie at the heart of the *Memoirs.* We will never know how much the letters Courtney sent Harley resemble those Hays sent Frend, because neither of
them kept their copies. That Hays's letters in the Godwin archive are virtually
identical to the ones Courtney sends to Mr. Francis suggests that the letters
she sends Harley mirror those Hays sent to Frend.[22] On that assumption, it is
not hard to see why Frend felt the need to cut off their relationship.

The first letter from Courtney to Harley is a rather straightforward declaration of love. When Harley does not clearly return her affections in his response, her biggest challenge becomes to find the reason that he does not love
her even though, as she puts it, "our principles are in unison, our tastes and
habits not dissimilar, our knowledge of, and confidence in, each other's virtues
is reciprocal, tried and established—our ages, personal accomplishments, and
mental acquirements do not materially differ." Confident that the problem
could not lie with her, because so many other people esteem and love her,
Courtney turns to philosophy for answers. If our ideas are formed from external impressions, she wonders "how, then, can I believe it compatible with the
nature of mind, that so many strong efforts, and reiterated impressions, can
have produced no effect upon yours?" Calling upon philosophy as her witness,

she declares that if she fails to secure his affections, "it will be a phenomenon in the history of mind!"[23] Nothing was too large to be relevant to her love.

Courtney's calls for truth resonate with her memories of her first encounters with the man "from whom my mind had acquired knowledge, and in whose presence my heart had rested satisfied."[24] She remembers listening to Harley's lectures and hearing "truths divine come mended from his tongue."[25] She thrills "with the energetic sympathies of truth and feeling—darting from mind to mind, enlightening, warming with electrical rapidity!"[26] When they were together, "every day brought with it the acquisition of some new truth," and she reveled in "'the feast of reason, and the flow of souls.'"[27] It is not difficult to recognize Courtney's descriptions as passionate, emotive versions of the truth Frend and Milner had argued about in the Senate House, but the points at issue between the two men are essentially irrelevant to Courtney's experience. As a woman, she has no place in their institutional context. Instead she has Harley, who holds the key to her experience of truth at least as much in his presence as in his teachings. "Will you no longer assist me in the pursuit of knowledge and truth?"[28] she cries when he begins to brush her off, and when his silence persists, the miserable young woman wonders whether she could still believe that "Truth and Good are one."[29] Whereas Frend's and Milner's discussion of truth was rooted in their relations with the divine, for Courtney it is tied to a man.

On an important level, Courtney recognizes this, and she lays out her position in a discussion of "independence" with Mr. Francis. When Courtney turns to her friend in despair, he chastises her for allowing her feelings to obscure "the first lesson of enlightened reason," which is that "the principle by which alone man can become what man is capable of being, is *independence*."[30] It is not hard to hear echoes of John 8:13, "and the truth will make you free," in Francis's call for independence. But what he does not notice as he translates Jesus's "freedom" into an eighteenth-century notion of independence is that he is essentially moving the meaning of the passage from the experiential and spiritual to the practical and political. This shift is all but invisible to Francis, because for a single man, like him, the "first lesson" of enlightened reason might well be independence. But for a single woman like Courtney, that kind of independence is simply not possible, and she knows it. "This is mockery!" she shoots back; "Why call woman to *independence*," when "the barbarous and accursed laws of society, have denied [it to] her?"[31] Courtney is well aware that a socially unconnected woman like herself could attain the kind of independence a single man could take for granted only through marriage to a respectable gentleman. "To have been the wife of a man of virtue and talents was my dearest ambition, and would have been my glory," she cries.[32] But the relationship between men and women was not at all reciprocal in the 1790s. The best Hays could do to grant her heroine the kind of power Harley had over her was for him to die in her arms.

The call for female freedom that lies at the center of Hays's novel was not widely heard. Hannah Lindsey might seem to be the perfect audience for Hays' arguments because she was every bit as forceful as Hays was and often criticized for it. Nonetheless, her marriage to a respectable gentleman protected her from the vulnerabilities that Hays faced, and she was unimpressed by the *Memoirs.* At least equally relevant to the reception of the novel was Hays's decision to share the raw intensity of her unrequited love. Placing the passionate letters she had written to Frend at the center of the work undercut her claims to be a reasoning person. From virtually the moment of its publication, her book was dismissed as the product of an unseemly woman.[33]

Even though Hays made her letters public, she did not reveal the object of her love, and her silence on this subject succeeded in keeping it a secret from contemporaries like Hannah.[34] Frend, however, certainly knew that it was he, and some sense of his reaction may be gleaned from the woman he eventually wed. His marriage to Sarah Blackburne was less to a passionate lover than it was to a family. All the evidence suggests that the Frend couple was devoted to each other, but there is no reason to believe that Sarah ever expressed the kind of passion that drove the pen of Mary Hays. At least equally to the point, Sarah was the granddaughter of Francis Blackburne and the niece of Hannah Lindsey. The marriage of Sarah Blackburne and William Frend marked the beginning of a new generation in the family of reason, but ten very eventful years were to transpire before that marriage took place.

# CHAPTER 8

## *Reasoning in Uneasy Times*

Notably absent in Hays's story of Courtney and Harley are any references to the banishments, treason trials, demonstrations, and starvation that were so central to London life in the 1790s. Frend, however, was living among them. As soon as the prosecutors in the treason trials failed to get convictions, he set out to reopen the case against his banishment in the House of Commons. In *A Sequel to the Account of the Proceedings against the Author of a Pamphlet, Entitled Peace and Union* he argued that the whole course of his trial was not only contrary to the letter and spirit of the law, but to "every former precedent."[1] Parliament did not see the situation as he did and his plea went nowhere.

Frend had a secondary goal in writing his *Sequel:* to maintain his position among the youth who had taken an interest in his case. "I am not reformed from those opinions, which my adversaries condemn; but, on the contrary am more tenacious of them than ever," he trumpeted.[2] He encouraged the students who had so vociferously supported his antics in the Senate House to continue to improve their minds and to be careful not to embrace any ideas or doctrines they had not carefully examined. He cautioned them not to allow themselves to be lulled into complacency and exhorted them instead to do whatever they could to be useful to society. It might not be easy, he admitted, but "upon the whole, the difficulties ye have to encounter may, to a philosophical mind, render your journey through life as desirable as that which seems to be strewed only with flowers."[3] With these words, he was doing what he could to maintain a position as a teacher and leader of youth.

It is not hard to see Frend's exhortations as a message to himself as well. His Jesus College fellowship was adequate for room and board but very little else. He prided himself on the way he had argued the legal points of his trial, and when he arrived in London he took rooms in the Middle Temple, a London law school that was just blocks from Lindsey's Essex Street Chapel. However, he quickly found himself prevented by what he called "an absurd regulation"[4] from becoming a barrister, and it was a challenge to know what else he could do. He might work as a kind of paralegal or offer assistance to those who "may wish to present the publick with an account of their voyages, their journeys, or of their reflections in any other mode of life."[5] Or he could make himself useful "in the drawing up of memorials, in lectures on elocution, in any other employment suited to a man of letters."[6] In the years that followed he undoubtedly tried all of these approaches, but none of them proved to be either secure or profitable. His banishment may have been a mild punishment for his day, but life as an academic exile in London was not easy. "How I got on is a wonder to myself," he later recalled.[7]

Although Frend began to find occasional work as a private tutor, he needed more than that. In an effort to reach a larger audience and generate more income, he moved into educational publishing. A group of Unitarians, including Richard and Maria Edgeworth and Anna Letitia Barbauld, had begun to write works to promote the development of reason in young children. In 1796, Frend joined their efforts with an elementary textbook entitled *The Principles of Algebra* "for the use of schools, academies, and colleges."[8] In *Principles* his goal was essentially the same as that of the other Unitarian authors: to introduce England's youth to the multifaceted powers of reason.

Frend's choice of subject may also be seen as a way to rescue reason from the forces that made its pursuit so dangerous in turn-of-the-century England. When he studied mathematics at Cambridge, the subject had always been secondary to the theology that mathematics modeled. In *Principles* he was shifting the center of gravity between the two subjects that defined certain knowledge in Lockean philosophy. As he did so, Frend was pointing the way toward an increasingly secular nineteenth-century world in which the purview of reason would shift away from theology and into mathematics.

The issues that attended the pursuit of reason in mathematics were no less complicated than they were in theology. Mathematics is truly a plural word that covers an enormous range of subjects. The form of reason that propelled Frend out of the Anglican Church and guided him through England's political and social landscape was pulled from geometry, but the subject of his textbook was algebra. This choice complicated Frend's mission considerably. The study of mathematics had been held up as the epitome of reason since at least the time of the Greeks, but their mathematics included only geometry and arithmetic. The Greeks said nothing about algebra, because they did not do it. The word itself comes from the title of a twelfth-century work in which Abu Ja'far Muhammad ibn Musa al-Kwārizimī used the Arabic term "al-jabr" to refer to a process of calculating by adding and subtracting quantities to opposite sides of equations. After centuries of travel through medieval Latin and early English, "al-jabr" emerged as "algebra."

Like its name, the subject itself evolved over time, and the recognizably modern study of algebra only emerged at the end of the sixteenth century in Europe. At its heart lay a very powerful system of symbols that could be manipulated to solve an enormous range of problems; by the time Frend was writing his *Principles,* Europe's mathematical practitioners had been using algebra to solve problems in astronomy, navigation, accounting, and engineering for more than two hundred years. At the same time, however, the subject did not fit comfortably with the classical mathematical studies of geometry and arithmetic. Geometry was the study of space, arithmetic was the study of number, but the subject of algebra was not clear. Algebraic manipulation was very useful in many areas, but Europe's intellectuals struggled to explain the power of a set of symbols that did not seem to have clear meanings.

Algebra's early users may be divided into two groups: those who used algebra for practical purposes like surveying, navigation, or commerce, and those for whom mathematics was a model for clear and distinct reasoning. The practical group eagerly adopted and developed the symbols that made their work easier. The more philosophically oriented tried to work out the implications of the symbols' power for their understandings of the nature of mathematical truth. All manner of thinkers, from Jesuits to lawyers to doctors to diplomats to court philosophers, viewed algebra through the lenses of their particular concerns and preoccupations.[9] As they did so, they used a variety of terms and phrases—"analysis," "algebra," "algebraic geometry," "calculus"—to describe the mathematical ideas they were exploring. None of these terms had universally accepted meanings; all were differently defined to fit the needs and intentions of the people who were using them. Looking back, it is truly impossible to find a single coherent mathematical order in the array of results that these men generated over the course of the seventeenth and eighteenth centuries.[10]

Frend's *Principles* fits easily and well within one of the areas of this theoretical landscape: a long-standing English tradition of algebra as "universal arithmetic." In this tradition, the English used algebraic symbols to describe processes and unknowns even as they insisted that, ultimately, the legitimacy of their results lay in the absolutely unchangeable numbers those symbols referred to. "A number may be greater or less than another number; it may be added to, taken from, multiplied into, and divided by another number; but in other respects it is untractable," is how Frend put it. Even if the whole world were destroyed, he explained, "one will be one, and three will be three; and no art whatever can change their nature."[11] This essential point grounded all of his efforts to keep his readers anchored in reality as they followed their reason through the conceptually treacherous world of algebra.

In *Principles,* Frend guided his readers from the most basic instruction in addition and subtraction through a chapter explaining the relations of geometry and algebra, to a consideration of ways to solve cubic equations. Throughout this progression, his intent was to demonstrate how to think with numbers. The first step entailed teaching students to add, subtract, multiply, and divide "without their slates,"[12] while making sure that they were always using the proper words to describe the operations they were performing. Once this conceptual grounding was clearly established, the second step was to introduce the basic symbols used to describe the operations they had been performing. As Frend's students made the move from mental to written arithmetic, it was centrally important that they read these symbols in ways that accurately reflected their meanings: for *3 + 4 = 7,* they should be reading "three and four equal seven"; for *8 – 5 = 3,* "from eight take five, the remainder equals three," and so on.[13] Frend took the same care with the words he used when writing his book. Thus, for example, he always referred to $a^2$ as "*a* to the second power"; and $a^3$ as "*a* to the third power," as opposed to "*a* squared" or "*a* cubed."[14] The issue, he carefully explained, was that using geometrical words in arithmetically based algebra risked confounding the meanings in such a way that the student would "pass through life without having a clear idea upon the subject."[15] For Frend, teaching students how to reason from clear and distinct ideas was the essential goal of all mathematical study.

The impossibility of negative numbers served as an essential marker for the "untractable" primacy of numbers in algebra. As Frend put it, a number "submits to be taken away from another number greater than itself, but to attempt to take it away from a number less than itself is ridiculous."[16] Trying to sneak negative numbers in with references to the practical mathematics of "book-debts and other arts" (i.e., accounting) was a sign of mental laziness. "When a person cannot explain the principles of a science without reference to meta-phor," he averred, "the probability is, that he has never thought accurately upon the subject."[17] Because Frend was intent on teaching how to reason cor-rectly, he insisted that whenever a student encountered "an impossible," that is, negative, value in the course of a problem, "he will impute it to the proper cause, either to an error in his mode of reasoning, or to false premises."[18] And so, every time a negative number seemed to be required to solve an equation, Frend found a way to achieve the same result by other, often more laborious but always numerically comprehensible, means.

Frend was far from alone in his determination to defend a meaning-based view of reason in algebra by eschewing negative numbers. The number-based, negative-avoiding algebra he was presenting was very close to that contained in all five editions of the standard Cambridge textbook, *The Elements of Alge-bra* by James Wood, which appeared between 1798 and 1830.[19] By all accounts it seems that Frend had succeeded in the task he had set himself; his *Principles of Algebra* was a clearly written, accurately presented algebra textbook.

But that was not enough. Publishing a mathematical book was very expen-sive because it required considerable specialized typesetting and huge amounts of proofreading. Frend had every intention of covering those costs by sell-ing his text to schools, but first he needed a patron to pay for publication. At this critical moment, Lindsey introduced Frend to Francis Maseres, a very wealthy man who had for years been a quiet but generous supporter of Uni-tarian causes. Maseres had also been deeply involved in mathematics for years. When Lindsey was following the literalist form of reason toward Unitarian-ism in the middle of the century, Maseres was following it into mathemat-ics. After graduating from Cambridge with first class honors in mathematics, Maseres published *A Dissertation on the Use of the Negative Sign in Algebra,* in which he insisted on holding the symbols of algebra to the same standards of clear meaning that characterized geometry. He hoped that his work would qualify him for the position of Lucasian Professor of Mathematics, but it did not. His insistence that all traces of negative numbers be completely removed from even the highest reaches of algebra was recognized to be at once epis-temologically principled and mathematically crippling, and the position of Lucasian Professor was awarded to Edward Waring.[20] Deeply disappointed,

Maseres left Cambridge for a career in the law. By the time Frend arrived in London in the 1790s, he was comfortably settled as the treasurer of the Inner Temple, a judge of the sheriff's court in the City of London, and cursitor baron of the Exchequer.

Maseres's deep commitment to literalist reason meant that he was one of Lindsey's early supporters, but he was never particularly interested in theology. He avoided the controversies that surrounded the Essex Street Unitarians, and when at his family home in Surrey simply attended the Anglican parish church.[21] When it came to mathematics, however, Maseres was a fanatic. For years he published paper after paper rooting out negative numbers so that algebra could claim to be as purely reasoned as classical geometry.

Publishing Frend's book seemed to be an excellent project for Maseres. But when he reached page 213 of Frend's 214-page text, he was appalled to find that although Frend had not actually used a negative number, he had used a technique that allowed for its use. Maseres quickly set him straight. "It is surely high time for every true lover of this science, who is zealous for the honour of it's [sic] purity and perspicuity, to exclaim [of negative numbers] as the good Archbishop Tillotson did with respect to the Athanasian Creed, 'I wish we were fairly rid of it!'"[22] he exclaimed, before adding a 300-page "Appendix" to Frend's textbook. Maseres's addition was anything but elementary. It was instead a highly technical piece that drilled into the furthest reaches of algebra in a determined attempt to eradicate negative numbers wherever they might be found. Including this mathematical root canal as an appendix was the price Frend had to pay for Maseres's financial support, though it also destroyed all hope of selling the work as an elementary textbook.

Maseres was an all-but-irresistible force for a poverty-stricken academic outcast. Frend had never seen any real dangers in the mathematics he had "sucked in with the first milk of alma mater"[23] until Maseres showed him that Cambridge mathematics was as insidiously infected with irrationality as was Cambridge theology. A newly enlightened Frend stood ready to use all of reason's principles—insistence on clear definitions, rejection of analogy, and devotion to historical purity—as weapons in the battle against negative numbers. His moment arrived in 1798, when the death of Edward Waring again opened the position of Lucasian Professor of Mathematics at Cambridge. Within weeks, Frend wrote a letter of application to the vice-chancellor of the university announcing that he had just "put into the hands of the printer a small work, intended as a second part of my principles of algebra" in which "the doctrine of negative roots or numbers less than nothing is wholly exploded."[24] With this work, Frend set out to bring reason to triumph in the world of mathematics and himself to triumph in the academic world of Cambridge.

Frend's route to this end lay through an attack on what is now known as the "Fundamental Theorem of Algebra," which establishes that every equation has as many roots as it has dimensions. The value of the theorem lies in the way it links a straightforward classification of equations to equally straight-forward ways of solving them. Simple inspection of an equation can quickly establish the highest power of $x$ within it, which is called its "degree." The next step is to cast the equation into a standard form, in hopes that a relatively straightforward set of procedures may be applied to solve the equation for $x$, that is, to find the roots of the equation. So, for example, using a modicum of algebraic manipulation, any equation of the third degree can be cast into the standard form, $x^3 + ax^2 + bx + c = 0$, which can be solved using clearly de-fined techniques. According to the fundamental theorem, there will be exactly three roots—that is values of $x$—for any third-degree equation. Those roots may be positive, they may be negative, several of them may even be equal to each other, but the number of roots will always be equal to the number of degrees of the equation. As its name suggests, the fundamental theorem is a wonderfully powerful tool, and continental mathematicians, including Jean Le Rond d'Alembert, Leonhard Euler, and Joseph-Louis Lagrange, devoted a great deal of energy to establishing its validity. Even as Frend was rejecting it, the young German mathematician Friedrich Gauss was publishing a dissertation in which he offered ways to prove it.[25]

Giving up the Fundamental Theorem of Algebra was a truly extreme posi-tion, but Frend saw no alternative. He did not deny the theorem's power, but he also saw that it required accepting "certain roots called negative," which in the seventeenth century had been "more properly" called "false roots, that is, no roots at all."[26] The issue was an ethical one: to teach students to reason with an algebra that included negative numbers would undermine their moral training. As he put it: "From false notions, falsehood must necessarily flow, if the reasoning employed upon them has been properly conducted. Hence the fundamental proposition, that every equation has as many roots, as it has dimensions, being built upon falsehood, must necessarily divert the mind to frivolous pursuits; though it may have been the means in some cases of conducting us to truth."[27] To counteract this ruinous tendency, it was nec-essary "to restore the analytick art to its true principles,"[28] that is, to find a way to classify and solve equations that did not assume the validity of nega-tive numbers.

In 1799, Frend took up this challenge in a second volume of his *Principles of Algebra,* with the subtitle *Or the True Theory of Equations Established on Mathematical Demonstration. Part the Second.* He knew from the outset that laying out a theory of equations in which all variables would have positive number referents would not be easy because "the forms of equations are more numerous than the leaves of autumn in Vallombrosa."[29] Nonetheless he saw it as his sacred duty to counter the "volumes upon volumes [that] have been written on the stupid dreams of Athanasius" and the similar volumes devoted to "the impossible roots of an equation of *n* dimensions." For Frend, the Trinity and the negative numbers were equally pernicious products of invalid reasoning, so over the course of 119 equation-filled pages he doggedly pursued his goal of classifying equations without using negative numbers. By the end, he could only claim to have taken the first steps down the path of reason in mathematics, but he nonetheless emerged triumphant. "The investigation of the properties of equations is endless" he exclaimed, and then followed this infinitude into the ranks of the angels. "By the class of intelligent beings next in rank above man, all these equations and all these curves are, perhaps, thoroughly understood, and the next class excels them as much as they do us. How great then must be that being to whom the thoughts of all these orders of beings are known at a moment's glance," he exulted.[30] Frend had been teaching mathematics for years, but not until he set out to remove the negatives from algebra did he find himself face to face with his God.

It is somewhat difficult to know how serious Frend was in offering his *True Theory of Equations* as the centerpiece of an application for the Lucasian Professor of Mathematics at Cambridge. The committee making the appointment knew as well as Frend did that mathematics was an essentially theological subject, and they were not about to appoint to the university's highest mathematical post the radical Unitarian who had been banished from Cambridge less than a decade before. Instead they made Milner the Lucasian Professor of Mathematics in 1799. There was, however, one person in Cambridge who was both willing and able to engage Frend's mathematics directly: Robert Woodhouse. Woodhouse's mother was the daughter of the Unitarian minister in Lowestoft, and after her son finished grammar school in North Walsham, he cultivated his Unitarian roots for a time at the New College in Hackney. However, he began shedding his Unitarian past when he entered Gonville & Caius College in 1791, and by 1795, when he took his degree as senior wrangler and first Smith's Prizeman, he was Anglican enough to be awarded the Gonville & Caius College fellowship that supported him for more than two decades. Woodhouse's conversion seems to have been a pragmatic move by a man who did not see the kinds of issues that divided religious groups as fundamentally interesting or important; his central concerns lay in mathematics, not theology.

In 1797, Woodhouse reviewed Frend's *Principles* and complimented him on the way he had "simplified the science of algebra" by "the exclusion of negative quantities," thereby removing "one of the principal difficulties that have perplexed and puzzled young persons in commencing the study."[31] He was less enthusiastic about Frend's *True Theory,* however. "The author of the present treatise has undoubtedly thought for himself; and his work deserves notice for its freedom from absurd notions and indirect demonstrations, and for the practical information relative to the solution of equations," he began, but that was as far as he was going to go with praise. He dismissed Frend's efforts to tie mathematical and theological reasoning as "flippant" and opined that Frend "would have done better if he had shewn that [the use of negatives] is unintelligible and useless in expediting mathematical reasoning."[32] Woodhouse saw no need to sacrifice algebra on the altar of reason.

Maseres emphatically did, and he was not about to let Woodhouse's challenge stand. He responded with a blistering letter that forced Woodhouse to concede that until the theory of negative numbers was better understood, it "had better *not* be taught" at Cambridge. Woodhouse's capitulation was only provisional though. He agreed that negative numbers should not be taught "at present," but he was absolutely clear that "since they lead to right conclusions, they *must have a logic.*"[33] Grappling with Frend's *True Theory* pointed Woodhouse down an intellectual path that was to fundamentally change the way mathematics was conceived in England. By the time De Morgan launched his career at the end of the 1820s Woodhouse's ideas were being developed to form a new view of reason and mathematics.

The *True Theory* marked the end of Frend's mathematical journey, however. In religion he had followed the virtuous path when he defended reason's monotheism; in mathematics he had clearly maintained and persisted in the essential point that the negative numbers violated the dictates of literalist reading that reason required. He was never going to moderate either of these positions, but he nonetheless faced the challenge of making his way in an increasingly troubled world. As the war with France wore on, England's economy suffered enough that the Bank of England began issuing paper money without the bullion to back it up. Frend was horrified. He denounced paper money as a "monster in nature,"[34] which undercut "the true point of mercantile honour."[35] In a natural, upright world, paper notes "wore an honest front, and represented labour and property," but notes without coins to back them acted like dreams "upon the distempered mind of the nation."[36] From Frend's literalist point of view, England's new paper money was as noxious as the negative numbers; neither had clear referents, and for him, the results were equally pernicious.

Even as Frend was defending it, reason remained under attack. In 1799, the Unitarian publisher Benjamin Flower was imprisoned for six months for publishing a work interpreted as libeling an Anglican bishop, and Gilbert Wakefield was jailed for two years for saying the English peasantry would not suffer if Napoleon took over the country. In the face of such pressures, the Essex Street community struggled to find its way. Lindsey was moving through his seventies and his energy levels were decreasing. Hannah, who was sixteen years younger, found it difficult to accept his retirement, and she and Disney were often at each other's throats. In Cambridge, their friends Hammond and Reynolds were both struggling against inflation. As Frend's gentlemanly cohort began to feel the strains of war, his views on social relations changed significantly. England's poor continued to suffer terribly, but he no longer counseled resistance. "It is the duty of the lower classes to support with patience the present evil, and to act with constitutional firmness for its redress," he wrote.[37] Six years earlier Frend had encouraged crowds of tens of thousands to "disregard" and "shew their contempt for" governmental decrees.[38] But subsequent years of relentless war had chastened him. Considerably less confident about the stability of his country than he had been in 1795, he was much more cautious about taking positions that might undermine it further.

Frend never abandoned the cause of reason, but by 1796, even his former students were deserting. Samuel Taylor Coleridge escaped into a world of poetry, while in 1798, Robert Malthus attacked Godwin's political optimism with an *Essay on the Principle of Population.* Frend was disappointed in Coleridge's defection and unimpressed by Malthus's arguments. "Evil could not originate in his [God's] fiat," he declared. If the natural processes God had put in place when he created the world "led to a plenum of human population at a remote time, the evils prognosticated would have their remedy."[39] Even as it seemed the world around him was crumbling into disarray, Frend remained convinced that the practice of reason would eventually solve all problems.

———•———

Through it all, Frend continued to contend with the social and political exile that hobbled his efforts to find employment. Lindsey tried to help by introducing him to influential and important people, but his interventions were not enough. "It is not merely the *odium theologicum,* but *publicum* also, far more pestilential at present, which has marked you out as an infectious person," he sadly explained.[40] This left Frend to cobble together a living with occasional tutoring and publication. He had emerged from Cambridge with a reputation as an exceptionally strong teacher and found positions teaching boys in the homes of Lord Oxford and the Duke of Cumberland. He was equally ready to work with young women, and surviving letters from 1806 and 1807 show him guiding the precocious young Annabella Milbanke, later Lady Byron, through Latin, Greek, English composition, and mathematics. But these were all temporary positions, and he remained stalled in his efforts to secure a steady source of income.

As Frend struggled through his trying times, Lindsey could only wonder how his protégé could remain "so happy and chearful."[41] To Lindsey's question, Frend might have answered that his faith sustained him. If so, the response would have been relatively new. For decades he had been discussing questions of religious doctrine without any indication that he actually saw himself and his own experience in Christian terms. From Tylden's near death in the Alps through his trial and beyond, there is no hint that the God and Jesus to whom he devoted so much scholarly attention affected him personally. The ecstatic conclusion to his *True Theory* is just the first indication of the ways Frend's difficult years were changing his religious life as much as they were his political views. In the midst of "the pain, the uneasiness, the disquiet of mind"[42] that dogged his perennially unemployed footsteps, he began to experience directly the healing powers of the God he seems previously to have known primarily as an intellectual exercise. His new religious sensitivity opened him to a whole new world of thought, feeling, and insight that lay beyond the pickiness of literalist reason. In the first decade of the nineteenth century, he began to supplement the understandings of God he had gleaned from close readings of the Bible with more directly experienced forms of insight that arose from reading the book of nature.

Frend's venture into natural theology began as yet another attempt to support himself. After the failure of his mathematical ventures, the community around Lindsey came together to rescue him. Lindsey offered his library as a place to write, Maseres put up the money for initial printings, and the publisher Joseph Mawman agreed to publish his elementary textbooks. This time, instead of writing for schoolboys, Frend followed the example of Richard and Maria Edgeworth's highly successful *Practical Education,* which focused on teaching mothers how to educate their children through materials to be found at home. In *Tangible Arithmetic,* Frend showed his mother-teachers how to introduce their young children to basic arithmetic using an "arithmetical toy" much like an abacus.[43] *Tangible Arithmetic* actually went through two editions, but, as Frend explained with a self-deprecatory grin, that was only because "the greater part of the first edition" had "been consumed by fire."[44] Considerably more successful was a series of astronomy texts entitled *Evening Amusements: Or, The Beauty of the Heavens Displayed,* which were directed to mothers and the teenaged children they were educating at home. In *Evening Amusements,* Frend returned in spirit to his childhood home in Canterbury, where his stepmother, Jane, had first awakened his love of learning.

Frend's choice of astronomy as his subject was timely. Ever since 1780, when William Herschel discovered a new planet, Uranus, a steadily increasing number of the English had become fascinated by his telescopic explorations. By the first decade of the nineteenth century, astronomical globes and telescopes were being manufactured for private use in the homes of those who could afford them. Frend directed his books to the ever-growing audience that was proudly displaying these instruments in their front rooms. In *Evening Amusements,* he set out to show his compatriots how to use their eyes, their telescopes, their globes, and their reason to understand the heavens above them.

Astronomy as Frend conceived it was no less reasoned than mathematics, but the variablity of the heavens made it more amusing. Their variety was evident not only in wonderfully unique events, like the comets that Herschel's assiduous sister Caroline began reporting in 1786, but was also to be found in the ever-shifting configurations of the familiar bodies in the solar system. Frend's *Evening Amusements* covered the whole of the nineteen-year lunar, or Metonic, cycle from 1804 through 1823. Each of the nineteen annual volumes contained thirteen chapters, the first introductory, the next twelve mapping the positions of the sun, moon, and planets month by month. The books proceeded cumulatively with each year building on those that had gone before. In the earliest ones, Frend focused his introductions on the basics of astronomy, from the shape of the earth to the motions of the fixed stars to the structure of the solar system, and he did not repeat these basics in the later ones. From a publishing point of view, the advantage of this design was that readers who came late to the series needed to buy the early volumes in order to catch up. From an authorial point of view, it meant that in the later books, he could freely expound on whatever insights he gleaned from observing night skies. Thus, for almost two decades, Frend's *Evening Amusements* served him as a kind of annual letter to an audience of women and young people.

All of Frend's efforts were directed toward leading his readers to a full understanding of their true place in God's universe. In this he was following the lead of Newton, for whom the distinction between theology and astronomy was vanishingly small. Newton repeatedly observed that the astronomical space he so completely understood with his mathematics exhibited the essential qualities of the divine. Like God, it was infinite, eternal, and ever present, or to use his words, "endures from eternity to eternity, and is present from infinity to infinity." The parallels were so powerful that in the *Principia* Newton described space as the divine "sensory" by which God thoroughly perceived and comprehended all things "wholly by their immediate presence to himself."[45] Later he drew back from the position that it was actually God's mind to the more moderate claim that God had "established" the infinite, eternal and ever-present space that shared his defining attributes.[46] In either case, Newton was clear that properly doing astronomy entailed matching our own thinking to the divine thought that is directly manifested in space.

When it came to the relationship of astronomy to theology, Frend was thoroughly Newtonian. The amusement that was the stated goal of his series was just a way to enable his readers to connect directly with the reason so clearly expressed in God's heavens. Although reason was for Frend the defining property of the human, he was convinced that without education it lay passive. In astronomy this was a real issue because "the heavens present two very different aspects to the active and the passive mind." Passive observers might find momentary pleasure in astronomical phenomena like sunrises or moonlight, but the active learner would experience them in ways that were far deeper, more satisfying, and more long-lasting. Frend therefore urged his readers to waken their "latent powers" by following their reason into the night skies where "your mind will naturally expand itself from the contemplation of created objects, to the power of him who called them into existence, and regulates their motions with the utmost exactness."[47] He wanted his readers to experience the heavens he was teaching them to observe, measure, and predict as immediate manifestations of the divine presence.

For Frend, the path to understanding lay through concrete interactions with everyday things. He told his mother-teachers how to model the astronomer's earth by twirling an orange scored with longitude lines and an equator between thumb and forefinger and how to illustrate the form of an ellipse by shaving the bottom of a conical sugar loaf.[48] He carefully connected the "parallel lines, and uprights, and declination and equator" of astronomy to the patterns of their weaving or embroidery and suggested that they might find ideas for ways to construct perfect squares or octagons or figures of any number of sides using the geometrical techniques developed for astronomical purposes.[49] Through it all, his ultimate goal remained always to open their minds to the divine order that reason could uncover in every aspect of the natural world. "Nature begins with the objects of sense," he explained, but those "senses are only inlets to the mind." To understand and properly fill their places in the universe, people needed to cultivate and apply their "superior powers" to the world they lived in.[50] Frend's readers could learn a great deal through observation, but understanding the ways God was manifest in the world entailed reasoning their way through those experiences.

Frend had no trouble following the mathematics of God's reason in heavenly space, but he did have some trouble explaining that God to the readers of *Evening Amusements*. He closed his 1804 volume with an outburst of wonder that from all of the myriads of astronomical objects "peopled with various beings, capable of enjoying the munificence of their Creator," God had chosen to send Jesus to the earth.[51] His conviction that the universe was filled with sentient beings was common in his day, as was the lesson of God's specific love that he drew from these teeming populations, but Frend was rather uncomfortable with the image of a universe in which human reason played such an insignificant role. In 1805 he tempered the humility of his 1804 conclusion with the recognition that "a man of understanding and virtue is a nobler object of contemplation than the beauties of an earthly landscape, or the glorious lights in the starry heavens."[52] Having thus extolled humans and the reason that defined them, he began to worry about the deism that might emerge from a too strong emphasis on human rational power. So, in 1806, he emphasized that the ultimate subject of astronomy was "the spiritual heavens" in which the Christian God "shines, as the God of love." Whatever might be found in "the book of nature written in the heavens" it would always be secondary to that to be found in the Bible "with which no other book will bear a comparison."[53] The movement of Frend's effort to articulate the nature of the God he perceived in the heavens—from wonder, through science, to a final resting place in the scriptures—followed a dynamic that was to be typical for the rest of his life. Although he was always eager to explore new ideas and approaches to the world around him, nothing would ever move him away from the scriptural literalism that had led him to Unitarianism in his late twenties.

In 1807 Frend turned his attention to a solar eclipse that would not be seen in England but which, he assured his readers, was going to be an impressive sight in other parts of the world. He guarded his readers against the ignorance that rendered eclipses terrifying by showing them that the darkness they created was a natural occurrence. He gave them careful instructions for tracing the orbits of the planets and the moon on a piece of pasteboard and then had them place different colored beads to represent the sun, moon, and planets in different configurations on these orbits. As he explained how to construct such a model, he emphasized how difficult it had been to arrive at the heliocentric system they depicted. Placing the earth at the center was considerably more intuitive than placing the sun there, but the price of uncritically accepting this approach was a horrifically complex system of planetary motions. Irregularities like retrograde motion, which earth-bound observers see the planets to be tracing, "are owing to the motion of the body on which we are placed," he explained.[54] But from the perspective of observers on the sun, those irregularities simply would not exist. He encouraged his readers to experiment with this kind of perspective-changing by considering the ways the rest of the system would look from the vantage points of different beads in their pasteboard models.

For Frend, practicing such perspectival shifts was not merely an intellectual imperative. What he called "the moral world" was filled with vastly more irregularities than were the heavens, and successfully negotiating it required constant practice in adopting multiple perspectives. Each individual perspective provides only a "partial view" of circumstances, he emphasized. Anyone who fails to recognize the individual limitations of partial perspectives would form "rash and erroneous judgments" of past and present events and would be correspondingly unable to think constructively about the future.[55] Working with others was the best way to get around the restrictions of the individual. Here Frend pointed to the astronomical community as a model. That the 1807 eclipse would not actually be visible to any of his English readers meant that he and they would have to rely on the reports of those for whom it was. In this the eclipse was typical. In fact, most things in the heavens, from the positions of the stars to the movements of the planets to the orbits of comets, could only be understood with the cooperation of observers scattered all over the world.

Frend was excited by this kind of international cooperation, even as he recognized its limits. He approached the heavens as a natural theologian for whom studying the heavens was an essentially personal experience of communion with the reasoning God who lay behind all of its manifestations. Truly understanding God's design required moving beyond the limitations of individual experience, but working from a communal perspective was not the same as giving up personal power; it was essential always to remember "that in this, as in every other science, we are not bound by any man's authority, but have full liberty for the exercise of our own judgement."[56] Frend was no more willing to cede his independence to an astronomical community than he was to a religious one. His goal in laying the mathematical foundations in the early volumes of *Evening Amusements* was to free all of his readers to follow their own reasoned paths to the God whose ways of thinking were evidenced among the stars.

Even as Frend was reaching toward the ideal of a multinational, multiperspectival astronomical community, he was creating a local one of his own. That sales supported four editions of his 1804 volume may be taken as an indication that he was beginning to overcome the *odium publicum* that had dogged his footsteps for so long. *Evening Amusements* began to be taken up by the kinds of boys' schools that had spurned his *Principles of Algebra*,[57] and the full series of nineteen volumes was reprinted in 1830. With *Evening Amusements,* Frend finally found a comfortable place for himself, his reason, and his God, and judging from the copies to be found in American libraries from New York to Wisconsin to California, at least two generations of English-speakers joined him there.

⸻

In 1802, Frend and his fellow radicals had rejoiced when the war with France they had so determinedly opposed was finally ended with the Treaty of Amiens. But when just over a year later war broke out again, Frend supported it. Some in his circle did not share his change of heart but most agreed that Napoleon represented a real threat to their country. Local militias sprang up all over England and began practicing their response to the French invasion they saw as imminent. Hannah Lindsey proclaimed herself to be "very happy that the country is so generally roused to a sense of danger from without & getting ready to meet it."[58] Frend had some kind of leg injury that prevented him from volunteering, but in a pamphlet entitled "Patriotism" and dedicated to "the volunteers of the United Kingdom," he voiced support for all the efforts to protect his homeland from the threat of Napoleon's marauding armies.[59]

By the time Frend was showing English mothers and children how to feel at home in God's universe, Napoleon was raising serious questions about their country's place in the world. In the early years of *Evening Amusements,* the French emperor was trying to squeeze their island into submission by cutting it off from all trade with Europe. This strategy was frighteningly effective, and in the month after the 1807 comet so excited Frend and his readers, the general's armies began marching through Spain to conquer Portugal, the last holdout against this blockade. By January 1808 this effort was successful enough that the English were able to trade only with Sweden and America, and neither of these relations seemed particularly stable. "This year opens with a deeper gloom even than the last," the noted Quaker William Allen wrote, while the Duke of Northumberland could only hope that "when things are at their worst they must mend."[60] Three months later, Frend joined them with a lament about the advent of a new power that "has sprung up from the waves; and now, standing on the shore, fills the world with terror."[61] The seemingly irresistible advance of Napoleon was forcing the English to call upon all of their spiritual as well as martial strength.

In the dark days of 1808, Frend began writing a political column for the Unitarian *Monthly Repository of Theology and General Literature.* The almost twenty years that had passed since he and his fellow radicals had looked with such hope to the French Revolution had fundamentally challenged their Enlightenment vision, and Frend struggled to understand its meaning in "A Monthly Retrospect of Public Affairs; or A Christian's Survey of the Political World." At the outset he tried the multi-perspectival approach and explained to the readers of the "Christian's Survey" that human generations were like planets that move always in harmony with God's plan, although "to those who live in one planet, the motions of the others are full of irregularity."[62] Over time, however, Frend found that perspectival adjustments were really not enough to explain the enormity of Napoleonic disruptions. As the dreadful years of Napoleon's rampage pushed him to confront the limits of his understanding, Frend turned to prophecy as "the only ground of belief of the rational believer."[63]

Frend's turn to the prophetic books of the Bible—primarily Revelations, Daniel, and Jeremiah—represented a dramatic change from his previous approach. Earlier, when he described himself as a "scriptural Christian," the scriptures he was referring to were the books of the Bible that contained the words of Jesus, that is, the four Gospels and Acts. When he discussed questions like those surrounding the nature of baptism or the value of communal worship, he would venture onto the somewhat shakier ground of the rest of the Old or New Testaments, but the prophetic books were something else entirely. They made no claims to be reasoned, and Frend was absolutely clear that they could not be read closely. But when it was impossible to make sense of his world either by "the principles of philosophy or the nature of the human mind," they offered comfort.[64] They were far too unreasoned to support any particular predictions, but they did provide the fundamental assurance that no matter how horrible the situation appeared from his earthly perspective, everything that was happening around him was actually in accordance with "the commands of the most high."[65] At times Frend allowed himself to wonder whether Napoleon was the "horrible beast" who roared until the nations trembled,[66] or the angel whom God commanded to "gather the vine of the earth, and cast it into the great wine press of the wrath of God,"[67] but he would quickly draw back from such idle speculations. He was convinced that it would someday become clear how everything was leading to "the establishment of that kingdom, which must finally be erected on the earth,"[68] but that day had not yet arrived. For the present, all Frend could do was allow his faith in this prophetic promise to support him as he tried to understand the meanings of the events that surrounded him.

Frend's responses to his world can seem incoherent. In some months Napoleon was seen as a great threat, in others he was a force for good; in Europe Catholicism had turned the people of Spain and Portugal into bigots, in Ireland it was the legitimate expression of an oppressed people. But he knew these were all incomplete readings of the true meanings of these events for the divine design. For him, the unfolding of that larger plan was structured by a number of basic themes of reason—disgust at superstition, support for education, hatred of intolerance—which were expressed differently in different situations. It was an enormous challenge to glean God's message in the chaotic world of Napoleonic Europe, but month after month Frend tried. Through it all he remained convinced that someday true meanings would emerge from the earthly chaos of national allegiances, religious doctrine, military victory, and crushing defeat that surrounded him.

Frend did not confine himself to international developments in his "Christian's Survey." Local events could be equally pressing. Even as he was applauding English attempts to strengthen Catholic Portugal against Napoleonic advances, he was also furiously reporting on the case of Reverend Francis Stone, who was being tried in a civil court for maintaining from his pulpit that Jesus was the son of Mary and Joseph.[69] Stone's plight served as a stark reminder that Anglicans could not be counted on to be tolerant, and although more than fifty years had passed since the case of John Jones had so terrified Blackburne, Unitarians were still vulnerable. That Frend no longer held a church position meant he was no longer liable to legal prosecution, but as Stone's case was moving through the courts, a different situation arose to remind him of how precarious his own position could be.

When Lindsey died in November of 1808, Frend shared his grief with the audience he had for years been addressing from the warm confines of his dear friend's study. In the December issue of *Evening Amusements* Frend wrote lovingly of the man who in private was always so "mild, gentle, affable, and courteous," but in public was powerful enough to establish societies through-out England that effectively called all men "to make the Scriptures the rule of their faith and actions."[70] Frend's celebration of the Unitarian movement Lindsey had initiated engendered an immediate backlash, and a number of readers cancelled their subscriptions. Startled and hurt, he responded with a defiance that was more reminiscent of the outbursts of his radical youth than of the gentle pedagogue he had become. He had already told his readers that a proper astronomical community was one that allowed "full liberty for the exercise of our own judgement,"[71] and he was certainly not about to adjust his message for the comfort of others. "It is the same God, whose powerful word operates both in the natural and the moral world," he insisted, and he was emphatically not going to "drive religion into a corner" just because he was doing astronomy. His readers were free to cancel their subscriptions, but he was equally free "to pursue his own plan in his own way, leaving to his readers to reject or assent to those reflections, which naturally arise in his mind in the course of his work."[72] Frend was never going to let others dictate his positions. God's reason continued to structure the heavens, and he faithfully guided readers through them until the end of the Metonic cycle in 1822.

# CHAPTER 9

## *Heiress*

By the time the uprising against Frend's Unitarianism broke out in *Evening Amusements,* he could actually afford to lose some of his readers. Two years before he published his obituary of Lindsey, a group of businessmen had offered him a position as the actuary of the Rock Assurance Company (the Rock). The legitimacy of insuring human lives was still rather questionable in the early nineteenth century, but the idea had been building in strength on the dissenting fringes of English society for several decades. Frend's essential qualification for the position was the endorsement of William Morgan, who had been remarkably successful as the actuary of England's Equitable Life Assurance Society (the Equitable). Morgan was himself a radical who had barely escaped arrest in the London Treason Trials, so the *odium publicum* that had so long frustrated Frend's efforts to find a position was not an issue. Becoming the actuary of the Rock promised Frend the kind of comfortable, steady income that had eluded him for years.

Frend nonetheless hesitated. Through all of his trials and tribulations, he had always remained a Cambridge-educated gentleman honorably supported by some form of church or aristocratic patronage. Becoming a businessman felt like enough of a comedown that six months after he accepted the position with the Rock he tried one more time to return to Cambridge, this time as a professor of theology. Only after learning that his application was being completely ignored did he identify himself as "actuary to the Rock Life Assurance

Office" in the author's byline of *Evening Amusements*. Taking the job at the
Rock meant that Frend finally had the means to follow through on the dreams
of marriage he had shared with his sister Mary more than thirty years earlier.
By all accounts he was a very attractive man, but in the decade after the Mary
Hays debacle he had held his charms in check. Now he turned his attentions
to Lindsey's niece and Blackburne's granddaughter, Sarah Blackburne.

Sarah's father, also Francis Blackburne, was the Anglican vicar of the small
Yorkshire parish of Brignal. In 1796, he and his stepsister Hannah agreed that
sixteen-year-old Sarah was ready to broaden her horizons by spending a winter
with the Lindseys in London. In her first venture out of the sheltered seclusion
of the Yorkshire Dales, Sarah was shy, taciturn, and cautious; "she vents her
mind slowly" is how Hannah put it, and "she does not easily find pleasure in
novelty so that there is no fear that her imagination will run away with her."[1]
Hannah nonetheless quickly warmed to the young woman, who had a way of
moderating what she herself recognized as her "irritable" temperament.[2] For
the next several summers, Hannah sent little gifts "to please and adorn"[3] her
niece, while eagerly awaiting her winter returns. In Sarah, Hannah had found
the daughter she never had, and under her watchful eye the young woman became a flowering favorite in London's Unitarian community.

For Sarah, visits to the Lindseys were religious as well as geographical journeys. Her father had given his daughter the same education in reason that his
father had given to Hannah. But in Brignal Sarah attended her father's Anglican church, whereas in London she went to the Essex Street Chapel. For the
young woman, the theological differences between the two churches were of
no more moment than were personal differences like those between Disney
and Hannah. Both congregations recognized the central importance of reason
to the relations between God and humans, and because she was a woman,
Sarah did not need to declare her theological convictions beyond that.

William and Sarah first met during one of Sarah's winter visits to London.
Their interactions became more prolonged after 1803, when Sarah moved to
London full time to help Hannah care for Lindsey, who was suffering a series
of strokes. Hannah was enormously grateful to the young woman who acted
as a "sheer anchor" through her husband's decline. When Frend secured his independence as actuary of the Rock, she did all she could to bring him into the
family as the husband of "invaluable Sarah."[4] Sarah's father was also pleased
to place his beloved daughter "under the care & protection of a man who can
appreciate her worth, and on whose virtue & fidelity I can rely with the most
implicit confidence."[5] Thus blessed, Sarah and William were married in the

Essex Street Chapel on January 16, 1808. Both of the Lindseys were undoubtedly there, with stroke-paralyzed Theophilus strapped upright in his chair and Hannah making sure he was comfortable. As they watched, William Frend and Sarah Blackburne effectively healed the split that had so painfully divided Blackburne and Lindsey forty years before.

Beyond this familial group, Frend's marriage came as a welcome surprise. "When a man is settled, his family becomes as it were a little world to him, where his principal cares and tenderness are centered," one of his friends enthused.[6] This description fit the Frend marriage very well. For more than forty years, Sarah and William focused their attention and tenderness on creating and maintaining the little world of reason in which they, and the children who were to join them, thrived.

William and Sarah welcomed their first child, Sophia Elizabeth Frend on November 10, 1808.[7] Our most direct window into the little world Sophia shared with her family is *Threescore Years and Ten: Reminiscences of the Late Sophia Elizabeth [Frend] De Morgan,* edited by her daughter Mary and published in 1895. The title's claim to be reminiscences accurately reflects the somewhat random and rambling nature of this piece, which Sophia wrote as she was nearing eighty. There are many reasons to question the historical accuracy of her remembrances of her childhood, but by the same token, it is not hard to recognize the importance to Sophia of the images that remained with her in her old age.

Sophia's childhood memories consist primarily of interactions with and impressions of the people she encountered with her father, who brought her with him whenever business or pleasure took him into the streets of London. Sophia remembered Coleridge "talking *to,* not *with* my father for many hours when I, being a child only, wished he would stop."[8] She remembered William Blake, "who had on a brown coat, and whose eyes, I thought, were uncommonly bright."[9] And she remembered John Quincy Adams comparing electoral politics in England and America over dinner in her home. Dining with a future president was just one of many memorable evenings that she spent listening to adult conversations. The growing girl drank in the ideas of those around her, but her learning was not free-form. Frend liked to describe his daughter as "an heiress of immortality,"[10] but she was at least equally an heiress of reason. From the moment of her birth, Frend devoted himself to showing his eldest child how to use her reason to understand the world she was growing into.

One example of his teaching may be seen in the story of the jailing of Sir

Francis Burdett, which takes up much of the first chapter of *Reminiscences*. Sophia's earliest direct memories of Burdett were of a man who liked children and welcomed her to play in the garden behind his house in St James Place; as she grew older she was drawn to the humanism of his campaigns against dueling, against flogging in the army, and in favor of prison reform. The story her father told concerned none of these. It focused on the challenges England's political radicals faced in the Napoleonic era before Sophia was old enough to remember anything. Frend told it to her so that she could understand reason's proper place in the political world.

The events that Frend wanted Sophia to know about began in the year before she was born. In 1807, he and his fellow radicals had succeeded in securing Burdett a place in the House of Commons on a platform promising to reform the electoral system that kept the landed gentry in power.[11] As the only radical member of Parliament (MP), Burdett was isolated, but events in his second year were nearly ideal for supporting his argument that the current system produced a government that was ineffectual and corrupt. The presenting issue was the fate of a military foray that set out to do battle with Napoleon in the Netherlands in the summer of 1809. The French learned of their plans and slipped away before the English troops arrived, leaving the English encamped on the low-lying mosquito-infested island of Walcheren. By the time they finally left in December, four thousand men had died and eleven thousand of those who returned to England were weakened by a particularly virulent combination of typhus and malaria known as "Walcheren fever." It is difficult to imagine more compelling evidence of military ineptitude than a disaster in an army that had not ever encountered the enemy.

The House of Commons initially resisted holding anyone responsible, but by February 1810, public outrage was loud enough that they finally agreed to conduct an inquiry. They insisted on keeping their investigation secret, however, and when some information was leaked they jailed the perpetrator, John Gale Jones. Burdett responded by writing an open letter to his constituents explaining what had happened. Crying that confidentiality had again been breached, the House of Commons voted to silence Burdett as well, by committing him to the Tower until the end of the parliamentary session.

This decision ignited a firestorm of protest. When Burdett declared he would not cooperate, angry crowds poured into the capital's streets while troops began setting up guns in London's squares. Sophia's father was a member of the inner circle of Burdett's advisors, and he related the situation in his "Christian's Survey." His April column began with the tensions between the unruly crowds gathered in front of Burdett's house in Piccadilly and the body of foot guards that was setting up cannons in Green Park. He then vividly described the drama surrounding Burdett's surrender and transportation to the Tower. Burdett's arrest was an essentially gentlemanly affair; after being carted to the Tower under heavy guard he was handed over to the constable, who politely walked him to the apartments that had been prepared for him. But the scenes in the streets were far less contained, and two members of the crowd were killed by the police. "The voice of thy brothers' blood crieth to me from the ground," Frend lamented; "both sides are our countrymen, and the military can derive no honour in such a conflict."[12] Frend had spent much of the previous decade exhorting London's crowds, but he would never do so again. The riots surrounding Burdett's arrest cemented Sophia's father fear of mob action.

Frend's new-found fear of crowds found its first public expression at the time of Burdett's release. Throughout the month of May, London's radicals protested the imprisonment of Burdett and Jones. When nothing came of their objections, they began laying plans to fill the streets with people to cheer Burdett's scheduled release. Supporters flocked from as far away as Scotland and Ireland, and when May 21 dawned warm and sunny, it seemed that all of London poured into the streets wearing blue in support of Burdett. Volunteer marshals carried white wands as they sat on their horses, but there was little need to control such good-humored crowds. The radicals had lost the argument against Burdett's imprisonment, but everything seemed to be coming together to make his release a singular triumph.

Things did not unfold as planned, however, and Frend was a major part of the reason. In the weeks before Burdett's release, Sophia's father had been making numerous trips to the Tower in order to impress the dangers of crowd action on the imprisoned MP. In the long run, he succeeded in persuading Burdett not to risk riding through the throngs that filled the capital. And so, while virtually all of London was waiting to cheer their hero through the streets, Burdett slipped out quietly through Traitors' Gate and allowed himself to be rowed with muffled oars out of town.

The organizers of the London gathering were caught completely off guard when Burdett did not show up as expected, and with each passing minute the people who had gathered to cheer his carriage became more agitated. Desperate to defuse the situation, the radicals managed to appease the crowds by making up a story in which Burdett had been *forced* to leave by boat and sending the newly released Jones down the route in a hastily procured cart. When

the sight of Jones's rickety conveyance with his name chalked on the sides succeeded in releasing the tension by amusing the crowds, the people who had
committed so much to bringing those crowds together heaved enormous sighs
of relief.[13] They were nonetheless deeply disappointed that the demonstration
that had begun so auspiciously ended with a whimper.

Burdett's failure to show himself to his hundreds of thousands of supporters marked a considerable downturn in the fortunes of the radical party; decades were to pass before they could again draw such crowds to their causes.
Nonetheless, Frend never admitted that rowing Burdett away was a mistake.
Sophia was only nineteen months old when these events took place, but her
father took great care to assure her that even the boatman endorsed the decision to take Burdett up the river.[14] Long after the issues had been settled and
all of the principal actors were dead, Frend's daughter was still using his re-
constructed conversation with a boatman as evidence that his political vision
was grounded in egalitarian reason.

The story of Burdett also supports a central insight that Sophia's daughter
Mary took from the experience of editing Sophia's *Reminiscenses:* "It would be
difficult to overstate the extent of her father's influence on the education and
character of my mother."[15] That influence was not the only one. Frend may
have carried her into his political world, but his daughter could never really
have a place there. Sophia was a girl, who belonged in the domestic sphere,
and there her mother reigned. When Frend entered his marriage, some combination of his Maddingly experience and textbook writing meant that he considered himself an expert on child-rearing; he loved teaching Sophia to read
using sets of ivory letters he had made, and he burst with pride when by the
age of nine she could read Hebrew passages aloud. Nonetheless, in the same
letter in which he reported these accomplishments, he acknowledged that any
efforts to move further into his wife's realm were met with firm resistance.[16]
When it came to the fundamentals of running their household, which included raising children and treating the sick, Sarah was in charge.

Frend's account of his and Sarah's disagreement suggests that Sophia's upbringing took place in the midst of a negotiation about how to define and
maintain the proper boundary between her mother's domestic world and the
political one that her father inhabited. The nature of the boundary may be seen
in a different side of the saga of Burdett that they related to Sophia. When
two exasperated marshals banged on the door demanding that she tell them
where Frend had taken Burdett, Sarah, who did not know the answer, shrank
in terror.[17] This response points to a reality that Sophia was going to live with

for the rest of her life; on the flip side of the story of her father's sagacity and power lay another story, that of a woman's proper place in a male-dominated world. In Sophia's version of the events, everyone agreed that whatever pressures they were under, the irritated men were wrong to have brought their masculine world to a woman's door. In adulthood she devoted a great deal of energy to finding places for the reason she learned from her father within the shelter of the domestic world she inherited from her mother.

During Sophia's childhood, the apartment in Blackfriars provided a stable platform for the development of a very bright and inquisitive little girl. Tucked among the tales of one rational thinker after another are the memories of a child with a lively imagination. Her father's mathematical sponsor, Maseres, routinely gave her family the key to the Temple Gardens, where she would lie on the grass to contemplate people and castles in the clouds. Or she would go into the Temple church, whose "dark nooks and mysterious corners furnished food for fancy," or weave stories about "the effigies of the Crusaders" that lay about "in the coarse long grass, most of them without their heads."[18] The streets of London could also provide food for her imaginings. She later well remembered one day when, her mind in a cloud, she stepped into the street right in front of the Duke of Wellington, who quickly "drew up" his horse and "quietly lifted his hat."[19] There was much that could be threatening in the exploding world of early nineteenth-century London, but Sophia's place in that world was safe and her mind expanded within it.

For more than ten years Sarah raised her ever-growing family in their rooms above the Rock, but her desire to get away from the huge and bustling city may be surmised from the pulls toward the countryside—glimpses of the Surrey Hills from the attic of the Rock, walks to the fields still to be found in the area of Kings Cross—that are part of her daughter's memories of the time. Sophia especially treasured summers in Clapham, which was rural enough to feel like country while being within commuting distance of the Rock. The town was also settled enough to support the constant demands of caring for the Frends' severely disabled fourth child and first son, Richard.[20] By 1819, Sarah's desire to escape the city finally prevailed, and the family moved their still growing brood to Stoke Newington. That Mary Wollstonecraft had run a school there and Richard Price had served as the minister of the Newington Green Unitarian Church made the community so appealing that Frend agreed to "vibrate every day like a pendulum"[21] between Stoke Newington and the offices of the Rock in London.

The Frend family moved into what had once been Daniel Defoe's rambling house. Sophia's account of its "clean, white, wainscoted rooms, with clusters of roses peeping in at the windows, its spacious recesses and unsuspected nooks and cupboards, and, above all, its large beautiful garden full of trees" rings true to the memory of an eleven-year-old. So does her description of their new neighbor, Anna Letitia Barbauld, who was well known as the Unitarian author of a number of successful books for children. What Sophia saw was "a very thin old lady" with an "ugly" hair style, and she proudly resisted the educator's suggestions for activities that "would be most appreciated by the young feminine mind."[22] It would seem that Barbauld was no match for the growing independence of Frend's eldest daughter.

The foundation for Sophia's sense of her own powers continued to be laid in the family, where Frend was teaching her to reason through mathematics, to worship through astronomy, to approach the Bible through its original languages, and to read widely in modern authors. "I am very glad to hear that you are reading Locke on the Conduct of the Understanding," her proud father wrote when she was on a visit to one of her Blackburne uncles in Somerset.[23] Frend went on to encourage his not yet thirteen-year-old daughter to "learn something every day & think often on what you have learned & thus you will gradually form your mind to embrace truth from whatever quarter it may come."[24] From Locke he wanted her to draw the essential message that the true value of understanding lay in the way it "distinguished us from the brute creation." For Frend this lesson was far too important to be confined to a single source, and even as he encouraged his daughter to focus on Locke, he pointed out that little distinguished the eighteenth-century philosopher from the psalmist who prayed, "give me understanding according to thy word."[25] The concept of reason Frend was teaching his daughter was manifest in all times and places, and recognizing it was essential to the full experience of being human.

Three years later, when Sophia traveled to visit her Blackburne cousins in Yorkshire in the summer of 1824, her mother's letters provide a warm and bustling picture of life with Sophia's siblings: Frances (14), Harriet (12), Richard (10), Alicia (8), Henry (5), and Alfred (2).[26] In Sarah's letters, friends got married, couples dropped in to visit, neighbors played practical jokes, a footman was hired, the master bedroom was recarpeted, Frend's brother Richard came to visit, and two-year-old Alfred began to talk. Out of doors, Frances was raising chickens, and Henry's little black bantam hen was "killed with the heavy cold rains which we had so much of."[27] The image that emerges from this, our only direct glimpse of Sarah's world, is of an active group cheerfully integrated into its surrounding community.

It was also a world that cared deeply about children, and parental advice flowed freely to the Frends' absent daughter. "Remember to grow straight & hold yourself up as your time is short for growing,"[28] her mother reminded. Her father gave her similar advice on comportment when she set off to her first ball. "The main point is to avoid affectation of every kind whether in walk, gesture or talk," he told her. "There is an awkward bashfulness & a bold look of self importance between which is the happy medium which distinguishes a well-bred woman." For the better part of a page he counseled her on when to bow, how to respond to a misstep in the midst of a dance, how to converse with a partner, but in the end, he assured her, "the great secret however is to carry with you a cheerful & innocent heart desirous of giving & receiving all the satisfaction which the amusement is capable of."[29] While their daughter was away, the Frends did all they could to ensure that their beloved child had the skills she needed to fit comfortably into middle-class society.

Knowing how to act properly in the social sphere was also essential to provide Sophia with the space she needed to think for herself. In his letter, Frend was at least equally concerned about her response to the teachings of an evangelical bishop as he was with the way she carried herself at a ball. The bishop's "zeal is commendable, his doctrine questionable," he warned. "Sanctification by faith requires only to be well defined & every Christian must adopt it," he explained to his far-away daughter. Frend then proceeded to search for the true definition of "sanctification" by tracing its origins from English through French to Latin and finally to the Hebrew word "Shdosh" which meant "a holy or separated people." Christians were separated from the world by their "faith in Christ" he went on, but that faith did not matter unless it was accompanied by action.[30] Frend's etymological analysis of "sanctification" is a classic example of his literalist reason in action, designed to arm Sophia against the bishop's questionable doctrines. He wanted to be sure that she carried herself properly in the presence of the churchman, but as she did so, it was at least equally important that she was properly following reason in her thoughts.

Frend was committed to showing his daughter the etymological minutiae of reason's linguistically rooted religion, but as Sophia grew older she began to see the limits to his kind of doctrinal focus. Stoke Newington was at least as much a Quaker as a Unitarian community, and though the two groups agreed on a number of social causes, efforts to work together were hampered by

Quaker fears of anti-Trinitarian belief.[31] As a result, although Frend proudly preserved an anti-slavery statement among his papers to show his commitment to the cause, he was never permitted to join in the Quakers' anti-slavery campaigns. As a woman, Sophia had more freedom than her father did. "Nobody was afraid of me," she grinned,[32] as she freely crossed doctrinal boundaries that sharply separated men from each other.

One of Sophia's boundary crossings involved working with the Quaker Mary Lister, who was such an indefatigable charitable worker that Frend described her as "the most determined beggar he ever saw."[33] In 1825, Lister founded an Invalid Asylum for Respectable Females, to give temporary shelter to women "employed in shops and in other dependent situations" who were "obliged by illness to quit their places."[34] Designed to help women who were too sick for convalescent care though not sick enough to need hospitalization, Lister's organization thrived until it was absorbed into the National Health Service in 1948. Sixty years after the Invalid Asylum first opened its doors to forty-six women, Sophia still took pride in the role she had played as a teenager in establishing it.

Sophia also benefited from her connections to another Quaker, William Allen, who was the epitome of someone who agreed with her father on the evils of slavery but could not work across the religious divide. Nonetheless, he happily invited Sophia to a series of lectures on chemistry, including experiments, where Sophia was proud to learn cutting-edge Daltonian atomic theory.[35] Sophia's report of Allen's class is the only hint that she ever received any kind of schooling outside of her home. In Sarah's letters to Yorkshire, she had cheerfully reported that Sophia's younger sisters Harriet and Alicia were happy in their school, but there is no indication that Sophia ever attended one. Instead, she read widely and drank up stories and impressions from the flocks of people who gathered under her father's roof "like martins in the summertime."[36] The result was an education that touched on virtually every subject imaginable without any of the discipline to be found in performing routine class exercises, or of the social integration with a group of peers that classroom study can impart.

An important part of the education Sophia received in her home focused on action in the world. She eagerly listened while Henry Brougham told of receiving wheelbarrows full of letters when he acted as defense lawyer for Queen Caroline,[37] while Robert Hibbert reached for the best approach to educating the slaves on his West Indian plantation,[38] and while George Birkbeck laid plans for the London Mechanics' Institute that later evolved into Birkbeck

College in the University of London. She also heard a great deal about the goings-on in the Astronomical Society and about the founding of the University of London. Frend's daughter was very proud of all the things her father did to support and develop these institutions, but none of them had direct relevance to her. She accepted that as a woman she would never become a lawyer, educate slaves, join a scientific society, or attend a university.

Sophia nonetheless saw herself as a reasoning person and was fascinated by the steady stream of "Jews, Turks, heretics and infidels"[39] who were drawn to her father's house by his principled open-mindedness. Some, like the retired Jewish merchant who regaled Frend "with the most wonderful and impossible astronomical and historical stories drawn from his favourite writers," were more entertaining than edifying.[40] Even more incomprehensible, though equally welcomed, was a man who showed up at the Frends' gate "dressed something like a Turk, in robe, scarf and turban," professing his belief in Christ, and insisting that the Second Coming was at hand.[41] Little came of Frend's efforts to make sense of these people beyond amusing Sophia and scandalizing the neighbors. When the minister of the Unitarian Church on the Newington Green laughed that "if there is a queer fish in the world, he will find his way to Frend's house," Sophia shot back "Pardon me, Mr. — —, I do not remember our having had the pleasure of seeing you there.'"[42] More than sixty years after delivering this response, Sophia still remembered being "frightened at her own temerity," but she did not regret it.[43] She could be as perplexed as anyone by her father's outlandish visitors, but she would never allow an outsider to criticize what went on in their household.

Sophia learned more from listening to another group, which was at once more respectable and comprehensible. She remembered these men as "peculiar people," all of whom "had a leading thought or special study," which they applied to a determined search for the single truth that underlay all the diversity that the early nineteenth-century English were encountering in their travels. Their various efforts reflected two of the essential characteristics of the reason that had for so long defined their host's pursuit of rational religion within the Christian tradition. All of them were convinced that the truth they were seeking had been known in the past before the vagaries of human history had obscured it, and all of them were equally convinced that the way to uncover that truth was through some form of etymology.[44] Language was only one of the sets of symbols that promised entry into the wisdom of the ancients, however. Those who gathered in Frend's house also struggled to make sense of the wide variety of objects and artifacts left by ancient peoples. Several were astronomers, who recognized the fundamental religious imagery to be found among the stars and constellations that had arched over all peoples throughout human history. Their approaches were different, but all were deeply involved in reading the past through its objects, its languages, and its peoples. For a teenaged Sophia, their visits constituted an ongoing archaeological, philological, and ethnological seminar that laid the groundwork for what was to become a lifetime of investigation into the deepest truths of human existence.

Among Frend's visitors was Thomas Taylor, an "earnest and energetic" Greek scholar whose "leading idea" was a Neoplatonic conviction that the philosophies of Aristotle and Plato were at base essentially the same. Coleridge dismissed him with the quip that in his work "difficult Greek is translated into incomprehensible English,"[45] but Frend greeted him with shouts of "Welcome, Jupiter Olympius!"[46] Another visitor to the Frend household was John Landseer, who went beyond Taylor's Western focus. In *Sabaean Researches,* he focused on the Babylonians and recognized that "the small and beautifully-graven cylinders known as the 'Babylonish gems,'" were actually seals, which could be read. This insight transformed objects previously seen as personal adornments or amulets into "historical documents as well as religious monuments" and thereby opened up whole new worlds to the gaze of reasoning scholarship.[47] Unfortunately, Landseer was deaf enough that all conversation with him had to be carried out in shouts, which were not always worth the effort because he was so "unwilling to listen to differences of opinion." One of his most determined interlocutors was John Bellamy, a self-taught Hebrew scholar who had produced a highly idiosyncratic translation of several books of the Bible. Bellamy's and Landseer's "hot and strong" arguments could be very exciting because, as Sophia remembered, "the Antiquary was deaf and Hebraist positive—so positive, indeed, that he sometimes startled us all by knocking his hand violently on the table and denouncing all former translators of the Bible as 'Fools! Blockheads! Asses!'"[48]

Godfrey Higgins was also there. He may have been quieter, but he was no less opinionated. In his 1829 work, *Celtic Druids,* he concluded that the Druids, Chaldeans, Brahmins, Magi, Pythagoreans, and Essenes were all representatives of a single ultimate religion, which had originated among the first people, who were "of one faith, one color, and one language, and spread over the entire globe."[49] Soon after publishing *Celtic Druids,* he became a Freemason because he found Freemasonry to be "particularly suited to reunite Jews, Christians, and Muslims in recognition of the essential unity of their religions."[50] Higgins was neither as loud as Landseer nor as explosive as Bellamy, but he was no less convinced that there was a single truth of the universe, which could be found through close readings of the relics of ancient civilizations.

In the midst of all of these enthusiasts, Frend remained absolutely steady in his focus on the Bible as the ultimate guide to this truth, but Sophia was not so set in her ways. She was uncritically fascinated by her father's many visitors, who were, in turn, charmed by her. Landseer delighted in telling her "how Joseph's signet, hidden in Benjamin's sack," was in fact a cylinder of the kind he was deciphering in the Babylonian context, and he carefully explained that what scholars had interpreted as pagan gods were really astronomical bodies.[51] Sophia was proud that Landseer trusted her to translate Hebrew texts that he believed referenced things like the stars mentioned in Job or the march of the Assyrian King Sennarcherib, but talking to Higgins was more fun. She spent hours considering possible implications of the "immense quantity of knowledge of a rather miscellaneous character"[52] that supported his various theories and eagerly perused the antiquarian books he recommended. With her father's visitors, Sophia began to put her education to use by deploying her reason to decipher and explore the deep truths that lay hidden in the world around her.

In 1827, when Sophia was nineteen, Frend introduced to his household yet another interesting person. Augustus De Morgan was twenty years old and had just finished his studies at Cambridge when Sophia's father first brought him to Stoke Newington. Frend introduced him as "a 'rising man' of whom great things were expected in science,"[53] but Sophia and her siblings saw a companion who shared their love of "fun, fairy tales, and ghost stories"; who played the flute beautifully and often; and who even showed her "a new figure in cat's cradle."[54] In later years, De Morgan liked to tease Sophia that Frend had just brought him home as marriageable material, but ten years elapsed between their initial meeting and their marriage, and for most of that time they related more as siblings than as lovers.

Frend, for his part, was very happy to adopt Augustus as a surrogate son. He loved his own sons, but Henry and Alfred were very young, and Richard was far too disabled to be able to follow in his father's footsteps. Moving into the role of Frend's son was also comfortable for Augustus, whose own father was away during most of his early childhood and died when he was ten. That his father was absent does not mean the young man was alone, however. He was the scion of a far-flung family whose members were scattered all over England's empire, and he entered Frend's household well equipped to engage in conversations with Landseer and Higgins. He was also primed for discussions with Frend, who set him on a path of exploring and defending reason that would occupy him for the rest of his life. Frend had come to reason through religion and politics; De Morgan's route lay through mathematics, but the two men had no difficulty understanding each other. Their different interests and approaches reflected the changes that had taken place in the more than fifty years that separated their ages, but reason's common ground supported them both.

# CHAPTER 10

# Son of India

Augustus De Morgan was born in the southern Indian city of Madura on June 26, 1806. His mother, who had been married for eight years at the time of his birth, was finding that life as the wife of John De Morgan was not easy. Her husband was a moody and restless officer in the Madras Army of the East India Company (EIC), and the tenor of her life with him is suggested by a list of the places where she gave birth to their children: John Augustus De Morgan was born at Madras on May 16, 1799; James Turing De Morgan was born at Masulipatam on June 4, 1800; Eliza De Morgan was born at Pondicherry on September 27, 1801; Georgiana De Morgan was born at Madras on February 15, 1805. A few months before Augustus was due, John moved his family yet again. This time he had been offered the choice of Vellore, which had a reputation as "a gay place," and Madura, which everyone agreed was "dull." John, who had not received a promotion that he had been expecting, "was in low spirits and sulky" and therefore decided to go to Madura.[1]

The pregnant wife who accompanied John to dull Madura was still reeling from a major tragedy that had struck her family two years before. Early in that year, she and John had followed the established tradition for the sons of EIC officers when they sent John and James, aged five and four, to England for an English education. The De Morgans carefully placed their children in eminently respectable hands—their guardian on the ship was the Honorable

George St John, third son of the tenth Lord St John of Bletsoe—but that was not enough to ensure their safety, and sometime in June 1804 *The Prince of Wales* went down off of the Cape of Good Hope.

Elizabeth was devastated by the loss of her first-born sons, and Augustus's birth, just two years after his brothers vanished, presented yet another major challenge. He was born at least half blind, with a right eye that was "only a rudiment, with a discoloration in the center, which shows that nature intended a pupil."[2] From this description it would seem that Augustus suffered from ophthalmia neonaturum, also called infant conjunctivitis, which is often caused by otherwise asymptomatic gonorrhea of the mother. Elizabeth, however, thought it was due to "sore eye," or "ophthalmia," which he had contracted from one of her Indian servants. Her interpretation placed the blame firmly in India, and supported her desire to leave.

As his mother was coming to grips with Augustus's visual handicap, his father was facing a different set of problems. As a colonel in the Madras Army, John De Morgan was in charge of a battalion of Indian soldiers, or sepoys. In the early nineteenth century, the EIC was in the process of trying to transform the motley crowd of native forces who had been fighting for them since the middle of the eighteenth century into an English army. Requiring that the sepoys conform to English standards of dress and routine turned out to be a particularly controversial aspect of the transformation. In May 1806 several were flogged for resisting orders that the Hindus remove caste marks from their foreheads and the Muslims shave their beards. This just stoked the fires of anger, and a month later tensions were running high enough that Augustus's father would crouch by his window during night-time sentry changes and listen for signs of planned sedition.[3] He apparently heard nothing in Madura, but on July 10, 1806, just two weeks after Augustus was born, the sepoys in "gay" Vellore rose up and killed as many as 130 British troops, including John's friend and counterpart, Colonel Farmhouse.

Some combination of Augustus's semi-blindness and the Vellore mutiny proved to be the final blow for the De Morgan family in India. As soon as Augustus was judged old enough to travel, Elizabeth and John boarded the *Jane Duchess of Gordon* with their three surviving children: Eliza who was just five, Georgiana who was not yet two, and four-month-old Augustus. The family arrived safely in England on April 12, 1807, and settled in Worcester, close to Mrs. De Morgan's sister, Honor Briggs. Elizabeth was vastly relieved to be home again on English soil, but little Georgiana survived her ocean voyage only to die in October. Family life nonetheless went on, and nine months later, on July 18, 1808, Elizabeth gave birth to another boy, whom she named after her husband's brother George. But John was not present for the birth of this namesake of his beloved brother. Three months before their new child was born, he had left Elizabeth to care for their living children and returned to India alone.

When John rejoined his family two years later, he proved to be a powerfully disruptive force. Throughout his stay, some combination of money problems and the patriarch's quarrelsome personality meant that the family did not stay long in one place. Elizabeth had lived close to her sister in Worcester for the whole time her husband was away, but over the course of two years, John moved them all over England: from Worcester to London, from London to Appledore, from Appledore to Bideford, from Bideford to Barnstaple, and from Barnstaple to Taunton. To four-year-old Augustus, his father was a rigidly religious disciplinarian and a harshly demanding schoolmaster. When John finally sailed to India in January 1813, his six-year-old son breathed a sigh of relief. The two never saw each other again.

Augustus found the story of his early life troubling, and it took him decades to come to terms with it. As an adult, he devoted several notebooks to locating himself in relation to his father's and mother's families. He approached the task mainly as a genealogist, carefully recording names, along with places and dates of births, marriages, and deaths. In addition, he talked to everyone he could find—including the captain of the ship that had carried him back to England—in an effort to fill in the stories he was too young to remember.

The longest and most detailed of Augustus's notebooks is devoted to piecing together the family history of the frightening and disruptive father he experienced for two and a half years when he was a child. John De Morgan was the only surviving son of a family of De Morgans who lived in India throughout the eighteenth century. The earliest representative that Augustus could find was a John De Morgan who married a Sarah Despomaire in 1717. Augustus was not clear how this John De Morgan related to the John De Morgan he claimed as his great-grandfather, and who was married in about 1733 to Ann Turberville.[4] The French overtones of all of these names—De Morgan, Despomaire, Turberville—suggests that they were Huguenots who came to India to escape persecution in France after the 1685 revocation of the Edict of Nantes, but by the time they were marrying each other in India their most central identity was as members of the unruly band of adventurers that constituted the eighteenth-century EIC.

Ann and John De Morgan had nine children, but only one survived to adulthood: Augustus's grandfather, Augustus, who rose to the rank of captain in the Madras Infantry. In 1769, he married Christiana Hutteman,[5] who promptly gave birth to two boys: George and John. Soon thereafter, giving birth a third time proved too much for her; both mother and child died within a year of each other. The widowed Augustus waited until John was five and George six before he judged them old enough to travel alone on a boat to England. Just one year later he was killed during the siege of Pondicherry. The death of this Augustus at the age of thirty-eight made him rather a hero in the next generation. His grandson and namesake carefully collected stories about the dashing man to record in his family notebook.

More immediately, Grandfather Augustus's death made valuable orphans of now seven-year-old George and six-year-old John, and when the news reached London, various would-be guardians entered into fierce litigation about "the keeping and investment of their money." This issue was so engrossing that several months passed before anyone thought about locating the actual children. By the time they did, the school where the two little boys were thought to reside had broken up, and no one knew what had become of the children.[6] Several months of police searching finally located the lost De Morgan boys "in the possession of a worthy old couple" who had taken them in from the streets. The De Morgan boys had seen little enough of their birth father and mother that they easily accepted the situation—in later years, John vividly "remembered going out every morning to fetch the beer for his <u>daddy</u> in the long coat"—but as soon as they were found, the authorities swooped them up and placed them in a grammar school at Lewisham Hill. The orphaned boys always knew they were De Morgans, and in 1789 eighteen-year-old George and seventeen-year-old John returned to their natal India to enlist in the Madras Army of the EIC.[7]

George and John De Morgan arrived in India just in time to fight the last major holdout against company rule: Tippoo Sultan, of Mysore. They both survived a major confrontation in 1790, only to plunge into another in 1792, in which George seems to have been killed almost immediately, though his body was not found. John refused to believe that his life's companion was gone, and he fought on ferociously for years in hopes of releasing his beloved brother from Tippoo's prisons. In 1796 John was transferred to fight in the war that drove the Dutch out of Ceylon, but still he hoped. It was not until Tippoo was completely defeated in 1798 that John was finally forced to accept that his brother was truly gone. As he was making this adjustment, he met Elizabeth Dodson, who was visiting her sister, the wife of a surgeon with the EIC. Within months, the two were married in Colombo, Ceylon. They moved to India soon thereafter.

The rest of John's life was played out during a period of considerable change in the EIC that was both home and country to him. In Grandfather Augustus's time, the EIC had basically been a loosely organized group of adventurers who were out to make their fortunes in a rich and exotic land. With the defeat of Tippoo Sultan, however, all of southern India had fallen into company hands, and the EIC was transforming itself into an occupying force committed to bringing English order to the subcontinent. That story developed against the backdrop of the contested interface between the EIC's ever more powerful bureaucratic civil service and its ever less autonomous army.

This transformation entailed curtailing the free-form processes that had enabled many eighteenth-century army officers to become rich. Already, in 1796, John had resisted a new policy that required officers to give up the immediate gratification of collecting shares of taxes levied on local bazaars in exchange for the assurance that they could retire on full pay after twenty-five years of service. His unsuccessful resistance to the Bazaar Fund may have been the first step toward identifying him as a man with "a certain litigious cast of conduct," who had difficulties agreeing "well with those above him."[8] He cemented this reputation in the fight over the Tent Contract System, which flared during the time he was moving his family to England. Under the old system, EIC officers acted as independent contractors for the maintenance of the tents and other materials necessary for their regiments. In 1807, the civilian authorities transferred these responsibilities to a departmental division of transport, with the promise that the moneys the officers lost would be offset by raises in their bonuses.

This was precisely the kind of infringement on officer autonomy that infuriated John De Morgan, and as soon as he returned from settling his family in England he joined those who were resisting it. In the summer of 1809, regiment after regiment exploded in open mutiny while their officers were engaged in a war of words with the governor. In August, the governor finally prevailed by removing twenty-five officers, including Augustus's father, from their positions. John had to come back to England because clearing his name required him to appear in person in the London offices of the EIC. In the two years that he was moving his family from place to place, Augustus's father was waiting for the outcome of a court-martial that was carried out behind closed doors. He was not privy to the deliberations that were determining his fate,

and it is not difficult to see his frustrated uncertainty behind the many moves
and moods that were so hard on Augustus.

Finally, after two years, John was exonerated and reinstated as an officer in
the Madras Army. Vastly relieved, he remained with his family long enough to
oversee a final move to Taunton in Somersetshire and to attend the christening
of a third son, Campbell Grieg De Morgan, who had been born on November
22, 1811. Having fulfilled these familial responsibilities, John set sail for India,
where he resumed his resistance to civilian direction in the EIC. His final
quibble was over an order to change the weight of the paper on which he wrote
official dispatches, but before it was resolved he developed a "liver complaint"
serious enough that he was sent back to England. Augustus's notebook narrative about his father concludes with the entry: "His only attempt at a diary
of this voyage is, in pencil '1816 Sept 27 left Madras in the Larkins E. J. Capt.
Durableton, sick with liver.' He died near St. Helena," on November 27, 1816.[9]

As an adult, Augustus devoted page upon page of his notebook to the attempt to disentangle the rights and wrongs of his father's career, but the picture that emerged remained frustratingly unclear. In the end, he took comfort
in recording that his father was a colonel when he died, and that even those
who spoke "plainly to me about him" agreed that John De Morgan "bore the
character of an honourable man and a good officer."[10] This gentlemanly set of
virtues was of paramount importance to Augustus. Recognizing them in John
De Morgan allowed him to find a resting place in the story of the troubled
and troubling man who was his father.

Standing as a counterweight to John De Morgan's angry isolation and
much more immediately present was Augustus's mother. When viewed
through her marriage, Elizabeth De Morgan can seem like a solitary soul,
abandoned by her husband to face the world alone. In fact, as she was the
fourth of sixteen children, she was never wholly alone. Her father, James Dodson, had supported his family with a job in the Customs House in London. As
his many girls grew up, he was well positioned to ensure them places in what
became known as the "fishing fleet," which carried eligible young women to
meet young men living in the East. The Dodson girls were very successful in
this venture, and eight of the nine married men who were actively engaged in
expanding England's empire. Sarah's husband was the surgeon general of the
EIC; Sophia's and Anna Maria's husbands were both admirals in the Royal
Navy; Honor's husband served as a doctor in Ceylon; Jane's husband rose to

the level of general in the EIC; Ann's and Georgiana's husbands were officers in the EIC civil service in Madras. Only the youngest, Henrietta, seems to have spent her whole life in England as the wife of a protector of the Ecclesiastical Courts.

This family provided the support that carried Elizabeth through decades in England without her husband. Hidden behind the diversity of their married names, the Dodson sisters formed a resilient web of family that covered the English world. They welcomed each other into their homes, helped with the numerous transitions of their peripatetic lifestyles, and acted as sponsors—that is, godparents—to each other's children. The number of these second-generation Dodsons was enormous. On his father's side, Augustus had no first cousins at all; on his mother's side he had seventy-two, ten of whom married each other.[11] The whole was so complicated that Augustus carefully preserved among his papers a second notebook in which one of his Dodson relatives carefully recorded the marriages of all the Dodson sisters, as well as the births, marriages, and deaths of all of their children. Augustus De Morgan proudly carried the name of his father's military family, but they were not present in his world. The extended family that interacted with him as a boy were the cosmopolitan Dodsons, whose vital statistics are recorded in this little book.

Yet another Dodson was important to Augustus's sense of self. Elizabeth's grandfather, James Dodson, is nowhere mentioned in the Dodson notebook, but he was a kindred spirit to his great-grandson. This Dodson was an intellectual, whose mathematical prowess elevated him from printer's son to master of the Royal Mathematical School and member of the Royal Society by the time he was forty. From these positions of relative stability, he began thinking about ways to insure his life so as to provide support for his wife and children in the event of his death. Somewhat ironically, he died before his scheme was implemented, but both the Equitable that blossomed under Morgan, and the Rock that grew under Frend rested on the actuarial foundations laid by Elizabeth's grandfather. There was, however, little room for the abstract world of Dodson's mathematics among the admirals and generals who married his granddaughters, so the first James Dodson had no place in the family stories that Augustus heard as he was growing up. But as he was forging his own path as a mathematics teacher, he was very proud of the great-grandfather who had preceded him.

Beyond routine recordings of his birth and marriage dates Augustus himself does not appear in either the De Morgan or the Dodson notebooks. He apparently wrote of his own life in a third notebook, which Sophia saw no reason to keep after she used it as a source for the *Memoir of Augustus De Morgan,* which she published in 1882. Like all memoirs written by grieving widows, Sophia's reconstruction of her husband's life is deeply colored by her personal lenses, but it was steeped in the family tradition of reason and conscientiously included only information obtained from sources she deemed reliable. When supplemented with the occasional comments Augustus made in letters to friends, her *Memoir* contains an image of the man he was to become.

One central issue shaped all of Augustus's life, but neither he nor Sophia was ever going to admit its importance. Although he could see very little, he learned early that it was unmanly to explain any of his decisions in terms of that handicap. By his own account, with his right eye he could see as much "as with any one finger—no more and no less," and with his left eye "without spectacles my reading distance would be less than six inches for moderate type, or my own handwriting. Four inches would be agreeable and convenient."[12] His eyesight thus made it unthinkable that Augustus would follow in any of his two families' many military footsteps.

Augustus's eyesight equally prevented him from traveling into the world over which his mother's family and friends ranged so freely. In later life, he liked to point out that when he first arrived in England he "had passed three fifths of my life on the water,"[13] but although he admitted that he remembered nothing about the trip, he declared that completing a journey of eleven thousand miles in which he crossed the equator twice was more than enough to last him a lifetime.[14] He was always proud to have been born in what he described as the "sacred city of Madura"[15] and loved listening to the tales of travelers who had been there, but he never returned. Instead, he lived in London, with its narrow streets and linear buildings. He always claimed that despite being "strictly unocular,"[16] he had no trouble seeing two dimensions, but outside of these urban spaces, he faced a visual wasteland. Trees loomed ominously massive, hedges ambled unpredictably, open spaces were "a *desolation,*"[17] and heights were sickening.[18] As he aged, it became so difficult for him to navigate unfamiliar landscapes that a week-long sojourn to the English countryside left him prostrate. He would never use his eyesight as an excuse, but he was never going to travel out into England's sprawling empire.[19]

Augustus's poor vision may have prevented him from roaming the physical world, but from an early age he traveled that world in books. It seems extraordinary, but the one myopic eye he could use was able to bear "any amount of work without fatigue,"[20] and after his father taught him to read, he never stopped. He began his lifetime of reading in the early nineteenth-century Gothic novels that Jane Austen spoofed in *Northanger Abbey,* and by the time he was twelve years old was well up on Walpole's *Castle of Otranto* and looking for more. When his mother took him on vacations to unfamiliar places, he eagerly looked for bloody armor and creaking skeletons; at night he lay awake so as not to miss processions of headless horsemen or of carriages driven by equally headless coachmen.[21] In his mother's house he listened with ghoulish fascination as his Uncle James Briggs claimed that Grandfather Augustus's head had rolled between his legs at the siege of Pondicherry. The same loquacious uncle then set his nephew to ponder the question of whether it was possible, as people had claimed, that his grandfather had foreseen his own death.[22] He may have had difficulties with his father, but from a young age the exploits of his family provided vivid grist for his imaginative mill.

The young Augustus approached his world in pictures as well as words. Whether because of or in spite of his dreadful eyesight, he had a visual imagination powerful enough to match anything he read or heard. He loved to draw, and throughout his life virtually any open surface was an invitation for illustration. His doodles are often silly, but there is an undercurrent of seriousness to most of them. So, for example, it is not hard to see the "Canonization of General St. Right-about-face" as a response to the stories about his grandfather he first encountered as a child (figure 4). Throughout his life, illustration remained a way for Augustus to access and explore ideas he was not ready to express in words.

For Augustus, drawing what was difficult to pin down with words was at least as useful in mathematics as it was anywhere else. His visual approach is central to the story of his first encounter with Euclid that Sophia included in her *Memoir.* In her telling, Augustus's immediate response to Euclid was to create elaborate images using ruler and compass, a process he described as "*drawing mathematics.*" Upon seeing what he was doing, one of his mother's friends quickly corrected the twelve-year-old boy: "That's not mathematics," he said crisply. This determined teacher then "rubbed out" all that Augustus had drawn, in order to focus the boy's attention on the words of "the first demonstration he ever knew the meaning of."[23] The clear lesson was that Augustus was not to use pictures to understand mathematics, but the many images that survive among the notes he wrote as an adult clearly show that he never stopped "drawing mathematics."

Figure 4: Augustus De Morgan, Cannon-ization of General St. Right-about-face.
Strong Room E, University College London Library Services, Special Collections.

Regardless of his approach to mathematics, Augustus's mother was devoted to the "quiet, thoughtful boy" who was "always speculating on things
that nobody else thought of, and asking her questions far beyond her power
to answer."[24] She encouraged her son in his voracious reading, smiled at his
drawings, and taught him how to play the flute so he could always have music
at his fingertips. She also did all she could to share the sense of belonging that
she had found in the Anglican Church. Most of Elizabeth's friends and family
were connected to the military, but the church was a major source of support
for the widowed young mother, and vicars formed another current in the river
of people that ran through her life. Elizabeth did all she could to introduce
her son to the churchmen who had offered her such solace, but her religiosity
went beyond the social in ways that were hard for him to negotiate. She practiced a deep evangelical form of Christianity that for her children translated
into a strict regimen of biblical memorization supplemented by two weekday
and three Sunday sermons every week.[25] It is not difficult to see Augustus's

Figure 5: Augustus De Morgan. Schoolmaster doodle.
Strong Room E, University College London
Library Services, Special Collections.

later rejection of the Anglican Church to be rooted in the soil of his mother's rigorous demands.

Augustus's early encounters with education were as negative as those with the church. He began attending school during the time his family was constantly relocating, and he described himself as the "victim" of an ever-changing series of teachers.[26] Twenty years after his last confrontation with a village school, he vividly captured his experience with the image of an enormous schoolmaster standing firm with a helpless little boy in tow (figure 5).[27]

The placement of this picture on the title page of a textbook is arguably as salient as the image itself. Although he spent his life as a teacher, Augustus always identified with the students under his care.

In January 1821, when Augustus was not yet fifteen, his mother sent him away to a school run by a Mr. Parsons in Redland, now a suburb of Bristol. There his visual impairment made him the butt of all sorts of gibes and practical jokes. The most egregious of his tormentors would hold a penknife to Augustus's blind-side cheek, and then call his name so that when he turned he stabbed himself. After a few iterations of this trick, which was fierce enough to draw blood, Augustus asked for help from his friends, who caught the penknife owner and turned him over to Augustus for a "sound thrashing."[28] In Sophia's telling, the moral of this story was that Augustus was growing into his birthright as a De Morgan who would defend himself. At another level, it is a story about learning that his eyesight could be a social as well as physical handicap. A large part of the education he received at Mr. Parsons's school focused on how to be a man, and part of what he learned was that he should never use his eyesight as an explanation or justification for his actions.

This tale of schoolboy bullying may also be read as a story of friendship. Whenever Augustus found himself among a group of contemporaries, he made friends that lasted his lifetime. At Mr. Parsons's school, the group that formed around him was drawn together by the rich world of his imagination. They were enthralled by his elaborate caricatures of classical scenes, drawn in the style of the political cartoonist James Gillray, and by late-night parties in which "De Morgan produced a match pistol and a tinder, snapped a spark and lit the candle," and proceeded to read aloud until all were too sleepy to go on.[29] More than sixty years later, the men who shared these adventures still vividly remembered their good times together.

Nonetheless, as an adult Augustus could find little positive to say about Mr. Parsons's school. Nighttime revels could be fun, but his daytime memories were of the mind-numbing expectation that he and his classmates learn by heart forty lines of Greek or Latin every weekday, with the whole two hundred to be repeated on Saturdays. Even the grin with which Augustus remembered the nimble subterfuges he and his friends developed to get around this grinding demand was not enough to erase its tedium.

Mathematics was different. Augustus arrived at the school knowing little enough that he had to begin at the most elementary level, but learning mathematics did not require any of the drudgery he found in the classics. Instead, as an awed classmate put it, he read his first text "like a novel," "dashed" into the next one, and was soon "out of sight as far as I was concerned." Decades after he had finished at the school, the wainscoting around the schoolboy's pew of St Michael's Church in Bristol still sported "the first and second propositions of Euclid and one or two simple equations" he had poked out using the "sharp point of a shoe buckle."[30] Apparently Augustus even drew mathematics in church.

The little piece of graffiti Augustus carved out while the required religious service droned on around him may stand as a marker that pointed the way forward from his childhood. He had long known that following in the footsteps of his father or his many uncles was not a real option, and living with his mother's religious demands was undercutting any attractions that might attend the life of a clergyman. He loved the imagined world of his drawings and the novels he read, but except in the dead of night that world was closed to him, and studying the classics in the light of day was simply awful. When Augustus left Mr. Parsons's school at the age of sixteen and a half, he was beginning to recognize that mathematics offered him a place to be and a place to roam.

# CHAPTER 11

# Reading Man

After two and a half years at Mr. Parsons's school, Elizabeth sent her eldest surviving son to finish his education at Cambridge. The Trinity College Augustus entered on February 4, 1823, had for centuries been vying with its immediate neighbor, St John's, for the honor of having the largest number of students and fellows. Its Great Court, complete with elaborate seventeenth-century fountain, was the biggest enclosed court of any Cambridge college. Through its west side lay the elegant colonnades of Neville's Court, first built to accommodate additional students in the early seventeenth century and later expanded with the addition of an elegant library designed by Christopher Wren. By 1823, the sons of England's burgeoning middle class had begun to flow into England's universities at a prodigious rate; between 1800 and 1820, the number of matriculants at Cambridge increased threefold, from 150 to 450 students.[1] On the order of 100 of the total were students at Trinity, which means that despite its impressive size, when De Morgan arrived there was no room for him in his college. So he lived in lodgings in the town until his final year, when a room opened up near what is affectionately known as "the windy passage" at the southwest corner of the Great Court.[2]

In addition to forcing many, like De Morgan, to live outside of their colleges, the university's phenomenal growth put considerable pressure on the organization of college studies. Admissions standards remained lax—most colleges required only a certificate of competence from a schoolmaster with a master's degree from either Cambridge or Oxford,[3] and many of those who entered never completed their degrees. The lifestyles of these so-called fast men continued to be dissipated and extravagant, but the growing middle class that was sending its sons to Cambridge did not relish this kind of debauchery. De Morgan was known as a "reading man," that is, a serious scholar. While other students were riding or rowing as a break from their studies, he was "foraging for relaxation"[4] among the books in Trinity's Wrenn library, and when he had enough of Trinity's tomes, he read all of the fiction he could wring out of the local circulating library.[5] He came into his own as a bibliophile while he was a student at Cambridge.

De Morgan read prodigiously, but he was never a hermit. He joined the Cambridge Amateur Music Union Society, and when his friends stumbled home after wild nights on the town, he would put down his books to soothe their headaches with his flute.[6] On other raucous evenings he and his friends dressed in their choristers' robes in order to sing Newtonian equations set to familiar tunes, which exercise proved rather more difficult than they had expected.[7] As he had done at Mr. Parsons's school, at Trinity De Morgan made friends to last a lifetime.

As her son was figuring out how to make his way, Elizabeth was pouring out her anxieties in letters from afar. "Dear *own* son of your father and mother," she gushed after he organized a budget, and then urged him to guard against "the sort of carelessness which brings many young men to ruin." Religion was another major theme, and she combed his letters to be sure he was not straying. All Trinity students were required to don their academic gowns and gather twice a day in the vaulted college chapel for a chanted service accompanied by an organ that was reputed to be one of the best in the country, but this was not enough to satisfy Elizabeth's evangelical soul. She exhorted him to supplement the required services with trips "to hear the Gospel on Sundays," "beseeched him" to read the tracts she sent, and emphasized again and again that unless he went "*by* yourself and *for* yourself to Christ for pardon and grace," he remained "liable to be hurled into everlasting torment by every little accident, every disease, nay, even by a crumb of bread going the wrong way."[8] Her son took this advice with a grain of salt. The sixteen-year-old who walked through the enormous east-facing Great Gate of Trinity College was ceasing to be his mother's Augustus and becoming De Morgan.

Elizabeth's first hope was that her son would focus on the classics, but De Morgan was no more interested in memorization at Cambridge than he had been at Mr. Parsons's school. He did poorly on the early exams in classics but soared on the later ones that focused on mathematics. "He is not only in our first class, but far, very far, the first in it,"[9] his tutor happily informed Elizabeth at the end of his second year. As De Morgan trooped to the required services in the Trinity College Chapel, he was much more likely to be impressed by the statue of Newton that Wordsworth described as "the marble index of a mind forever / Voyaging through strange seas of Thought alone"[10] than he was by his mother's admonitions.

The mathematical seas that De Morgan sailed were different from the ones that Newton had charted in the late seventeenth century, however. Newton as a person was not the wise, transcendent figure of the Trinity statue, and before his death he had fought bitterly with Gottfried Leibniz over which deserved credit for discovering the calculus.[11] This quarrel had a profound effect on mathematical posterity. By the time of Newton's death, the mathematical world was divided between the English, who accepted his view of the subject, and the continental thinkers who followed Leibniz's lead. Throughout the eighteenth century, the continental and English mathematical communities worked independently of each other until the differences between them were so enormous that it was all but impossible for the two groups to understand each other.

When De Morgan arrived at Cambridge, this situation was slowly beginning to change. A major force behind this movement was the same Robert Woodhouse who so thoughtfully responded to Frend's *Algebra* in 1803. Woodhouse's review of Frend's book was just one of a long series in which he considered not only English works, but also French ones by men like Sylvestre-François Lacroix, Joseph-Louis Lagrange, and Pierre-Simon Laplace. Woodhouse's engagement with French mathematics in the early 1800s was highly radical in an England that was reacting to Napoleon, but a deep chasm had separated French and English mathematics even before the revolution. Woodhouse was totally uninterested in the priority dispute that was the accepted origin of this controversy: "The truth seems to be that Newton and Leibnitz arrived at the same discovery, each proceeding by his own proper route of ideas," he said crisply.[12] Woodhouse's Cambridge education had brought him to the apex of the mathematical world following the English approach; he was now deeply interested in exploring the alternative routes that continental mathematicians had been following.

The continental route of ideas that Woodhouse was exploring may be called "analysis," to distinguish it from the English tradition of "algebra" or "universal arithmetic." The difference is that instead of piously taming algebra's symbols to the service of arithmetic, continental analysts were adventurously following those symbols into a whole new world. The results were staggering. As they pursued a form of reason that was more rule-based than subject-focused, enormous new mathematical vistas opened before them.[13]

The *Geometry* of René Descartes formed the basis of the analytic tradition. Developing that geometry entailed moving one step beyond an algebraic equation, like $x^2 = 16$, to recognize it as a specific instance of a more general relationship between two variables, which could be written as $x^2 = y$. The next step was to see that the equation $x^2 = y$ is just one of many that could be formed between $x$ and $y$; other examples would include $x^3 = y$; $x^3 + x^2 = y$; $x^2 + y^2 = r^2$. What Descartes did in the first half of the seventeenth century was to explore possible connections between these kinds of general equations and geometrical figures on a plane. To do so he would designate one line in a drawn figure as the x-axis and another as the y-axis. This approach was soon regularized into dividing the plane into quadrants with a horizontal x-axis and a vertical y-axis, and thus assigning an $x$ and a $y$ value to each point on the plane so that figures would correspond to equations of $x$ and $y$. Working within this "Cartesian" plane allowed mathematical thinkers to move easily from geometrical figures to arithmetical equations and back again.[14]

Moving geometry onto the Cartesian plane changed the subject beyond recognition. Using equations to describe spatial figures suggested that geometrical curves should be defined as those that could be described by equations, rather than as those that could be constructed with compass and straightedge, which was the Greek definition. In some cases, the two were easily seen as equivalent; a circle could be expressed with the equation $x^2 + y^2 = r^2$. Other cases were more ambiguous: the parabola traced by the equation $x^2 = y$ could not be drawn with compass and straightedge, but it could be found among the Greek conic sections. Other situations were considerably more confusing. When the sine, which the Greeks defined as the ratio of opposite side over hypotenuse in a right triangle, was moved to the Cartesian plane it became the general equation sine $(x) = y$, which was represented by a curve that undulated over and under the x-axis infinitely in both directions. Following the trace of a point on a circle that rolled along the x-axis generated a curve known as the "cycloid," which seemed to have all the characteristics of a curve but could not be written as an equation at all. Coming to terms with the wonders of the mathematical world opened up by Descartes's *Geometry* engaged European analysts for centuries.

Newton and Leibniz were both acting as analysts when they shared the honor of "discovering" calculus in the late seventeenth century. The calculus that they each developed was focused on one of the central challenges of Descartes's *Geometry*—the way it connected the continuous world of geometrical space to the discontinuous world of numbers. The issue may be seen in the move from the discrete, arithmetic values of $x^2$ to the continuous curve of $x^2 = y$. Hidden within this apparently simple move is the startling reality that there are an infinite number of values between any two values of $x$ (and between any two values of $y$). This infinity exists as much in the gap between $x = \frac{1}{2}$ and $x = \frac{1}{3}$ as it does in the gap between $x = 2$ and $x = 3$. A central challenge of Cartesian geometry was how to understand all of the teeming arithmetical infinities that are somehow embedded in simple curves.

The calculus that Newton and Leibniz developed at the end of the seventeenth century was born of the effort to understand the interface between discrete arithmetic and continuous geometry. What we now call the "derivative" is a precise expression of the slope of a tangent to a single point on a curved line, and the "integral" is a precise expression of the area of a space bounded by curved lines. Each of these values had always been easy to find in figures bounded by clearly defined straight lines; the challenge was to find their values when their boundaries became curved and the neatly defined distances between their points disappeared. Over the course of the seventeenth century, analysts managed to find the derivatives and integrals of specific curves, like the parabola, the sine, and the cycloid. Newton's and Leibniz's triumph lay in creating a single method to cover all of those specific cases. The calculus they developed could be reliably used to solve problems generated within the new world of Cartesian geometry, where equations from the discrete world of arithmetical numbers defined curves in the continuous world of geometrical space.

Developing a clearly defined set of techniques for calculating the derivative and the integral of curves on the Cartesian plane was an enormous achievement, but Newton and Leibniz were still faced with the challenge of explaining and justifying the validity of their techniques. The two men approached this task in radically different ways. Newton's calculus was grounded in the physical vision of his *Principia*. He transformed the Cartesian plane into a two-dimensional mirror of the absolute time and space that contained his cosmology by treating the horizontal axis as a measure of absolute time and the vertical axis as a measure of absolute, linear space. From this perspective, geometrical curves became the tracks left by quantities, which he called "fluents," that were moving through a continuously flowing mathematical time. The calculus that Newton presented in the *Principia* was the science of motion within this world of absolute, true mathematical time and space. In it, the slope of a curve, which we would call its "derivative," was the "ultimate velocity" of a "fluent" at a particular instant. Explaining how anything could be said to have a velocity at a particular moment—that is, could be moving without a change in time—was all but impossible. Newton nonetheless tried, in famously opaque passages like the one in which he explained that the derivative is to be understood as "the ultimate ratio of evanescent quantities," that is, as "the ratio of the quantities not before they vanish, nor afterwards, but with which they vanish."[15] Newton represented this "ultimate velocity" or "fluxion" (or "derivative") by placing a raised dot over the moving variable or "fluent" to produce $\dot{y}$.

Leibniz took a radically different approach to defining the derivative. He lived in an essentially noncontinuous, atomistic world. For him, then, the spaces between discrete numbers on a continuous curve were filled with infinitesimally small quantities, which he called "differentials" and represented as $dx$, $dy$, and so on. Leibniz's differentials were so small that no amount of addition or multiplication could ever render them finite. However, since they comprised a universe of tininess in which relations were structured just like those among finite arithmetic numbers, the result of dividing one of Leibniz's infinitesimals by another one was a finite number. Thus, for example, if $dy$ were twice as large as $dx$, then $dy/dx$ would be the finite number 2. Leibniz represented the derivative as the ratio of two discrete infinitesimals, $dy/dx$.[16]

The different sets of symbols Newton and Leibniz produced reflected the differences in their efforts to explain themselves. Newton's dot notation was as opaque as his explanations of it, and his symbols served simply to mark the place of the underlying ideas they were covering. Leibniz's $dy$s and $dx$s operated in the opposite way. They were widely recognized as meaningless, but they were powerfully effective not just for solving problems but also for opening new areas of study. Leibniz's algebraically nimble $dy$s and $dx$s could skip through all kinds of situations that were all but inaccessible to Newton's $\dot{y}$s. They opened up whole areas—from multivariate calculus to differential equations—that continental analysts were eager to explore.[17]

When presented this way, Newton's symbols seem clearly inferior, but they were actually very well suited to the context in which Frend and his compatriots learned them. For those to whom the calculus provided a glimpse into the mind of the Creator, the clumsiness of the dot notation could be seen as an advantage that required investigators to keep their focus on the ideas behind the symbols or the words. All of the objections of incomprehensibility that Frend leveled against negative numbers were equally applicable to Leibniz's infinitesimal *dx*s and *dy*s, but the nature of "fluents," "evanescent quantities," and "ultimate velocities" were wonderful subjects for rumination on "solitary walks."

Over the course of the eighteenth century the different symbols that Newton and Leibniz used for the derivative became the marker of a fundamental set of differences that distinguished continental and English views of the nature of mathematics. When it came to justifying themselves philosophically, the English had the clear advantage, but ignoring the issue allowed the continentals to be far more creative. *Allez en avant; la foi vous viendra!* (Forge ahead, faith will follow!) was the battle cry of generations of continental thinkers for whom the English conceptual focus was an impediment to mathematical progress.[18]

De Morgan came of age in a different world, however. By the time he was in college, the French Revolution had fundamentally changed thinking about mathematics in France. After the revolution, mathematical thinkers from all over Europe were brought together to teach their subjects in the newly formed École Polytechnique, an engineering school in which the value of reason lay in solving problems, not in reading the scriptures.[19] Establishing a clear foundation for the calculus became a central concern for those who were trying to teach the subject to a new generation. In 1810, Sylvestre-François Lacroix adopted a hybrid approach in his *Traité élémentaire de calcul différentiel et de calcul intégral.* There he used Leibniz's *dy/dx* notation for solving problems, while using Newton's ideas of motion in absolute time and space for explaining its results. It was not a totally satisfying position, but it worked, and more than one hundred years of successful mathematical development supported his decision to trust to the basic solidity of the calculus without worrying too much about the strength of its foundations.

In England, things developed differently. In 1803, Woodhouse, who was closely following French foundational discussions, published *The Principles of Analytic Calculation,* in which he enthusiastically embraced the Leibnizian notation, with little concern for its ultimate legitimacy. Mathematics, he declared, was essentially an inductive science. Just as scientific laws were formed by generalizing individual experiences, mathematical laws were found by generalizing individual results. It is more natural to treat "the steps by which we ascend to expressions, more and more general, merely as so many improvements in the language of Analysis," he explained. For him, the move from writing *(x + x + x)* to writing *3x* was an inductive one; as was the move from *(xxx)* to *x³*. All of these symbolic shifts could be useful as a way to streamline thought, but their validity rested always on the specific cases from which they were generalized.[20] When approached in this way, the successful development of *dy/dx* symbology was itself enough to legitimate its use.

Woodhouse's *Principles* put him in the somewhat odd position of entering a French discussion by means of an English book. But he was much more interested in getting his countrymen to follow him to France than he was in getting the French to engage his arguments. Introducing his English colleagues to the Leibnizian *dy/dx* notation was essential to this part of his project because it was the only way that they could read French mathematical papers. Surely, he argued, it was "desirable to have the same notation universally adopted, in order to facilitate the communication of science between different nations."[21] He tried to be diplomatic as he pushed his compatriots to adopt the alien symbols. The "method of fluxions, or of limits, or the infinitesimal calculus" was not "absolutely erroneous,"[22] he explained, but Leibniz's *dy/dx* symbols were less "ambiguous," "more readily extended," "most readily apprehended by the eye," and even less subject to "typographical errors."[23] Whatever one's view of the foundations of the calculus, Woodhouse was convinced that using the *dy/dx* notation was the best way to bring together the worlds of English and continental mathematics that had been divided for far too long.

Woodhouse's *Principles* had no immediate effect on the ways mathematics was taught at Cambridge. Achieving that change would require a kind of political maneuvering that he was incapable of. The credit for actually introducing the *dy/dx* notation to the English goes to a group of young men who arrived at Cambridge eight years after a disappointed Woodhouse had abandoned mathematics for astronomy. Charles Babbage, John Herschel, and George Peacock were devoted to bringing French mathematics to England and were the founding core of what they called the Analytical Society; the slightly younger William Whewell (pronounced Hew-well) joined them in 1812. De Morgan arrived at Cambridge a decade after the graduation of its members had dissolved the Analytical Society, but their ideas had a profound effect on his experience there. Some were tutors at Trinity College; others he met only after he graduated. However he first encountered them, each and all were to be his intellectual companions for life.

Babbage was the fiery rebel of the group. His father was a merchant banker whom his son described as "a tyrant in his family."[24] Perhaps because the two got along so poorly, the young Charles was remarkably free of concerns about conforming to the middle-class world in which he was brought up. Whereas very little was expected of Babbage when he entered Cambridge, a great deal was expected of Herschel. The son of England's celebrated astronomer, William Herschel, John had studied with a private tutor who ensured that he had already mastered everything in the curriculum before he arrived at Cambridge, so he sailed through the university with the reputation of being a genius. Despite his name, Peacock was the little brown bird of the group. Neither the offspring of a well-to-do businessman nor the scion of a powerful scientific family, he was the diligent son of a clergyman, whose efforts at Cambridge were rewarded with a variety of scholarships that enabled him to afford his tuition.[25] The youngest, Whewell, was the brilliant child of a Lancashire carpenter who scrimped and saved so that his boy could receive a Cambridge education. The tenor of the Cambridge this group entered was much the same as it had been when Frend was expelled almost twenty years earlier; England was still locked in war with France, Isaac Milner had just been once again elected to serve as the vice-chancellor of the university, and fears of French ideas remained strong. At Cambridge these young men were drawn together by their determination to open their university to the ideas being developed in the larger world around it.

The precipitating event that led to the formation of the Analytical Society occurred in fall 1811, when an evangelical student movement to distribute Bibles drew a number of the principals in Frend's trial out of the woodwork. The opening round was fired by Frend's cousin, Herbert Marsh, who in the almost twenty years since he had recused himself from his cousin's trial, had become the Margaret Professor of Divinity at Cambridge and a pillar of the Anglican establishment. In September, Marsh preached a sermon in St Paul's Cathedral in London in which he insisted that England's children should learn their religion from the Anglican Prayerbook, and Frend gleefully rose to the challenge. Here comes the old argument, he scoffed, in his "Christian's Survey"; if England's children are brought up reading the Bible, "the church will be overthrown and with it the state."[26] Rather ironically, Milner agreed with Frend on this point. The only problem Milner saw in the students' proposal was the danger of allowing them to do anything on their own initiative, and so he set out to ensure that the organization be formed by the administration. Marsh objected strenuously. Thus, throughout the fall of 1811, the entire university was gripped by a struggle over Bible distribution, which was finally resolved with the formation of the Cambridge Auxiliary Bible Society.[27]

This discussion of Bible reading was the epitome of the Cambridge parochialism that Babbage, Herschel, and Peacock found so stifling, and when Babbage found the "the walls of the town were placarded with broadsides" focused on how best to distribute the Bible, he counterattacked. Holding up Lacroix's textbook as a work "so perfect that any comment was unnecessary," he proposed forming a society to translate it into English. Babbage's sacrilegious call struck a chord among his peers, and in 1812 a small group of Cambridge students came together to consider French mathematics. Their goal was to advance "the Principles of pure D-ism,"—that is, the Leibnizian $dy/dx$ notation—"in opposition to the Dot-age of the University,"—that is the Newtonian dot notation.[28] The young men of the Analytical Society were determined to move English discussions of reason out of the church and into mathematics.

It was a heady ambition. Month after month, the Analytical Society gathered for boisterous discussions of the nature of reason in mathematics.[29] By the end of a year, Babbage and Herschel were ready to spread their ideas to a larger audience by publishing the *Memoirs of the Analytical Society of the Year 1813*. It was a horrifically difficult task to print a book filled with continental mathematical symbols, but they pushed through to the end. The result was a work that virtually none of their countrymen could read, so it fell stillborn from the press. "The fire of enthusiasm spreads only where it meets with inflammable matter to receive and cherish it," Herschel sighed.[30] Clearly the Analytical Society was not going to get anywhere until there was a larger group of people able to follow their ideas.

Changing the curriculum at Cambridge was the obvious next step toward accomplishing this goal, but neither Babbage nor Herschel was interested in taking on this project. Both of them finished at Cambridge in 1813—Herschel as senior, or first wrangler, Babbage without a degree because he had managed to offend the university examiners enough that he was barred from taking the Tripos. Babbage was not particularly concerned; by 1814 he had married the love of his life and was living on his father's money in London. Herschel was also in London, circling law, chemistry, and astronomy as possibilities for a life's work. Their efforts to change the ways mathematics was understood at Cambridge were just the first step of a larger program to open their whole country to new ideas about the nature of mathematics, of science, and of knowledge, that each of them was to pursue in rather different ways for the rest of their lives.

This left Peacock to push through the reforms of the Cambridge curriculum. With no family wealth to set him free, after graduating just below Her-

schel in 1813, he remained as a fellow of Trinity College, and by 1816 had succeeded in publishing an English translation of Lacroix's *Traité élémentaire* for
the benefit of Cambridge undergraduates. Making Lacroix's textbook available in English may have been a first step toward educating Cambridge students in French calculus, but at least equally important were the practical
steps Peacock took to change the symbology of questions on the Tripos examination. The year after his translation appeared, he served as moderator of
the exam and posed the questions using the *dy/dx* notation, with translations
into the Newtonian dot notation at the bottom of the page. A huge uproar
greeted this move, but his quiet persistence managed to wear the opposition
down. By the time De Morgan arrived at Cambridge in 1823, the questions on
the mathematical Tripos were routinely posed using the continental notation.

Peacock's translation of Lacroix provides an interesting perspective on the
issues and discussions that had so energized discussions in the Analytical Society. Although Babbage may have proclaimed Lacroix's text to be "so perfect that any comment was unnecessary," the work that Peacock published
was accompanied by copious revisionary footnotes. Peacock's explanation for
these reflects his continued respect for conceptual objections to Newtonian
mathematics. "Our notion, indeed, of a ratio, whose terms are evanescent [the
derivative] is necessarily obscure, however rigorously its existence and magnitude may be demonstrated; and its introduction into all our reasonings in
the establishment of this Calculus is calculated to throw a mystery over all its
operations, which can only be removed by our knowledge of its more simple
and natural origin," he insisted.[31] And so, in footnote after footnote, he offered
the neo-Baconian generalizing of Woodhouse's *Principles of Analysis* as a cor-
rective to Lacroix's neo-Newtonian approach.[32] The result is strange but may
be seen as an accurate reflection of its moment. At the beginning of the 1820s,
no one—whether French or English—was entirely clear about how to understand the nature of the mathematics that they were pursuing.

Although all of this took place before De Morgan arrived at Cambridge,
Peacock and Whewell were both fellows of Trinity College when he was a student there. Whewell was a powerful personality whose dinner conversations
held the whole college enthralled. Peacock was quieter, but a superb teacher.
Thomas Thorp was De Morgan's assigned college tutor, but as the young man
began to distinguish himself on the mathematical exams, Peacock began to
take an interest. Studying mathematics with the older man opened "a new
life" to De Morgan.[33] From Peacock he learned both the Newtonian and the
Leibnizian approaches to the calculus as well as the tensions between them.

He studied the Analytic's translation of Lacroix, complete with its amending footnotes, and then plunged into the works of Laplace, Legendre, and other Frenchmen, including Voltaire. All around him "every high undergraduate, and every mature B. A." was trying "to settle irrevocably the true foundation of the Differential Calculus," but De Morgan "made up my mind that they ought to wait and that I would wait."[34] Wrestling with mathematics was opening whole worlds to him, and he was much more interested in tasting a variety of ideas than he was in choosing among them.

De Morgan's enthusiastic engagement with his studies was setting him up for a fall, however. Even as he was embodying the phrase "reading man" that was the mark of the Cambridge scholar, changes in the university were crushing the practice. One of the most striking manifestations of the huge increase in student numbers can be seen in the development of the Senate House examination, or Tripos, that was the gateway to an honors degree. In the fourteen years before De Morgan sat for the exam, the number of students who took the Tripos almost doubled.[35] The pressures attendant on examining and ranking more than a hundred students meant that the interactive oral process that had in Frend's day led to class rankings gave way to a considerably less personal one. In De Morgan's day, the Tripos entailed five consecutive weekdays of written work, with set questions, written answers, and a clear set of marks allotted for each answer.[36] Each day was divided into three two-hour blocks beginning at nine in the morning and ending as late as ten in the evening. At eight o'clock on Friday morning, a list was posted in the Senate House that arranged the students in a set of "brackets" according to the marks they had earned so far. Faced with a new and clearer indication of their ranked positions, the students entered a final day of what seemed like "mortal struggle,"[37] which ended at five o'clock on Friday afternoon. The Tripos was well on its way to becoming the most gruelingly competitive examination in the Western world.

In the months leading up to these events, virtually everyone in Cambridge got into the act of predicting and betting on who would be senior wrangler. When the name of the student who had thus triumphed was announced somewhere between ten and midnight on the last Friday of the exam, whole colleges, from servants to masters and everyone in between, poured onto their greens to cheer their winners and mock the defeated. Similar scenes surrounded the posting of the rest of the list in the Senate House at eight the next morning. "The moment the doors fly open, in rush, towards the Tripos, hundreds of Gownsmen and Townsmen of all ranks and conditions, to read the fate of themselves, their friend, or their masters."[38] The graduation ceremony began just two hours after the initial pandemonium surrounding the unveiling of the Tripos results. With everyone dressed in elaborate gowns that celebrated their status in the community, the senior wrangler marched first, "taking precedence even of the very nobility,"[39] followed by the rest of the candidates grouped by college. After hours more of elaborate ceremony, they had all received their degrees, and "both the fortunate and unfortunate," sat down to a feast.[40]

De Morgan saw himself as above all of this drama. When his college tutor, Thorp, wrote a gloating letter calculating the odds of his becoming senior wrangler, he flipped the letter over and drew a series of gaping mouths, the largest of which features a protruding tongue with a fly crawling up it (figure 6).[41] De Morgan's self-confidence extended beyond private expressions of disgust with his tutor's attitude. It also determined his response to one of the effects of the growing power of the Tripos. By the time he arrived at Trinity, Cambridge's most academically motivated students had begun turning to private tutors to prepare them for the Tripos, but he was not interested in working with someone whose focus was on ways to solve mathematical problems quickly and accurately. What went on in their classrooms had nothing to do with the wide expanses of "Metaphysics, Mental Philosophy and even Theology"[42] that he was finding in mathematics.

De Morgan's mother was appalled, and letters flowed out exhorting her son "not to disregard good advice," "not to be so willful."[43] She wrung her hands when she heard of him "as a man who reads much, but who is not likely to *do* much, because he will not conform to the instructions of those who could assist him."[44] De Morgan nonetheless stood firm. He was completely confident that what he was learning through wide reading and independent study would allow him to succeed on the Tripos without the help of a private tutor. Many agreed with him, and when he began the Tripos it was widely predicted that he would emerge as senior wrangler.

De Morgan and his supporters had underestimated the fundamental nature of the changes involved in making mathematics a competitive sport, however. He may have had no trouble thinking through all of the questions he was given, but he was not as lightening quick at solving problems as were those who had trained with private tutors. When the "brackets" were published on Friday morning, everyone could see that he was not going to finish first, and on that Friday night the senior wrangler was declared to be Sir Henry Percy Gordon of Peterhouse. The next morning, De Morgan's disap-

Figure 6: Augustus De Morgan. Mouths. Manuscript letter,
Thomas Thorpe to Augustus De Morgan, MS 913A/2/9. By permission
of the Senate House Library, University of London.

pointment was further compounded when he found that two of his Trinity classmates, Thomas Turner and Anthony Cleasby, had also finished ahead of him. Through all of the marching and festivities of graduation, he was merely fourth wrangler.

De Morgan took his fourth-place finish as a body blow. For the rest of his life he railed against the Cambridge system of "CRAM," and obsessively tracked the careers of the three men who finished above him in support of his conviction they were no better than he. His more immediate response was to shake the dust of Cambridge from his feet and repair to London. He never returned to his Alma Mater, even for the shortest of visits, but his Cambridge education was not defined by the Tripos. His four years of reading at Trinity had transformed him into one of England's most powerful rising intellectuals. He was only twenty years old when he turned his back on Cambridge in January 1827 and headed into the rich new world of science and mathematics that awaited him in London.

# CHAPTER 12

✦

# *Laying the Groundwork*

Within days of his disappointing finish on the Tripos, De Morgan had joined his mother, sister, and two younger brothers in their house at 25 Hatton Garden in London. In the almost thirty-five years since Frend had arrived in 1793, the city's population had risen from under a million to more than one and a half million. Death rates no longer outstripped birth rates, but London remained huge, and overcrowded. This did not bother De Morgan, who was always deeply disquieted by what he experienced as the profligate disorder of the countryside. For the rest of his life, even as England's capital city continued to grow in size, filth, and social disorder, De Morgan happily made his home there.

Comfortable though he found London to be, the young man still had to find a way to earn a living. His mother had hoped that he would emerge from his years at Cambridge as a priest, but by the time he graduated he was quite clear that he would not serve as an Anglican minister of England's established church. What he could do instead was not immediately obvious. His eyesight disqualified him from the military careers pursued by so many of the men in this family, and he was advised that his limited "toleration for ignorance and folly" meant he would not be a good doctor.[1] That left the law, and soon after he arrived in the city, he began dutifully trudging the mile from his mother's house to study at Lincoln's Inn.

De Morgan's daily walk may have pleased his mother, but the traditional study of law was emphatically not what had drawn him to London. He was looking for intellectual excitement, and he was in a very good place to find it. More than ten years after Napoleon's fall, England's capital city was a roiling cauldron of social, political, and cultural change. Of particular interest to a brilliant young man just escaped from the confines of Anglican Cambridge was a new and powerfully emergent scientific scene.

De Morgan's initial entrée into the London scientific world was through one of his mother's friends, William Stratford. As a naval officer retired on half pay, Stratford was an archetypical representative of one of the major components of the English world of post-Napoleonic science. No longer at war, England's naval officers were beginning to turn their attention to the lands they were visiting, the seas they were sailing, the winds they were riding. Years of experience with navigation meant that this group tended to see their world in terms of physical variables like gravity or terrestrial magnetism, and they began bringing home quantities of measurements. Once back in London, they joined forces with a group of mathematical enthusiasts who were ready and willing to find order within these numbers. Among them were businessmen who eagerly turned mathematical skills honed in the pursuit of profits to calculating orbits, improving lunar tables, and determining positions on earth.[2] In addition there were the young Cambridge graduates John Herschel and Charles Babbage, full of energy and eager to open their country to new ways of thinking. Stratford had no trouble seeing the promise in Elizabeth De Morgan's eldest son. Within weeks of the young man's arrival in London in 1827, Stratford was drawing him out of the stodgy world of the law and into the company of these mathematical practitioners.

Intellectually powerful though they were, these mathematically inclined enthusiasts did not fit comfortably into the scholarly landscape of London. In that city and the England that it centered, scientific power had, since the time of Newton, been concentrated in the aristocratically dominated Royal Society. Its president, Joseph Banks, was himself increasingly crippled with gout, but he traveled by means of a gargantuan correspondence with naturalists all around the world, encouraged young adventurers in their travels, and welcomed scientific travelers either to breakfast with him in London or to join international gatherings at his house in Surrey.[3] Warmed by his sociability, the Royal Society flourished.

Large it may have been, but Banks's Royal Society was not all-encompassing. It was always a reflection of the interests of his wealthy leisured world, in which there was no place for the nitty-gritty practicalities of precision measurement and mathematical calculation. From Banks's point of view, mathematics was "little more than a tool with which other sciences are hewd into form." Even Newton, he insisted, "owed his immortality to his discoveries in Natural Philosophy not mathematics." In Banks's Royal Society "a *general* Acquaintance with the Sciences and classical Learning are of much more Consequence" than was any "*profound* Knowledge of the Mathematics,"[4] and when he became president in 1778 one of his first moves was to remove mathematicians from its ranks. A group of furious mathematical practitioners began to meet separately in a society in Spitalfields, but by the early nineteenth century they had been all but overcome by "a sense of loneliness and desertion." Mathematics was "at the last gasp and astronomy nearly so,"[5] but there was no hope of redress in Banks's Royal Society.

In the 1820s, the frustrations of this group came to a head over the *Nautical Almanac*, an annual publication that consisted of highly accurate tables of measurements of lunar positions taken at the Royal Observatory at Greenwich. The historical roots of the *Nautical Almanac* lay in the 1760s, when Neville Maskelyn had vied with John Harrison for the prize, established by the English Board of Longitude, for solving the problem of measuring longitude at sea.[6] Both men approached the challenge by finding ways a traveling ship could ascertain its distance from Greenwich—Harrison by constructing chronometers, Maskelyn by focusing on lunar observations. In 1765 Harrison was awarded the prize, but Maskelyn was named Astronomer Royal.

That the Astronomer Royal was the champion of lunar calculations points to practical realities. Harrison's chronometers were wonderfully accurate, but they were far too expensive for use by the vast majority of England's seafarers. The country's navigators were therefore left to compute their longitude from the lunar tables of the *Nautical Almanac*. Accuracy was of paramount importance because an error in the third or fourth decimal place could result in a possibly fatal mistake of over thirty miles at sea. From 1765 until his death in 1811, Maskelyn employed a hardy band of "calculators," mostly men but some women, to compute the values for these lunar tables. He was a meticulous overseer who devised a complex system of checks and balances to ensure that the *Nautical Almanac* was free of dangerous mistakes.[7]

Although producing the *Nautical Almanac* entailed precisely the kind of painstaking calculation and nit-picking attention to detail that Banks disdained, Maskelyn was a member of the wealthy upper class who managed to maintain good relations with the Royal Society. But after Maskelyn's death in 1811, Banks shrugged the production of the *Nautical Almanac* onto a sprawling Board of Longitude, composed of men like the secretary of the Treasury, the judge of the High Court of Admiralty, and a sprinkling of Cambridge academics, including Isaac Milner. By 1818, this group's *Nautical Almanac* was so error-filled that Frend was warning the readership of *Evening Amusements* to watch for mistakes. When the 1820 edition was followed by a list of 160 errors, he asked his readers to warn any friend "going to sea" that he should "take the list of errata with him; and to correct his volume throughout by it before he uses it."[8] Frend then led his audience deep into the crabbed world of astronomical computation so that they could fix the errors themselves. He did not apologize for the difficulties they would face. He would certainly prefer that the *Nautical Almanac* be error-free, but the work required to fix its flaws offered a wonderful opportunity for the exercise of reason that he was eager to share with sailors all over the world.

Frend was not the only person concerned about the mistakes in the *Nautical Almanac*. In 1818 it was denounced on the floor of Parliament, and over the course of the next several years, its most flagrant problems were brought under control by a reconstituted Board of Longitude under the direction of Thomas Young. Even as he improved the product, however, Young remained a Banksian conservative who fiercely resisted attempts to extend membership in the Board of Longitude beyond the aristocratic purview of the Royal Society.[9] This exclusion infuriated the motley crew of businessmen, mathematics teachers, naval officers, and engineers who were deeply invested in the *Nautical Almanac*. In January 1820, Francis Baily, who was taking up astronomy after a highly successful career as a member of the London stock exchange, gathered a dozen astronomical enthusiasts, including Herschel and Babbage, to consider forming an Astronomical Society. By the time of the group's second meeting on February 8, Herschel had written and Baily had published an "Address" laying out goals and a vision, and a designated committee of eight had drawn up a set of rules and regulations. At its third meeting on February 29, eighty-three people—Frend among them—had signed up as members of the fledgling society.

The political implications of declaring their independence from the Royal Society became immediately clear as the Astronomical Society set out to find a president. When, on February 29, they unanimously agreed to offer the position to the Duke of Somerset, Banks implored Somerset not to accept. "He apprehends the ruin of the Royal Society," the duke explained, before noting that he was "strongly attached" to Banks and withdrawing not only from the presidency but also from membership in the newly forming group.[10] The Astronomical Society refused to be shut down and responded by putting John Herschel's father, the eminent astronomer, Sir William Herschel, in the chair.

This episode was just an initial skirmish in a battle between the Astronomical Society and the Royal Society that raged throughout the 1820s and beyond. Baily was the Astronomical Society's major champion. As he was founding the society, he was fashioning himself as a scientific man about town by joining the Royal Society, the Linnean Society, and the Geological Society. He understood power, and having securely planted himself in every one of London's major scientific organizations, he began to use his positions in support of the Astronomical Society. The more he battled, the more engaged he became in what for him were the inseparable goals of pursuing astronomical science and strengthening the Astronomical Society.[11]

One of the group's first stated goals was creating a star catalogue "upon a scale infinitely more extensive than any yet undertaken and that shall comprehend the most minute objects visible in good astronomical telescopes."[12] This was an enormous project, but not an empty dream. In the same year that De Morgan began attending their meetings, Stratford and Baily were triumphantly publishing an astronomical catalogue that contained the positions of 2,881 stars.[13]

Stratford's and Baily's catalogue rested on a combination of extensive observations using the best telescopes and most precise measuring instruments available—and prodigious amounts of calculation, which the authors did themselves.[14] They approached this challenge in the spirit of an Astronomical Society member who wrote of the "peculiar charm" to be found in harnessing reason for the "well-directed pursuit of facts."[15] From this perspective, when Maskelyn outsourced the calculations for the *Nautical Almanac,* he was destroying an essential value. "Paper astronomers" who merely calculated motions and positions "without any knowledge of what they are intended to represent and actually effect" were not exercising their reason.[16] For the enlightened men of the Astronomical Society, attaining precision in both measurement and calculation served as a contemplative practice, a welcome opportunity to exercise their reason as they pursued their scientific goals.

Babbage early emerged as a dissenting voice against this practice of reason. Even as Stratford and Baily were exalting in the practice of reasoned calculation, he was already envisioning a computer that would calculate astronomical tables mechanically. This project was always more dream than reality, but his vision of a dehumanized mathematics fundamentally challenged those who valued the practice of mathematics as one of the highest forms of human reason and understanding.[17]

De Morgan emphatically rejected Babbage's program, but the focus on precision that characterized the Astronomical Society posed problems for him as well. Although he certainly could do it, he was never particularly intrigued by calculation; his visual impairment rendered the telescope "an insidious foe," which he avoided looking through "on principle";[18] and the society's other precision instruments were equally out of his reach.[19] But for De Morgan, the precision they afforded was just the prelude to huge areas of knowledge that were being opened by the Astronomical Society's "well-directed pursuit of facts."[20] So, for example, Baily's first foray into astronomy was actually designed to address a historical question. In 1811, he had begun scouring astronomical records in hopes of dating precisely the solar eclipse predicted by the ancient Greek philosopher Thales. When he found that extant lunar tables were not precise enough to do the job, he proceeded to compute "all the solar eclipses during a period of seventy years, six centuries before the Christian era."[21] It was a typically enormous project, but at its end Baily emerged with a date: September 30, 610 BCE.[22] In his hands, astronomical precision had the power to open the ancient world.

Western civilization was not the only one whose past could be illuminated by astronomy. Non-Western civilizations had developed under the same skies and were thus equally open to astronomical reasoning. H. T. Colebrooke was just returned from thirty years in India when he was elected to be the second president of the Astronomical Society. For him, mathematical astronomy served as an entrée into an enormous world of ancient Indian history and agricultural practice.[23] For members of the Astronomical Society, examining the world through the lens of mathematical astronomy was a uniquely powerful way to understand the cultures of other times and places.

De Morgan may have been unable to see the stars, but insights to be gained by those who could drew him into the Astronomical Society. Baily's work on Thales led directly into the kinds of antiquarian interest he had pursued so happily in the Trinity College Library, and Colebrooke's interest in India fed his deep and abiding love of all things relating to the country where he was born. The opportunity to meet and work with Herschel and Babbage was another enormous draw for the young man who had heard so much about them at Cambridge. At the Astronomical Society, their expansive vision of science was bearing fruit in efforts to educate observers, form a library, "stimulate astronomical research by offering prizes,"[24] and bring in foreign astronomers as associates. The results were electric. "What may not be expected from so liberal an association?" an early member of the Astronomical Society enthused. "Happy the country where the love of science alone causes so many men of enlightened minds to combine in such an object! Happy, also, those who dwell there!"[25] De Morgan certainly fit the description of someone who was "happy to dwell there." He was elected a fellow in 1828, made a member of the Council of the society in 1830, and named honorary secretary in 1831. For the next thirty years some combination of these positions kept him at the center of the dynamic world of London's astronomers.

William Frend was among the first people De Morgan encountered in the growing Astronomical Society. Frend had long been friends with Baily and regarded Babbage and Herschel with indulgent affection. De Morgan may have at first seemed to him to be just another Cambridge intellectual, but their friendship quickly blossomed into something more. De Morgan immediately grasped Frend's interpretations of reason in both theology and mathematics, and from then on, the two men recognized each other as soul mates. Within weeks of their first meeting, Frend was urging Stratford to bring the young man to his rambling home in Stoke Newington. De Morgan fit in beautifully there. At the time of his first visit, Frend's oldest son, Richard was twelve, but he was sadly disabled. Frend's two other sons were just seven and five, which meant they were still too young to engage in their father's world of reason. For Frend, a brilliant Cambridge graduate around the age of his older daughters was a welcome addition to the household.

De Morgan not only filled the gap created by the distribution of sons in Frend's family, he also contributed the expertise of one of his seemingly innumerable Dodson uncles to the ongoing discussions. Lieutenant Colonel [later General] John Briggs was the husband of a younger Dodson sister.[26] At the time Augustus was making the transition from Cambridge, John and Jane Briggs were returning to England after twenty-six years in India. De Morgan was fascinated by his newly discovered uncle's firsthand knowledge of India. The defining mark of Briggs's erudition was linguistic. Within two years of arriving in India at the age of fifteen, he had learned both "the Hindustani language" and Persian, and this was just the beginning. Throughout the of rest his time in India, he was constantly trying to learn the local languages and dialects of the areas where he was working. Briggs's linguistic efforts were essential to his larger goal of interacting respectfully and constructively with his Indian colleagues.[27] In addition, his study of languages supported a deep and abiding interest in the history and culture of the world in which he was living. His expertise underlay the four-volume translation of the history of India by the sixteenth-century Persian historian Ferishta that he produced during the six months of his homeward voyage,[28] and ensured his welcome into the Royal Asiatic Society upon his arrival in London. His erudition also proved fascinating to the ongoing seminar that flowed through the Frend household in Stoke Newington.

Briggs's interests fit well with the linguistic views of reason that characterized Frend's colleagues, but his deep knowledge of Eastern languages pointed toward a world far larger than that bounded by their Hebrew, Greek, and Latin. His study of languages was part of a well-established tradition among Indian scholars that had grown up in parallel with the Unitarian linguistic tradition of the 1780s. Throughout the 1770s and 1780s, as Lindsey, Frend, and their fellow Unitarians were pushing into ever more ancient languages to find the true expression of Jesus's religion, linguists in India had been following the same form of reason into fields at once more ancient and more geographically broad. In 1786, William Jones had argued that Greek, Latin, and Sanskrit were so clearly and closely related that "no philologer could examine them all three without believing them to have sprung from some common source, which, perhaps, no longer exists."[29] From this perspective, antiquity stretched much farther back than the time of Christ, and religious possibility extended far beyond the confines of the Western tradition of Jews, Christians, and Muslims.

Briggs followed the focus on meaning that characterized this kind of linguistic thinking into astronomy and emerged convinced "that all the Pagan Pantheon will be found to be symbols of the planets and constellations."[30] This approach transformed basic observational astronomy into a form of language that, if read correctly, would be an invaluable guide through all of the complexities of all the world's religions and ancient civilizations. Interpreting the heavens in this way was certainly less exact than reading them through the mathematics of the Astronomical Society, but for Briggs, it promised a much richer understanding of the deep currents of meaning that swirled through all of the world's religious expressions.

Briggs's firsthand knowledge of India brought a whole new level of expertise to discussions in the Frend household. De Morgan was completely fascinated, but he was never going to venture further along the paths that his uncle was blazing. He hated learning Greek and Latin in school, and he had absolutely no desire to add Hebrew, Persian, Sanskrit, or Arabic to his repertoire. Even as he learned what he could from his Uncle Briggs, he would never

claim expertise of his own. Sophia was equally intrigued, but she had a different sense of her limits. Her father had brought her up to trust herself, and this
teaching was never tempered by the discipline of school. The conversations
that coalesced around Briggs served to deepen her conviction that "much of
modern [religious] doctrine has gained something of its form, at least, from
ancient symbolism."[31] For the rest of her life, Sophia remained open to the
deep meanings that were waiting to be found in the world around her: in the
stars, in ancient artifacts, and in the words of the people she encountered.

Within less than a year of De Morgan's arrival in London, some combination of the Astronomical Society and the Frend household had become his
intellectual family. But neither could rescue him from life at the bar. For that,
he had to find a paying job. In December of 1827, he responded to an advertisement calling for applicants for professor of mathematics at a radical institution that was taking shape on the outskirts of London. If he could land that,
he could escape the fate of becoming a lawyer.

The institution that accepted De Morgan's application called itself the
London University, but that seemed a rather pretentious and misleading name
for a joint stock company that had been refused a royal charter and therefore
did not have the power to give degrees. The London University did, however,
have a powerful vision. In the same way that the Astronomical Society was
challenging the Anglican establishment within the scientific community, the
London University was challenging the Anglican establishment within the
educational world. It was to be an enlightened institution in which legitimacy,
whether of knowledge or of the people who pursued it, was to be grounded in
reason, rather than in religion or social class.

Frend was as involved in the founding of the London University as he was
in the Astronomical Society.[32] The initial impetus for the project came from a
poet, Thomas Campbell, who returned from a trip to Germany determined to
create a religiously inclusive university for the people of London. This was just
Frend's kind of project, and Sophia remembered gatherings at her house from
the time she was about twelve. By 1825, the idea had gained enough momentum to move the project beyond the Stoke Newington home. On February 9,
1825, Campbell published a letter in *The Times* calling on the Whig MP Henry
Brougham to take action. Brougham was the perfect spearhead for Campbell's project. Within weeks of the appearance of Campbell's letter, Brougham
began exploring the possibility of receiving a royal charter "to establish a town
university." When that approach failed, he applied to the government. Parlia-

ment was reluctant and did what it could to create obstacles, but there was too much momentum behind the idea.[33]

Even as Parliament was thrashing out the terms of its existence, three of the richest supporters of the London University were buying an eight-acre tract of vacant land on the northern outskirts of the city. There, despite having raised rather less than they had hoped with their stock sale, the group threw themselves into the construction of a building more "suited to the wants, the wealth, and the magnitude of the population for whom the Institution is intended, than one commensurate with its present means."[34] The building they so determinedly erected sent a powerful message about the university it was to house. Its classical style stood in marked contrast to the churchly towers of Cambridge and Oxford. And whereas those Anglican colleges were surrounded by protective walls, the London University opened directly onto the street. This was a significant statement. There were to be no religious tests for entry or exit from the institution. It was to be equally open to Anglicans, Christian dissenters, Jews, and Muslims.

The enormity of the challenge involved in creating a university to fill the building's promise was evident from the moment it was proposed. Campbell's determination not to create any "barrier to the education of any sect among His Majesty's subjects"[35] proved very difficult to conceive in the religiously rigid England of the early nineteenth century. The issue quickly coalesced around the question of who would teach theology. After months of wrangling, it was finally agreed that the best way forward was to maintain "an expressive silence upon points of theological difference."[36] Institutionally this policy was carried out by making the new university nonresidential. This allowed the institution to focus on the students' "secular education alone" while leaving their religious instruction "to their parents and to themselves."[37] Various schemes were launched to ensure religious instruction was available in the neighborhood of the university, but no religious practice of any kind was to be supported within its walls.

Frend was deeply disappointed by this outcome. When he heard that "divinity, as it is called, is to form no part of the studies of the new university," he returned in memory to classes in which Paley would approach a passage from the New Testament "with the same freedom, as a passage from Homer, Herodotus, or Aeschylus,"[38] and the students were free to interpret what they were reading "each according to his own manner."[39] For Frend, this openness to difference was an essential hallmark of truly reasoned discussion, and studying with Paley had shown him that it could be found in theology as well as in the classics. The key to Paley's success was that his courses were grounded in a search for truth as it was described in John 8:32: "And ye shall know the truth and the truth shall make you free." This truth was not the same as the one that supported the arguments about who should occupy the theological chair at the London University. Frend marked the difference as one between "truth" as it was understood in the sciences and reflected in theological squabbling, and what he designated as "the truth," which Jesus was referring to in John. Frend recognized that "freedom from errour of any kind is a pleasing state to the mind," and that it was "highly desirable & and advantageous to possess it."[40] But, he insisted, this scientific truth was different from and essentially secondary to the truth that Jesus proclaimed, which was more experienced than known. The experience of the truth is what Locke was pointing to when he wrote of the certainty that could attend mathematical understanding. For him, as for Frend, the route to it lay through reason, but they both recognized that such efforts could take one only so far. Frend himself had reasoned from the Bible for decades before he closed the second volume of his *Principles of Algebra* with an inspired testimony to the immediate experience of the truth that emerged from his mathematical difficulties.[41]

For Frend, this kind of experience provided the solid ground for truly reasoned understanding. "Truth and freedom are inseparable," he insisted, and all talk of liberty was in vain unless it was founded on the truth.[42] He recognized that the pursuit of science was an admirable endeavor, but scientific understandings of truth were not enough to support a university in which people of all religious persuasions could learn together, connected "by the ties of universal love."[43] Frend's decades of conversation with members of the Jewish community were grounded in a mutual recognition that they were all reaching toward the truth. He looked forward to the final coming of Jesus when "the names of Jew & Christian will cease to exist," but he recognized that, in his present world, that was a state of bliss "of which we cannot now form any conception."[44] In the meantime, creating a truly diverse university in this imperfect world required recognizing the primacy of the truth that transcended all religions.

Frend was an aging man when the London University was founded, however, and his world was being replaced by a new one. The ideal of the London University was of a place "without religious rivalship,"[45] that is, one in which theology was simply left out of discussion. In this way, the institution could embrace religious diversity but without a trace of that diversity, or of religious thought at all, within its classrooms. To make this possible entailed erecting an impenetrable wall between religious understanding and academic reason, between the truth and everyday truth. Figuring out how to define and negotiate the parameters of this division engaged the administration and faculty of London's secular university for decades.

Mathematics played an essential role in these considerations. Once theology was eliminated as a subject, mathematics rose to an essential position as the study that would bring students to their full potential as reasoning human beings. By the end of 1826, the university began looking for the right person for the position of professor of mathematics. They tried to recruit Babbage and Herschel, in hopes of increasing the prestige of the new university, but both turned them down. A sizeable list of other candidates applied, but the committee was dissatisfied enough that on December 22, 1827, they readvertised the position. De Morgan sent in his application on the same day. All of his Cambridge tutors wrote letters of support, and two months later, on February 23, 1828, the Council, which was the governing body of the university, "resolved that Mr. De Morgan be appointed professor of Mathematics."[46] Just a year after his Tripos defeat, the reading man was vindicated.

At least that is how De Morgan saw it. At the time, however, becoming professor of mathematics at the London University was a rather tenuous proposition. Certainly it was not at all clear to De Morgan's mother that it was a good idea for him to give up his hopes for a brilliant legal career in order to teach in what was commonly known as the "godless institution of Gower Street."[47] Babbage had already declined the position because the salary was small, and the university had "no dignity to confer as yet."[48] Even Frend wondered. When he wrote to offer his congratulations, he declared himself "confident of the advantage which the institution will derive from your service," but he was not "quite so clear" whether it would ultimately be an advantage to De Morgan.[49] The new professor himself did not have any doubts: "My choice will be to keep to the sciences as long as they will feed me,"[50] he declared. Becoming the professor of mathematics at the London University ended the nightmare of finding the law standing between him and the mathematics he so loved.

It was not just mathematics that attracted De Morgan to his new position. Still smarting from his Tripos defeat, he could now teach the subject as it should be taught. Not everyone saw this as an advantage; the reason Herschel gave for turning down the position was that he had "a positive dislike" of teaching.[51] But for De Morgan, there could be no higher calling than bringing students to their fullest humanity by teaching them to reason through mathematics.

Just before classes began, De Morgan clearly laid out this position in his introductory lecture to the mathematical classes at the University of London on November 5, 1828. His starting point was the primacy of reason to the human condition. It is universally recognized, he there proclaimed, "that God had distinguished man from the brutes by what is denominated the gift of reason," and "the success of every individual in the world must depend on his power of reasoning on the events which immediately concern him, and the circumstances in which he is placed."[52] The power of reasoning was, however, not innate. According to Locke, humans are born with wide open minds, but a developing mind was not really "like a blank sheet of paper fit and ready to receive any impression and retain it for its own use."[53] Before it could reach its full reasoning potential, the mind had to be "prepared": its "powers of arranging and combining" had to be developed.[54] Fully reasoning people were not born; they were the products of a well-designed education.

Having thus laid out the basic educational challenge, De Morgan turned his attention to the subject most fit for the task. Mathematics, he explained, was a language that was uniquely appropriate to the development of reason. Two essential properties distinguished mathematics from other spoken languages, making it uniquely suited to this educational task. On the one hand, mathematics was a "compendious language"; on the other, it was a language that "has been from its peculiar structure a never failing guide to new discoveries." These two characteristics combined to create a study that was uniquely able to "call into exercise some of the powers that most peculiarly distinguish man from the brute creation."[55] For De Morgan, learning to think using this language was the most effective way to develop the powers of reason that were so essential to the formation of a fully human being.

De Morgan's focus on language positioned him neatly in the Lockean tradition of reason that had so inspired Lindsey and Frend. The young man's challenge was to locate the linguistic approach to reason that had propelled them into the Unitarian Church within the secular London University. His description of mathematics as a "compendious language" made this transition possible. From a modern point of view, calling mathematics a language is to recognize the analogy between mathematical and linguistic structures and thereby open both subjects to formal structural analysis. But that is not what De Morgan was doing. His mathematics was not just a "language"; it was a *compendious* language, which made all the difference. According to the

*Oxford English Dictionary,* "compendious" means "containing the substance within small compass, concise, succinct, summary; comprehensive though brief." When De Morgan used the word to characterize mathematics, he was focusing attention on the particular way mathematical words, terms, and symbols relate to their referents. As he did so, he was transferring the meaning-focused view of reason that eighteenth-century Unitarians had honed on tight readings of scriptures into mathematics, where it could neatly fit within the secular vision of the London University.

There were other advantages to moving the development of reason from the realm of natural language to mathematics. In vernacular language, words have ambiguous and overlapping meanings, but in mathematics, every word and symbol has a singular and precise meaning. De Morgan offered Locke's *Essay Concerning Human Understanding* as a model of "correct application of terms," but he pointed out, Locke could only write as concisely as English allowed. Thus, over the course of a single paragraph even this careful writer used the words "solidity, resistance, hardness and impenetrability" to refer to the same thing. These four words, De Morgan went on, are so alike in meaning that they are often used "indiscriminately one for the other," but they do not mean precisely the same thing.[56] It could be very difficult to capture the differences among them in simple words, but those differences are nonetheless real. This kind of imprecision is a characteristic of all spoken languages. Mathematics, however, is different. In mathematics, as Locke had recognized, "The idea of two is as distinct from the idea of three, as the magnitude of the whole earth is from that of a mite."[57] In the compendious language of mathematics, meanings might be very close to each other, but each is absolutely distinct.

Whenever De Morgan referred to mathematics as an "exact science," it was this conciseness of meaning he had in mind. The kind of precision that was so crucial to the calculations of the Astronomical Society and the kind of exactitude with which a given problem could be solved could also be part of mathematical practice, but it was not the heart of the subject he proposed to teach at the new London University. There he was touting a mathematics in which symbol and meaning were so tightly bound together that arguments within it were completely free of any difficulties that are "independent of the reasoning itself."[58] For him, mathematical thinking was the purest form of reason itself because there was essentially no distinction between its symbols and its meanings.

De Morgan's understanding of mathematics as a compendious language was essentially the same position that had led Frend to eliminate the nega-

tive numbers from algebra, and the younger man's first published work clearly
reflects his mentor's influence. In De Morgan's translation of the first three
chapters of L. P. M. Bourdon's *Elements of Algebra,* "designed for the use of
students in and preparing for the University of London,"[59] he immediately
confronted the problem of negative numbers. Whenever these offending un-
clarities appeared in the course of a problem, De Morgan carefully explained
that they had no clear meaning and were therefore mathematically illegitimate. Nonetheless, he had to admit using them could generate correct results. Because his goal in teaching mathematics was to convey the precision of
reason, he could not accept them without comment, but as soon as he moved
into the world of natural language to explain their status he became muddled.
His struggles to explain how negative numbers could be at once mathematically meaningless and pragmatically useful, which for him meant at once invalid and legitimate, render this work all but unreadable.

De Morgan's problems with negative numbers did nothing to undercut
his commitment to the value of mathematical reason, however. He summarily
dismissed the value judgments of those looking for practical utility in their
studies and respectfully left the challenge of understanding the physical world
to natural philosophy. For mathematics he assumed a different mission, at
once circumscribed and huge: the mission of a mathematics teacher was to
bring students to their full humanity by enabling them to "distinguish right
from wrong."[60] The key word in this phrase is "distinguish." In De Morgan's
view, people made bad decisions because their reasoning was not sufficiently
precise, not because they were evil. As he put it in a later publication, "Evil
is often chosen in preference to good, not from any lack of desire to do what
is right, but from a want of means to distinguish clearly, in difficult circumstances, where the proper course lies."[61] The essential problem, De Morgan
went on, is that people's efforts to follow ethical rules are continually muddied by "the vague and erroneous use of words."[62] The "potentate" can incite
crowds, secure in the conviction that most of his audience is unable to listen
critically to what he is saying. As a result, De Morgan went on, "At this moment we see hundreds on the verge of crime and misery, because they cannot
see through the misapplication of a few words."[63] For him, the best possible
way to avoid these problems was to teach young people the compendious language of mathematics in which they would directly encounter the precision
of reason in its purest form. Having learned how to make sharp distinctions
there, they would always be ready to think clearly and distinctly in all other
area of their lives.

One more issue faced the young professor who was offering mathematics as an alternative to the theology that had no place in a secular university. In response to those who would argue that "mathematical demonstration contributes to the formation of a self-sufficient and arrogant spirit,"[64] De Morgan explained that studying mathematics was an essentially humbling activity. Part of the excitement of the subject was the way it led to new discoveries, but it was nonetheless the case that "every one who is acquainted with the mathematics is aware that the higher he advances, the more widely does the horizon open around him, that for one difficulty which he is able to conquer, a hundred remain to baffle his utmost efforts." Even considering all of the advances that had been made in mathematics since the time of the Greeks, "human knowledge is still as limited in its extent, as small in its proportions as are the atoms in a sunbeam in comparison with the sun itself."[65] This was just the kind of thinking that had led Frend to close his *Algebra* with rapturous contemplations of the divine mind that could understand it all, but De Morgan could not take that step in the London University. Nonetheless, the mathematics he was pursuing in his secular university was large enough to encompass all of the experiences of truth to be found in sacred texts.

The subsequent decade was a particularly dynamic one for De Morgan, who was just twenty-two years old when he laid out his view of mathematics as the highest form of reason at the opening of the London University. In 1831, the Astronomical Society he served as honorary secretary was elevated to the Royal Astronomical Society and placed in charge of the *Nautical Almanac*. In 1838, the London University he so eloquently addressed in 1828 finally gained legitimacy as University College London (UCL). In the same year, eighty-year-old Frend happily passed the torch of reason to a new generation when De Morgan married his daughter Sophia. De Morgan himself was to change in many ways, but the view of mathematics as the compendious language of ever-expanding reason that he first outlined in his introductory lecture was to remain the essential platform that grounded his thinking for the rest of his life.

# CHAPTER 13

# Unitarian Women

The opening of the London University occurred midway between the passage of two parliamentary acts that moved all of England in the same religiously diverse direction. On May 9, 1828, King George IV signed the Supplemental Test Act, which lifted the restrictions on dissenters contained in the Test and Corporations Acts; on April 13, 1829, he signed the Roman Catholic Relief Act, which lifted the additional strictures that had been directed against Catholics. These strokes of the pen brought down the legal barriers that London's Unitarians had fought against thirty years before, and Frend was thrilled. "What a glorious day for this, I may call it United Kingdom!" he crowed,[1] in the belief that accepting the legitimacy of mutually contradictory creeds would push all Christians to embrace the truth that essentially joined them.

Frend experienced the passage of these acts as the culmination of decades of struggle for religious freedom, but their passage at the end of the 1820s may be equally seen as a first step toward accepting the legitimacy of the new middle class that was leading England's Industrial Revolution. In the first part of the nineteenth century, provincial industrial cities like Manchester, Birmingham, and Leeds were expanding even faster than London, and with their rise the center of wealth and power was moving away from the landed aristocracy that gathered within the established church. Many of the new industrialists were dissenters, whose political legitimacy was established in the religious acts, but that legitimacy meant little if they did not have representation in Parliament. It took several more years, but on June 7, 1832, King William IV signed the Representation of the People Act, commonly called the Great Reform Act, which allowed the middle-class inhabitants of cities like Birmingham, Manchester, and Leeds to vote. This political victory thrilled Frend almost as much as did the religious victories of the late 1820s, but at the same time he was deeply concerned and confused by social changes that accompanied the rise of the middle class. He never doubted that giving voice to its members was essential to the political health of his nation, but he was at a loss to understand how to respond to the problems of the poor that seemed equally intrinsic to the Industrial Revolution.

Lady Byron, the widow of the Romantic poet Lord Byron, was a major ally in Frend's efforts to understand and ameliorate the problems of England's industrial poor. The two had first met in about 1806, when she was Annabella Milbanke, and her parents, Sir Ralph Milbanke, 6th Baronet, and his wife, the Honorable Lady Milbanke, hired Frend as her tutor. The two of them got along very well, and for several years Frend led his adolescent pupil through geometry, algebra, physics, astronomy, Greek, Latin, Hebrew, and English composition.[2] As she grew up and he moved into his position at the Rock, their correspondence tapered off and this phase of their relationship ended.

Frend sent his former pupil a couple of congratulatory notes upon her marriage to the dashing Lord Byron in January 1815, but their correspondence did not resume in earnest until the following spring when that marriage was on the rocks. Nothing in the young woman's education had prepared her for life with someone as mercurial as Byron. Through all of the ups and downs, she clung to the times—fleeting but real—when Byron was kind to her, but as she became more visibly pregnant, such moments were increasingly rare. After her daughter, Ada, was born in December, the young mother began to fear for their safety. In January 1816, Lady Byron took her newborn child home to her parents. One look at their miserable daughter persuaded Lady Byron's parents that she must never return to her husband, but she herself remained deeply conflicted. Despite—if not because of—all she had suffered, she was still completely entangled with her husband and in desperate need of a friend who would stand by her side as she tried to separate from him.

In spring 1816, William Frend became that friend. The man who had once taught her biblical languages so that she could learn about the God of the Bible now encouraged his young friend to go to that God for solace. "You are walking in the light and have nothing to dread,"[3] he soothed; "when the groans of the heart are felt & language ceases to be necessary," you can be "assured that God heareth you."[4] In a later letter he supplemented this advice with one of his favorite astronomical images: "There is only one way of seeing these things. Place yourself in the Sun & all the irregularities which we think we observe in the planetary motions disappear."[5] In Frend's letters, Christian religion and Copernican astronomy combined to form a loving world where his former student had a place, and reading them supported Lady Byron through "the most afflicting trials."[6] Finally, in May 1816, Lord Byron resolved the issue by signing a deed of separation and leaving the country.

As the immediate intensity of Lady Byron's problems began to fade, she turned her attention to raising her child, and her correspondence with Frend again dwindled. But by the end of the 1820s, Ada was growing up and Lady Byron reached out to Frend for advice on finding a tutor for her. From that point, their conversations widened to include all of their common interests. This part of Frend's correspondence with Lady Byron is our richest source for the ways he moved his eighteenth-century understandings into the post-Napoleonic world.

Frend's and Lady Byron's letters in the late 1820s and early 1830s focus largely on ways to respond to the problems of the poor. Times had changed since Frend had listened so sympathetically to the St Ives market women in 1793. In that earlier world, a very small number of landlords owned the vast majority of land, pieces of which they let to tenant farmers who paid their rents with a portion of their produce. These farmers, in turn, hired laborers on a yearly basis to work the land. Frend's market women were most likely the wives of these laborers and were supplementing their income with spinning. They would have been living in cottages that were surrounded by enough land for the maintenance of small personal gardens and not far from larger stretches of common land where they could gather wood and graze livestock. Their lives would have been far from luxurious, but they could usually cobble together a subsistence from some combination of these resources.

By the time Frend and Lady Byron resumed their correspondence in the late 1820s, the circumstances of England's rural poor had deteriorated significantly. An unintended consequence of the effort to improve English agriculture through enclosure, which Frend had recommended to Hammond and Reynolds in the first decades of the century, was the systematic reduction of the common lands that had supported the workers' most basic needs for grazing and wood. Laborers began working for money rather than subsistence, and most were not paid enough to live on.[7] When there was enough food to go around, it was possible to let this situation ride, but in poor harvest years, it was not. In 1827 the harvest was excellent, and the English countryside quiet. In 1828, however, the harvest was not good, and in 1829 it was even worse, until by March 1830, the ever-warm and welcoming Frend was "full of embarrassments" trying to extricate himself from his efforts to help a struggling family "half of which is under my roof."[8] In fact, all of England was beginning to feel the effects of the new social order.

After the Frends had somehow disentangled themselves from the pauperized family that had found its way into their home, they joined their Quaker neighbors in an organized effort to distribute soup, coal, bread, and clothing to the starving poor in their neighborhood. Frend was heartened that the group succeeded in dispensing "some hundred quarto of soup"[9] within just four days of launching their effort, but the problems of England's poor were far too large to be solved by such piecemeal efforts. By the following November, when Frend visited his brother in Canterbury, he found himself in the midst of the Swing Riots, in which angry mobs were destroying threshing machines, setting destructive fires, and going house to house demanding higher wages, lower rents, lower tithes, lower taxes.[10] Frend searched for solutions to no avail. A month later, even as he celebrated Christmas in Stoke Newington, his mind remained "a good deal oppress'd."[11] Trying to find constructive ways to relieve the sufferings of the poor of England was to occupy his thoughts for the rest of his life.

Lady Byron was contemplating the same issues in the fashionable resort town of Brighton. The city was familiarly known as the "Old Ocean's Bauble," but behind its glittering facade lay fetid slums where the poor, the drunk, and the unemployed crowded together in filthy hovels.[12] Lady Byron was acutely aware of this juxtaposition of poverty and privilege. At the end of the long lean winter of 1828–29 she wrote to Frend: "If it not be consistent with the most practicable views of benevolence to 'sell all I have, and give to the poor,' I cannot feel myself acquitted of a violation of the precept, unless I study how I may empty all I have in the service of the poor."[13] Lady Byron never lacked money, but she did know what it was to suffer. Almost as soon as she arrived in Brighton, she began working with a local physician, Dr. William King, to improve the situation of those who were struggling amidst the riches of her town. King's efforts had already earned him the title of the "Poor Man's Doctor,"[14] but when the two first met his vision was expanding beyond medicine. He wanted to establish a Mechanics' Institute in Brighton.

The idea behind King's new venture came from the same Henry Brougham who had worked so effectively in support of the London University. For

Brougham, establishing a university to teach London's middle classes was just one part of a larger project of educating England's population. In 1825 he laid out an educational vision for England's working people in *Practical Observations upon the Education of the People*. His central proposal was to create a host of working-class libraries, or Mechanics' Institutes, to benefit those who were not "so entirely occupied with labour as not to have an hour or two every other day at least, to bestow upon the pleasure and improvement to be derived from reading—or so poor as not to have the means of contributing something towards purchasing this gratification."[15] Brougham saw these libraries as providing a great deal more than gratification or entertainment. Their mission was to help people use "the knowledge gained at schools, for their moral and intellectual improvement."[16] In Mechanics' Institutes, England's working-class people would be able to enjoy all the benefits that accrue to those who are able to exercise their reason.

Political economy took pride of place among the topics Brougham advocated for working-class readers. As they were being buffeted by fundamental economic displacement "expounding to them the true principles and mutual relations of population and wages" would be "wholesome for the community."[17] Understanding science would in a similar way promote their comfort and stability within God's world: "The more widely science is diffused, the better will the Author of all things be known, and the less will the people be 'tossed to and fro by the sleight of man, and cunning craftiness, whereby they lie in wait to deceive.'"[18] In Brougham's view, enabling the laboring poor to reason about their world the way the educated middle classes did was the best way to ensure the stability of society. His ideas resonated so powerfully with the middle-class reading public that his tract went through twenty editions in the first year of its publication.[19]

The importance of individual responsibility lay at the heart of Brougham's vision. Over and over in his *Observations* he insisted that the working-class libraries he was proposing must be organized in such a way that they were independent—both of one another and of their upper-and middle-class sponsors. The essential form of independence in Brougham's model was financial. Although he allowed that "the rich" might seed such institutions by giving "a few books as a beginning," they must then withdraw to let the institutes run on the "weekly or monthly contributions" of the working-class members.[20] Otherwise, "they enforce the appeal to gratitude by something very like control; and they hurt the character of those whom they would serve."[21] Brougham's insistence that each Mechanics' Institute be member-controlled was as much the product of experience as of ideology. Earlier in the decade he had worked with George Birkbeck to create a Mechanics' Institute that opened its doors in London in 1823. However, even as Brougham and his solvent supporters, including William Frend, were contributing their time, money, and expertise in support of the project, the institute's working-class members were fighting back against what they saw as middle-class meddling. They, as well as a host of modern scholars, saw efforts to control in Brougham's plans to educate the people.[22]

There was, however, a downside to turning such institutes over to their working-class members. Without the financial support of people with means, most of them failed. In this, Lady Byron's experience in Brighton was typical. After she and King and his other supporters had rented the building and amassed the books, the Mechanics' Institute's working-class members did not have either the time or the money to keep the project going. And so, as its financially solvent sponsors retreated, the Brighton Mechanics' Institute collapsed.

Dr. King and Lady Byron were not so easily diverted from their mission to help the poor. As the Mechanics' Institute crumbled, they began to embrace the ideas of a different visionary. Robert Owen had in the 1810s successfully managed the New Lanark Mills on a cooperative model. In Owen's view, a competitive environment created combative workers whereas a supportive one would create sympathetic and happy ones. His success in building such a community among the workers in the profitable New Lanark Mills gave his ideas considerable credence, so Lady Byron and Dr. King turned their attention to forming a Workers' Cooperative in Brighton. Frend was highly skeptical from afar, but his reservations did not quell Lady Byron's hopes. "Were you brought into contact with the members, you would be convinced that their objects are laudable, & tending to the public good,"[23] she insisted. Further, it was simply inconceivable that such hard-working, rational people would not succeed.

Two years later, Frend had his own direct experience of a workers' cooperative which had been founded by a group of Quakers in the seaside village of Hastings. The focus of this group was a store that featured local fruits, vegetables, meats, and cheeses as well as "articles of clothing of a coarse, substantial kind."[24] Frend, predictably, found the energy of the "many intelligent working men" associated with the Hastings Cooperative Association irresistible, and he immediately set to work to rationalize their business plan.[25] After a couple of years of support from the Frends and their like, the Hastings Cooperative Association took the further step of setting up a communal library, and Lady Byron stood at the ready to send books.[26] There is no reason to believe that its success was due to its business practices, however. As had been the case with the Brighton Mechanics' Institute, the success of the Hastings Cooperative Association depended on the money provided by its the middle-class sponsors.

———

The Frend family's trip to Hastings in 1831 was prompted by a personal tragedy. In the spring of that year, their eldest son, Richard, died. Frend had seen his disabled son as "soul without body," someone whose "misshapen form was forgotten in the powers of his mind." Surely "our poor boy" would not have lived so long had it not been for Mrs. F's years of "utmost attention," he wrote to Lady Byron.[27] In Frend's world, it was inappropriate to succumb to sadness at the death of someone who was now with God, and after pouring out his sorrow in this letter, he did not mention his son again. But Richard's passing marked the end of an era. Just four months later the Frends moved away from their Stoke Newington home.[28] Their first port of call was Hastings, but after a few months of quiet in that "primitive little fishing town," the family relocated to London.[29]

The move was a shock to Sophia, who had loved the sheltered quiet of life in their rambling house and surrounding gardens. "The greatest grief of all was to bid farewell to the large spreading oak in the field," in the branches of which she had read an enormous variety of books. By the end of her life she had come to see that her reading "was far from being a systematic course of intellectual training," but still, she liked to think that "the beautiful sunrises which I sometimes watched threw a light over it."[30] It was a wrench to leave that bucolic world behind.

It was also very exciting. Sophia had just turned twenty-three when her family settled into a brand-new row house on Bedford Place, a one-block street between elegant Russell and Bloomsbury Squares and just a few blocks from the London University. It was a quiet street, and the house had a secluded yard behind, but the neighborhood was nonetheless embedded in the urban crush of London. As soon as they arrived, Sophia and her sisters began venturing forth on carefully chaperoned expeditions into the exploding metropolis around them.

On one memorable evening Sophia, Frances, and Harriet were escorted by "three or four very earnest, if rather Utopian reformers" into the middle of an experiment in cooperative living that Robert Owen was leading in an enormous rotunda on Grays Inn Lane, within blocks of their house. The scale of this project was huge: the space included several large rooms suitable for lectures, balls, and concerts, and a school for children was in the offing.[31] The Frend sisters went there to attend a ball, where they were quickly enveloped in a group of two thousand attendees that included a wide variety of people from tradesmen to servants to "tidy laundresses and others lower in the social scale."[32] Frend had raised his children to see all people as equal in their reason, but his daughters were much more amused by their variety. At one moment they were entertained by a "long-nosed" eccentric who declared himself "a creature of circumstances."[33] In the next they listened in amazed amusement as a lady garlanded with any number of odd ribbons and beads explained the philosophy of the gathering. "'Appiness is our aim, and we encourage recreation calculated to dewelop the 'ole individual while his intellectual capacities are enlarged. But this is 'indered by superstition and ignorance, so we must put down kings and priests, and get rid of all class distinctions,"[34] she proclaimed. These were just the kinds of equalizing ideas that Frend had been exploring all of his life, but his daughters were startled to hear them from the people gathered in Grays Inn Lane. Looking back over a lifetime, Sophia was willing to allow that Owen's "system had a foundation of truth,"[35] but when confronted with it directly, she and her sisters protected themselves with laughter. They were not ready to listen to claims of equality coming from the mouths of people with lower-class accents and ill-fitting clothing.

Even as Sophia giggled with Frances and Harriet, she recognized that rejecting reason's egalitarianism called for an alternative view of the human condition, and she had one ready to hand. Four years before the Frends moved to London, they had joined the crowd of seven hundred who flocked to a series of lectures on the new psychological science of phrenology, which claimed it was possible to read the internal depths of human minds from the external traces of the human skull.[36] The subject was originally developed by a Viennese physician, Franz Joseph Gall, who saw himself as a follower of Locke. Yet the characteristics he found on human skulls went beyond Locke's intellectual faculties of reason, memory, and imagination to include character traits like inquisitiveness, secretiveness, or combativeness. In Europe, Gall's ideas were firmly and effectively resisted by the Viennese court, where he made personal enemies; by the Catholic Church, which denied that human powers were located in the physical brain; and by the scientific establishment, which questioned the validity of his experimental work. In England, in contrast, Gall's ideas were promulgated by his charismatic assistant Johann Gaspard Spurzheim, and they thrived.[37] The power of Spurzheim's delivery may be seen in

Sophia's enthusiastic response to the lectures she and her family attended in 1827. In fact, phrenology shaped her views of the world for decades.

An important element of Spurzheim's success in England stemmed from his insistence that phrenology was a practical subject, one that his listeners could adopt, practice, and develop on their own. The Sophia who arrived in London in 1831 was the perfect audience for this do-it-yourself phrenology. The possibility that one could discover a person's true character through an examination of her or his skull opened up a whole series of interactions considerably more entertaining than those available in the average parlor or dancing party. It gave her an excuse to accept an invitation from Colonel Briggs that she go with him to Camberwell to read a skull purported to be that of Oliver Cromwell; her conclusion was that "benevolence and conscientiousness were large; hope and ideality smaller."[38] Equally amusing was having the Irish political leader Daniel O'Connell, often known as "the Liberator," remove his wig so she could read his skull. In this instance, Sophia was not above playing to her audience, and "seeing the Liberator's love of approbation not small," she endowed him with "many saintly and heroic qualities." At the same time, however, she was quietly recording "the relative sizes of all the largest and smallest developments in the head," which provided ample material for analysis and discussion when the Liberator wasn't listening.[39]

Interestingly, phrenological readings were successful even when wrong. Sophia particularly remembered a family visit to the home of "Mr Holmes, a German, living in London and lecturing on phrenology."[40] Holmes kept a considerable collection of "skulls of criminals, idiots and other abnormalities,"[41] as well as a number of the plaster casts he had made of living people. One of these casts was of Augustus, who had already told Sophia how uncomfortable it had been to have it made; now she was amazed to see Mr. Holmes shake his head and sadly pronounce it the head of a man with wonderful endowments but "no power to make them active."[42] Listening to Mr. Holmes declaim on Augustus's character was precisely the kind of experience that made phrenology so enticing. That his pronouncements were open-ended meant that his audience could push back, pointing out features he had not mentioned, questioning his interpretations of those he had. The silliness of Holmes's conclusions led Sophia to see how difficult it could be to read the head of an unknown person "in any but the most archaic fashion,"[43] but she nonetheless persisted in reading the skulls of the people around her.

Whatever Sophia made of a phrenologist's powers, the central message remained: people's minds and characters were as different as were their faces.

That these differences were part of the human body served as a powerful counter to the Owenite idea that a person's character was "moulded from without, instead of having its origin within the soul,"[44] and it neatly explained the failure of Owen's many efforts to create perfect societies on the assumption that "anyone might attain perfection in any line if properly educated."[45] Phrenology's multifaceted view of humanity fit Sophia's experiences in London considerably better than her father's monochromatic emphasis on egalitarian reason. It convinced her that "only those schemes of reform which allow for an aristocracy of Nature, which in itself implies a giving by the strong to the weak, have a change of ultimate success."[46] This adjustment did not change her commitment to helping the poor, however. In her new London neighborhood, she found herself in the midst of a "knot of philanthropists" and was eager to join them.[47]

Lady Byron was a major source of support for Sophia's charitable efforts. The two had briefly crossed paths when Frend was counseling Lady Byron through her separation, but Frend's eldest daughter was only seven at the time and Lady Byron far too engrossed in her own life to notice the child. By the late 1820s, however, Sophia was quite grown up, and she began to appear at the edges of the letters between Frend and Lady Byron. By the early 1830s, the two women came to realize how much they had in common and began to correspond directly. Unfortunately for the historian, Sophia later asked Lady Byron to destroy her letters, even as she carefully preserved the ones Lady Byron sent to her. Thus her voice survives only as filtered through Lady Byron's. Even though indirectly, these letters provide a revealing view of the development of Sophia's thinking in the formative years of her twenties.

Sophia was far too restless to be confined to her mother's domestic world, and Lady Byron served as her portal to the larger female world beyond. The young woman had already begun her charitable work in the sheltered suburban world of Stoke Newington; after her move to London, Lady Byron introduced her into the rarified world of London's high-society philanthropy. By the time the Frends were settling into London, several years of throwing money into failing cooperative societies had led Lady Byron to the conclusion that working men were essentially incapable of improving their condition, and she was turning her attention to more malleable children. Her inspiration was the Swiss educator Philipp Emanuel von Fellenberg, who had set up a successful school on his estate in Hofwyl near Berne. Deeply impressed by the reports of Fellenberg's school, Lady Bryon, in spring 1834, rented the requisite land and buildings and hired a schoolmaster for a "labour school" for about thirty boys.[48] The students in the Ealing School divided their days between study and agricultural labor in the hopes that they would emerge as healthy and industrious schoolmasters.

The school thrived until 1852, but it was too far from London for Sophia to play an active role in it. Lady Byron directed the younger woman's attention instead to the Children's Friend Society, a group that proposed sending the thousands of vagrant children who lived on the streets of London to work in the colonies. As the group began actually moving children off the streets of London, its support base mushroomed. In 1833, the society opened a school to prepare street boys for the journey in an abandoned silk mill in Hackney Wick. In 1834, the crown princess, Victoria, opened the Royal Victoria Asylum for Girls in what is now Walpole House on the Chiswick Mall. Lady Byron made sure that Sophia had a place on the Victoria Asylum's management committee.

Working with "gutter children" was a perfect fit for Sophia. More than fifty years later, she could still be moved by the memory of how quickly children responded to the basic comforts of warmth, clothes, and "wholesome food,"[49] but her official place was not among them. The charge of her management committee was "to procure and prepare the house and the matron, or matrons, to collect money from friends, and to find and recommend little destitute children for admission."[50] The last of these duties—finding girls to fill the Victoria Asylum—was actually the most difficult because the overwhelming majority of London's street children were boys, whose female counterparts were taken in as servants of all varieties. This meant that while the boys' school in Hackney Wick was filled to overflowing, the Royal Victoria Asylum had trouble finding girls to occupy its beds.[51] But when it came to raising funds, Sophia's committee was exceptionally successful. The same Amelia Matilda Murray who had supported Mary Lister's work in Stoke Newington was in charge of the girls' hostel, and she was a formidable social presence. By the time Murray became maid of honor to the young Queen Victoria in 1837, Sophia was part of the management of by far the largest and most successful charity in all of England.[52]

Sophia and Lady Byron's charitable activities were engrossing, but their intellectual friendship was even more so. In the 1830s, Lady Byron began inviting Sophia for long visits to her home in Fordhook. Sophia experienced these visits as "short peeps into paradise,"[53] and they were similarly wonderful for Lady Byron. She found in Sophia "a pleasant resting place" for her thoughts, "like a green bank, over which flit lights & shadows in rapid succession, but where no dark cloud casts a gloom."[54] Theirs was an asymmetrical relationship—Lady Byron was always Sophia's patroness, both because she was sixteen years older and because she was a Lady. Nonetheless they were devoted to each other.

Sophia and Lady Byron were particularly connected by their shared Frendian upbringing, but they were also part of a larger group of Unitarian women who were moving beyond his ways of thinking. In a letter to Sophia in which Lady Byron was wrestling with Frend's reading of John 8:58—"Before Abraham was, I am"—she also reported on a discussion of Maria Edgeworth's novel *Helen* at Miss Joanna Baillie's house. Edgeworth and Baillie were both prominent Unitarian writers—Edgeworth near the end of her career, Baillie at the beginning of hers. *Helen,* the just-published Edgeworth novel that they were discussing that afternoon, provides a useful entrée into the structure and preoccupations of the wider world of Unitarian women in which Sophia and Lady Byron were embedded.

A defining feature of that world was a clear separation between male and female spheres of influence. This position flew in the faces of radical women like Mary Wollstonecraft and Mary Hays, who had shared their views of womanhood with Sophia's father in the 1790s and beyond. As reflected in *Helen,* the movement away from those egalitarian views was self-conscious. In the novel's first section, Edgeworth's wise mother figure, Lady Davenant, expounds at length on how mistaken women were to aspire to masculine powers. In her youth, Edgeworth's character explains, she tried to emulate Mme de Stael by forming a salon but was rescued from her folly by a respected male friend who told her that the men in attendance were laughing behind her back. Being dismissed as silly by men was quite enough to show her that aspiring to be a politically powerful woman was "unsuited to the manners, domestic habits, and public virtue of our country."[55] The message Lady Davenport passes on to Helen, as well as to her readers, is that the only proper place for successful women is in a world of female sociability.

In the female world of Edgeworth's *Helen,* conversation is the major source for insights into truth. Whereas Frend searched for truth in close readings of the Bible and De Morgan found it in mathematics, Lady Davenport and her friends look for it in the words they speak and hear. In their world, the worst possible sin is to hide true meanings or intentions through obfuscation of any kind, and the action in the novel is driven by the dreadful consequences of social lying. The eponymous heroine is a transparently open young woman who is drawn into an ever-increasing mass of social confusion by the many lies told by her sweet but flighty best friend, Cecelia. This character's earliest untruths are the whitest of lies, offered in a warmhearted effort to help Helen overcome her social awkwardness. Over time, however, Cecelia's fibs develop lives of their own. They become large enough to crack the relations between Helen and her fiancé, and they lead him into an almost fatal duel. When Cecelia finally confesses to the lies that are ruining the lives of both Helen and her fiancé, her own marriage begins to break apart. It takes some time for Cecelia's husband to recognize that justice requires him to forgive his contrite wife. With this forgiveness, in the final sentence of the book, both Cecelia and Helen can be "perfectly happy in Love and Truth."[56]

The capitalized final words of Edgeworth's *Helen* point to the basic structure of the principled social world of reason to which Unitarian women, including Sophia and Lady Byron, aspired. In that world, the warmth of love and friendship must always be tempered by the demands of truth. Maintaining the proper balance could be delicate, however. On the one hand, fibs and white lies were completely unacceptable; things should always be called "by their right names," and "truth must be told, whether agreeable or not." On the other hand, "whoever makes truth disagreeable commits high treason against virtue."[57] One of Lady Byron's letters to Sophia neatly illustrates the kinds of dilemmas this ideal could create in practice. She admitted she had "made a false excuse to a person in order to withdraw her eyes from a paper that would have given her great pain & caused injury to another." She then explained that her behavior was the result of a previously held belief that "benevolence might modify truth," but now she realized it would have been better simply to stop the person from reading the paper while declaring she could not give a reason.[58] Without Sophia's answering letter, we don't know what she made of this particular incident, but the delicate negotiations of truth-telling remained a major preoccupation for her as well.

At least as important as maintaining truth in personal relations was the search after what Frend had earlier designated as "the truth" and which Sophia and Lady Byron called "Truth" with a capital *T*. This Truth was a constant topic of conversation at Fordhook.[59] Arriving at Truth was the goal of what Lady Byron described as her "dear Metaphysico-Theology";[60] all of her and Sophia's biblical researches were directed to this end. Sophia for her part was trying to formulate some kind of reasoned understanding of its basic structures from her readings of contemporary authors like Dugald Stewart. The Truth that these women were reaching toward was too large to be captured by reason alone. It was a matter of the heart, and much of their effort to articulate it was poetic. Sophia apparently submitted a number of poems to the Unitarian *Monthly Repository* that had published her father's "Christian Survey," but none was published. However, some of those that did appear may give a fuller sense of what it was Sophia and Lady Byron meant when they talked about Truth with a capital *T*.

A characteristic example from 1832 is an extended poem in blank verse that describes an inspired dream of divine perfection. For a page and a half, the dreamer joins a throng of diverse people moving toward a temple that rises "high in pillar'd pomp." Once in the temple, all clamoring ceases as the crowd raptly gazes upon an "Angel form" "dazzlingly bright" "with snowy wings outspread." And then,

> The spell that bound that deathly stilly throng
> In silence, was dissolved, and there uprose
> One universal, rending, deafening shout—
> One word was all I heard—that word was *Truth*
> And I awoke! Awoke unto a world
> Where yet the angel form is veiled in clouds,
> Oh! God our Father, when wilt thou send down
> The blessed light from Heaven to pierce the gloom?
> Thou wilt in thine own time,—thy will be done.[61]

Sophia's and Lady Byron's letters drip with the yearnings toward the bliss of perfect understanding reached for in this poem. All of their intellectual endeavors were fundamentally shaped by the challenge of recognizing and embracing the transcendence of a Truth that might be glimpsed in a dream but could not be grasped or held.

The Truth that Sophia and Lady Byron were seeking was easily translated into the truth that Frend saw lying behind all Christian creeds. But even as their intellectual father figure was supporting their efforts in terms of his eighteenth-century version of reason, Sophia and Lady Byron were being challenged to adjust his vision to fit their nineteenth-century lives. A major catalyst for these changes was the arrival in England in 1831 of the Bengali intellectual Rammohun Roy.[62]

Roy was the son of a prominent Bengali family who came of age in the final decades of the eighteenth century. A pleasantly soft-spoken young man, he developed close relations with a number of British civil servants in Bengal and then moved to Calcutta, where he began to flourish as both author and publisher.[63] His political message was liberal; he argued for constitutional reform in India and for freedom of the press, the rights of women, the importance of education, the humane treatment of prisoners.[64] An important part of his power lay in his ability to adapt these messages to different audiences by expounding them in Bengali, Persian, and English. Over the course of the 1820s, his Asian audience grew until he was ready for the larger horizons of England. When Roy arrived there in early April 1831, his liberal political positions resonated powerfully with reform movements that were already underway. He was greeted by enormous crowds in Manchester and Liverpool, happily perched among the foreign dignitaries at the coronation of King William IV, and was eagerly sought after by intellectuals from Macauley to Brougham and beyond.[65]

Sophia first encountered Roy in late May, when he spoke at the Unitarian chapel in Finsbury and again when he dined at her father's house three months later. She found him "a very fine-looking man," whose "dark sparkling eyes, olive-brown skin and black beard, with picturesque Eastern turban and robe, made him a very striking object,"[66] and his conversation was even more intriguing than his looks. Although he had been raised as a Hindu and seems to have adhered to this tradition throughout his life, he had learned of Christianity from Unitarians in India and had been following the *Monthly Repository* for years. Much of his charm for England's Unitarians lay in the way he argued for a rational religion from his Hindu perspective.

Roy had laid out his basic message in his first book on religion, which he published in Persian with an Arabic introduction and intended for an Islamic audience. There he proposed that a dynamic between two themes lay at the heart of all religious traditions. On the one hand, he saw that "turning generally towards One Eternal Being is like a natural tendency in human beings and is common to all individuals of mankind equally."[67] On the other, he recognized that "the inclination of each sect of mankind to a particular God or Gods, holding certain especial attributes, and to some peculiar forms of worship or devotion" was a tendency supported by "habits and training."[68] For him, true religion was rooted in the common ground that lay behind these impulses.

Roy's message of an essential monotheism that lay behind all religious difference resonated powerfully among England's Unitarians, but they were not the only ones who found a reflection of themselves in conversation with this colorful man with a foreign accent. In Roy's two and a half years in England, the same man who pleased the Unitarians by "dwelling on the unity of God and the moral teaching of Jesus," charmed the Quakers by "acquiescing readily in their doctrine of the reception of the Spirit," and won over more orthodox dissenters and churchmen by "accepting salvation by faith, meaning that entire trust in the Divine goodness which shows itself in all the actions of life."[69] At first his universal appeal struck his English hearers as laudable. Over time, however, his ability to appeal to so many different groups who saw themselves in fundamental conflict with each other made him seem more sycophantic than wise; by the time a fatal attack of meningitis tragically ended his life in fall 1833, many had come to see Roy as just a "seeker for popularity."[70] Lady Byron and Sophia did not agree with this critique; they saw Roy's ability to appeal to many different groups as a reflection of the universal Truth that he was espousing.

Frend agreed. He had no difficulty fitting Roy's rational religious message into the reasoned Christian worldview he had been developing for decades. The Indian's "One Eternal Being" coincided with Frend's determined monotheism, and Roy's references to specific cultural traditions resonated with Frend's concern with the creedal overlay. Talking to Roy strengthened the older man's determination to strip away all the cultural confusions that obscured the truly universal message Jesus had taught in the Bible.

The younger generation heard something more radically challenging. Sophia and Lady Byron understood Roy to be saying that the Bible represented the cultural side of the Christian religion, that the whole of biblical Christianity was a particular religion supported "by habits and training." Neither woman was ever to abandon her fundamental Christian commitments, but both recognized that their Christianity had always to take its place within a larger world of religious seeking after Truth. How this search was to be accomplished was certainly a challenge. Expanding religious understanding beyond the reasoned reading of the Bible they had learned from Frend confronted both Lady Byron and Sophia with vast new worlds of uncharted territory.

Lady Byron made the first move. In August 1835 she told Sophia she was going to the Sunday services with a friend who attended a local evangelical Anglican chapel. When Sophia objected, Lady Byron located her defense in the "Truth" that lay beyond mere words. The Evangelicals, she acknowledged, were mistaken in many things; in particular, they confused "Wonder for Veneration," and "Fear for Humility." However, she went on, they were "superior to any religious sect I know, in carrying the sense of their Christian calling into everything" and in the way "their piety irrigates the whole surface of life, instead of being left to flow through one or two great channels."[71] Drawing a distinction between the everyday practice of religion and the theoretical arguments of theology, she proposed to "take my theology from Unitarians—my Practical religion from Evangelicals and Quakers."[72] Therefore, Lady Byron concluded, she could both attend Anglican services with her friends and neighbors and be true to her Frendian Unitarian upbringing.

Sophia was appalled as she watched her friend slip into prevarication, if not outright lying. If the Evangelicals "were fully aware of the opinions you entertain, surely you would not be admitted to their communion" she objected, but Lady Byron was unmoved.[73] Her evangelical friends had at least some inkling of her Unitarian leanings, she responded, and even if they didn't, those views lay on a creedal level where it was egotistical to insist on them. What is more, if she were to try to proclaim her theological views, she was unclear how she could talk about them when words meant such different things to different people. To declare oneself a "Unitarian" was to become "an Athiest," "an Infidel," "a Deist or Philosopher," "a Radical or Revolutionist," "a rational believer," or "a 'Christian Hero,'" depending on who was listening.[74] In short, she concluded there were so many contradictions in the word it was not worth worrying about its true meaning. Sophia was aghast. She called Lady Byron a "hypocrite" and fought back so fiercely against the claim that words and the theology they expressed did not matter that an exasperated Lady Byron asked "to postpone the argument a little longer."[75] Five days later she took her case directly to Frend.

In the letter she wrote to Sophia's father, Lady Byron argued that "in the present state of knowledge" direct opposition to religious error was more likely to strengthen "fallacies & absurdities" than it was to weaken them. Her view was that the clergymen who rattled on about the Trinity knew "very well that they are talking nonsense,"[76] but they were not about to stop and, in the final analysis, it didn't matter. What did matter was not the words that were spoken but rather the ability of the hearers to understand those words. No one can hope to " know Religion 'pure & undefiled' until they are so themselves," she explained, and once they have achieved that condition "the 'traditions of men' will not prevent them from coming to the Truth."[77] This being the case, the best she could do was to hold out the "hand of fellowship"[78] to the people attending services around her, whether those services were Anglican or Unitarian. For her, the Truth lay beyond those doctrinal differences.

Lady Byron's letter struck at the very heart of the practice of reason that had defined Frend's life—his Unitarian conversion, his banishment from Cambridge, his fifteen years of exiled unemployment, and his final work as a businessman. For a full two weeks he "revolved the subject over & over in my mind" before finally figuring out how to respond. In the end, he chose not to defend reason directly but rather to recognize how differently reason manifested itself in people's lives. His first example was Robert Tyrwhitt, who had lived his whole life at Jesus College but "never went to Chapel or to any church except St Marg[aret]s where the liturgy is not used." His second was Baron Maseres, who "did not disguise" his Unitarianism when in London, but who "when resident in Surrey, occasionally attended" the local church. The third was himself. Frend was unapologetic about having lived his life as "an Obtrusive Unitarian." He believed that to worship in an Anglican church was to be like the biblical apostates who worshipped Baal, but he also recognized that it was not his place "to judge another's servant." As long as she was "thoroughly convinced" with a mind that was not "biased by any extraneous circumstances," Lady Byron should proceed in her own way, knowing that she was safe with the God "who knoweth the secrets of all hearts & who judgeth not as man judges."[79] Frend never regretted the life he had led, but he also recognized that in matters of religion everyone had to find the right path for him- or herself.

Frend's loving response seems to have served the purpose of patching up Lady Byron's quarrel with Sophia, and the two remained friends even as Lady Byron moved into the Anglican fold. Frend's daughter was not, however, persuaded by Lady Byron's arguments. Reasoning from both the words written in sacred texts and the words spoken by trusted people remained central to the search for Truth that was to be a major focus throughout her life.

Sophia never abandoned the semantic view of reason she had learned from her father, but Roy had a powerful effect on her. She saw that the Indian's ability to communicate effectively across so many cultural and religious boundaries lay in his knowledge of the "one great spiritual truth which was expressed in different forms by all." In the 1830s, she still interpreted Roy's ability to speak "the Truth" biblically, and attributed it to his having embraced a "doctrine, which is a literal practical truth, taught by Christ himself—'Seek ye first the kingdom of God, and his (its) righteousness, and all these things shall be added to you.'" By the end of her life, however, she was equally comfortable recognizing the power of Roy's religious roots in "the East, where inspiration is still believed in, and a certain amount of preparation for it is practiced in many of the native religions."[80] Sophia had been listening to conversations about India for years, but her direct experiences with an Indian led her to new understandings of human difference.

Sophia's move to a more ecumenical view of religion greatly expanded the sources available to her religious searching. The Bible remained centrally important, but she was willing to supplement what she found there with input from different cultures all over the world. At least equally important was her recognition that the many truth-telling friends, neighbors, servants, and children who populated her woman's world might themselves become powerful sources of insight and instruction. She was often challenged to know how to apply the definitional rigor of William Frend's biblical religion to the multitudes of people and texts she encountered, but she was never daunted. Her efforts opened her to vistas of Truth that she found vastly greater, more exciting, and more satisfying than those her father had known.

# CHAPTER 14

———◆◆◆———

# *Gentleman of Reason*

De Morgan is nowhere to be found in Lady Byron's correspondences with Frend and Sophia. In the period when they were caught up in the problems of the poor, he was engaged in issues more immediate to himself. It was good to have an income that allowed him to move out of his mother's house, but a couple of years spent teaching elementary mathematics to classes of teenaged boys left him ready for something different. In the summer of 1829, De Morgan's Uncle Briggs provided the change of scene he needed by transporting him to the international world of Parisian science and mathematics.

Briggs's initial move to Paris was sparked by his experiences living in England in the volatile years preceding the passage of the Great Reform Bill of 1832. He rented a house in the countryside just in time for the Swing Riots, and when he escaped to London was robbed — first by his footman and then by the footman's accomplices.[1] By spring 1828 he had had enough and moved his wife and five children to Paris, where his teenaged daughters could attain continental polish and he could work on his book in peace. He was warmly welcomed by Jules Mohl, who was "so interested in Oriental literature that he settled in Paris where the greatest advantages and facilities for such studies were then to be found."[2] In addition, he carried an introduction to the naturalist Baron Cuvier, who led him into the world of Parisian natural history.[3] For Briggs, life in Paris was a great improvement over life in London.

Briggs wanted to share the excitements of scientific Paris with his nephew

and so invited the young man to become his research assistant and amanuensis. De Morgan leapt at the chance. Throughout the summer of 1829, he
gathered material for his uncle's book, squired his cousins to evening events,
and eagerly explored the world of Parisian astronomy, physics, and mathematics. As Briggs was pursuing paleontology with Cuvier, De Morgan was
meeting the physicist Ernst Chladni, the astronomer Jean-Baptiste Biot, the
statesman Victor, duc de Broglie, and the mathematician Jean Nicholas Pierre
Hachette.[4] When he visited, the community was divided between participants
in the Bourbon restoration and committed remnants of the Enlightenment
that had fallen with Napoleon. For a fascinating summer, De Morgan soaked
up their perspectives before returning to his London teaching duties.

The following summer was different. At the end of July 1830, political resistance to the monarchical regime exploded into revolution. Briggs saw the
trouble coming and managed to get his family out of Paris just two weeks before the street fighting that overthrew the government. In August, De Morgan moved the other way to join the crowd of English political tourists who
flocked to Paris to see the sights of the overthrow. He emerged deeply impressed. "From all I have heard since my arrival, I feel that our language has
no words at all to express the degree of admiration with which I regard the
conduct of the French people. I had not the slightest idea that large masses of
people ever acted together for good," he exclaimed.[5]

Within weeks of his return to England in September, De Morgan's scientific community exploded into its own revolution, which pitted the aristocratic science of the Royal Society against the scientific community of the
rising middle class. The precipitating cause of what Sophia later called "the
great battle of 1830" was an 1828 parliamentary decision to ignore the Astronomical Society and leave production of the *Nautical Almanac* under the aus-
pices of the Royal Society. After a year of simmering fury over this lack of
respect, Babbage published a passionate attack in which he lambasted the
elitism of English science. In a world with vanishingly few salaried positions
and virtually no government funding for research, the pursuit of science was
closed to everyone except for the very rich, he raged. Babbage's book was both
ferocious and personal enough that when Herschel read a draft, he wanted to
give his friend "a good slap in the face." Babbage was not a temperate man,
however, and he clearly saw the strength of his position in the tense political
and social world that was England in 1830. "I will make them writhe if they
do not reform," he growled and published his book unchanged.[6]

The response was immediate, and England's scientific enthusiasts were

soon deeply divided between the "Declinists," who saw elitism as destroying English science, and the "anti-Declinists," who defended the status quo.
Within weeks, the president of the Royal Society, Davies Gilbert, found his
position so uncomfortable that he resigned it in favor of the Duke of Sussex.
This effort to place a prince of the blood in the chair of England's premier
scientific institution added fuel to the fire of the Declinists, who drafted a reluctant Herschel to stand as the alternate candidate. As Herschel's response to
Babbage's original draft suggests, he was not a man who enjoyed a fight and
was relieved when the duke won the election. Soon thereafter he escaped the
politics of English science by means of a five-year trip to map the southern
skies from the Cape of Good Hope.

Herschel's withdrawal did not solve the problem of the social elitism of the
Royal Society, however, and within months of their failure to capture its presidency, England's scientific reformers were hard at work creating an alternative.
The British Association for the Advancement of Science, or BAAS, committed
itself to spreading science throughout the British Isles by bringing England's
men of science together to present and discuss their ideas in large and inclusive annual meetings to be held in different towns. Any doubts about the viability of this radically open approach to science evaporated with the success of
the first meeting, at which more than two hundred paying members gathered
in York in the summer of 1831. Just four years later, the BAAS boasted more
than two thousand members, with special sessions for women drawing additional crowds, and Britain's cities were eagerly competing for the honor and
economic boost of hosting its bustling meetings. Public interest in science in
England burgeoned.

The creation of the BAAS also brought the people who had previously
been scattered among the Geological Society, the Linnaean Society, and the
Astronomical Society together as a group. At the 1833 meeting of the BAAS,
William Whewell suggested that the men gathered there recognize their identity by calling themselves "scientists," but they were far too culturally conservative to adopt such a new-fangled, term for themselves. Samuel Taylor
Coleridge suggested the term "clerisy" to designate what he saw as "a sort of
national church of intellect," but the members were unwilling to accept the
neo-religious overtones of this designation. In many ways, the most accurate
descriptive phrase for the group was coined more than 150 years after they first
came together. Jack Morrell's and Arnold Thackray's phrase "gentlemen of science" highlights the social concerns that underlay their resistance to Whewell's new-fangled coinage. They were for the most part "cultured" Anglicans,"

who followed a latitudinarian theology and a politics of "moderate reform." They saw themselves as highly respectable, but very few of them could claim to be gentlemen in the most traditional sense of the word. Their challenge was to create a social space for themselves in which scientific prowess guaranteed the privileges and respect traditionally accorded to the landed gentry.[7]

De Morgan's Cambridge education, membership in the Astronomical Society, and determination to devote himself to a life at the forefront of intellectual change would seem clearly to locate him among Morrell's and Thackray's gentlemen of science, but equally important factors marginalized him. The professor of mathematics at the "godless" London University was not a liberal, latitudinarian Anglican but a rebel who had left the Anglican Church entirely; the young man who responded to the French revolution of 1830 with dispatches hailing the power of the French *citoyens* had moved far beyond of the politics of "moderate reform"; and the reasoned, mathematical thinker who could barely see the natural world around him was never involved in either the founding of or the subsequent meetings of the BAAS. These differences were disqualifying enough that De Morgan did not merit a single mention in Morrell's and Thackray's *Gentlemen of Science,* but he was just as determined to be granted the respect due to a gentleman as were those who did. As they were forging positions for themselves as gentlemen of *science,* De Morgan was creating a place for himself as a gentleman of *reason.*

The most straightforward way to designate the difference would seem to be calling De Morgan a "mathematician," but that would not address his problem. The term "mathematician" had a much longer history than did Whewell's sibilant "scientist," but it carried at least equally powerful negative social connotations.[8] De Morgan had early confronted these objections in his own family's reactions to his great-grandfather, James Dodson. De Morgan remembered being told "when I was very young, on my making reference to him in company, that it is 'not usual to <u>cry stinking fish</u>,'" that is, to draw attention to one's rotten relatives. Reflecting on this incident as an adult, De Morgan concluded, "The reason of this loss of his name is, I fully believe, the <u>gentility</u> of his descendants, who, as I have observed, think [of] the Mastership of the mathematical school of Christ's hospital as <u>low</u>."[9] The biography of Dodson preserved among De Morgan's papers drips with fury against those who would judge him, and by extension his great-grandson, to be less worthy of respect because a teacher of mathematics.

De Morgan made a similar point in a tongue-in-cheek paper written sometime in the 1830s in which he set out to prove that "A. De Morgan is not a gentleman." The premise of the proof that followed was that to be a gentleman, a man must either "be free of labour"—which clearly did not describe De Morgan—or "carry on some employment the need whereof floweth from the loss of paradise by our great grandfather Adam." De Morgan defended this definition by pointing out that military officers, lawyers, clergymen, physicians, landowners, and surgeons were all considered gentleman because of their professions, but none of those professions would be necessary in the paradisical world of Eden before Adam ate the apple. Being a teacher of mathematics, however, was "an employment which would have found place even though man had continued perfect; for they would have sought, both for use and enjoyment, every way in which the mind might add to its forces." From this De Morgan concluded with a flourish, he lacked "an essential of gentility" and was therefore not a gentleman.[10] There is no reason to take this analysis seriously, but that he took the trouble to develop it is evidence of how important the designation of "gentleman" was to him.

Even as De Morgan brooded in private about the ways he did not meet the traditional criteria of gentility, in public he fought fiercely for the esteem accorded to those who did. His search for gentlemanly respect began with a battle over the way professors were treated at the London University. When he first accepted the position as professor of mathematics there, he was happy to declare his independence from both Cambridge and the Anglican Church and to devote his efforts to educating students outside of the establishment. This commitment left open the larger definitional problems that faced him as a professor in the newly defined university. It was not at all clear what it meant to be a professor in a nonresidential secular university. Guaranteeing the respectability of that role was one of the major challenges that faced De Morgan and his professorial colleagues at the London University.

Looking to the examples of Oxford and Cambridge was not particularly helpful in this effort. Professorships at the Oxbridge Universities had long been treated essentially as sinecures; there is little evidence, for example, that Isaac Milner did any original mathematical work during his twenty-two years as Lucasian Professor of Mathematics. This was slowly beginning to change in De Morgan's post-Napoleonic world. Robert Woodhouse practiced astronomy after being named Plumian Professor in 1822, and Adam Sedgwick pursued geology after being named Woodwardian Professor of Geology in 1818. No professor was ever expected to engage in regular classroom teaching, however; that role was reserved for college tutors who were also acting in loco parentis. The closest anyone at Cambridge came to the twice a day lecturing that De Morgan faced were the private mathematics coaches whom he had scorned when he was a student. He did not object to teaching; on the contrary, all the evidence points to his having been an inspired teacher who was deeply committed to his students. Nonetheless, even as he threw himself into his position, he dressed himself elaborately every day in answer to those who saw him and his colleagues as workers in a "Cockney University" established to serve "dustmen," "coal-heavers," and "sweepers."[11]

Dressing like a gentleman was not enough to establish the status of a professor of mathematics at the London University, however. Structural issues bound up in the university's nontraditional organization raised questions about the power and privileges of all of its professors. From its inception the London University had been set up as a joint stock company. The initial capital was raised by selling £100 shares; the governance of the university was through a council of twenty-four men elected from among these shareholders. The result was a purportedly academic institution being run by a council of people who were hoping to make a profit.

This state of affairs caused problems from the beginning. Even before Augustus signed his initial contract, he and his colleagues had to negotiate their relations to a council divided between those like Frend, who saw the university as depending on the professors, and those others who believed the professors depended on the university. Initially, the second of these groups prevailed, and the members of the faculty were asked to sign contracts accepting the authority of the Council over their work. It was only after they balked at this request that the professors were allowed to hold their positions as long as they adhered "to the constitution as set forth in the Deed of Settlement."[12] The skirmish over the initial contracts was thus resolved in favor of professorial autonomy.

The underlying tensions between the Council and the professoriate nonetheless remained. The grandiose building on Gower Street had effectively eaten up the London University's initial capital, leaving the Council with few resources to pay expenses. In the initial business plan for the university, both the return on the stockholders' investment and the salaries of the professoriate were to be financed by student fees. But there were problems here as well. In much of their original thinking, members of the Council had taken the University of Edinburgh as their model, but when it came to student enrollments, the analogy did not hold; whereas enrollments in Edinburgh averaged about 2,000 students per year, enrollments in London hovered at only about

650.[13] Until and unless it attracted more students, the London University was in a financial bind.

Initially, this was not a problem for the professoriate. When they were hired, the Council had guaranteed each of them a salary of £300, which was comparable to what they would have earned as tutors at a Cambridge College.[14] Unlike the situation with college tutors, however, these funds were guaranteed for only three years. After that the professors' salaries were to come from the fees students paid to take their individual courses. This arrangement was not an issue for De Morgan because his mathematics classes lay at the heart of the curriculum; in his first year he lectured to almost one hundred students and had even more the second year.[15] However, his more specialized colleagues could not hope to generate classes of this size. So, for example, Robert E. Grant, who taught anatomy in the medical school, could only hope for classes of between thirty and forty. These numbers were not enough to generate an adequate income.

In the summer of 1831, as the end of the academic year approached and professorial salaries were no longer guaranteed, tensions between the faculty and the Council came to a head. The presenting issue was the case of Granville Sharp Pattison, a professor of anatomy. Pattison was both a poor teacher and an academic traditionalist who so infuriated his radical colleagues that they organized to have him fired.[16] Nonetheless, when the Council began to look into the complaints that had been raised against him, De Morgan joined a number of other professors who objected to the firing.

De Morgan's defense of Pattison was grounded in the man's status as a professor and gentleman. In an impassioned letter to the Council, he pointed out the ways their policies were undermining the professoriate in relation both to the students and to society at large. When it came to the students, he argued, "No man who feels (rightly) for himself will face a class of pupils as long as there is anything in the character in which he appears before them to excite any feelings but those of the most entire respect."[17] He was clear that the students' respect for the professoriate would be shattered as soon as they knew that the Council could fire members of the faculty at will. London University professors needed to be accorded the same respect given to "the clergyman, the lawyer, the physician, the tutor or Professor in the ancient Universities." In his eyes, what all of these groups had in common was that they were "secured in the possession of their characters," which meant that "nothing but the public voice, or the law of the land, can touch them." The only way to ensure that De Morgan and his colleagues were equally respected was to make them equally free. In practical terms, this meant that professors' positions could not be terminated except "by death, voluntary resignation, or misconduct."[18] The professors at the London University were gentlemen, who were every bit as honorable as were those who managed them. The Council had no right to fire them without cause.

In the event, De Morgan's eloquence went for naught, and the Council dismissed Professor Pattison without presenting any evidence that he had failed in his "adherence to the constitution as set forth in the Deed of Settlement." On the contrary, Pattison was assured by the Council that "nothing which has come to their knowledge with respect to his conduct has in any way tended to impeach either his general character or professional skill and knowledge."[19] This statement acted as a red flag to De Morgan's bull; in the Council's effort to give Pattison some support for future job prospects, he saw "distinctly laid down the principle that a Professor may be removed, and, as far as you can do it, disgraced, without any fault of his own."[20] He resigned on the spot.

Frend firmly supported De Morgan in his resignation and offered the encouragement that there was still "ample field for your exertions" in the law.[21] De Morgan appreciated the support, but his experience at the London University did nothing to weaken his determination to "keep to the sciences as long as they will feed me."[22] His salary as a professor had been low enough that from the beginning he had been supplementing it with freelance work: teaching pupils in his lodgings and consulting for insurance companies. At one point it seemed that he might land a full-time position with the Amicable Insurance Office.[23] But even as that possibility was falling through, a new and in many ways far more interesting path was opening before him. Departing the London University freed De Morgan to devote himself fulltime to educating the population of England in a different way.

———•———

The cause that drew De Morgan in as he was leaving the London University was an offshoot of the vision of working-class libraries that Frend and Lady Byron had pursued so actively in Brighton. Even as Brougham was laying the groundwork for Mechanics' Institutes in his *Practical Observations upon the Education of the People,* he had also been recognizing that working-class reading groups and libraries needed appropriate reading materials in order to succeed. He immediately rose to the challenge: "I am not without hopes of seeing formed a Society for promoting the composition, publication, and distribution of cheap and useful works,"[24] he wrote in his *Observations.* In November 1826, Brougham chaired the first meeting of the Society for the Diffusion of Useful Knowledge, universally known as the SDUK.

Brougham was as much a businessman as he was a visionary, and his SDUK was to be a profitable venture. Within two years of the first meeting, he had connected with the publisher Charles Knight, who shared both his vision of providing readings for the poor and his hopes for making it profitable. For almost two decades, Knight, Brougham, and the board of the SDUK engaged in a complex process of trial and error to find the right balance among their vision of working-class edification, their public's interest in their efforts, and the economics of book publishing in the 1830s and 1840s.[25]

The SDUK's mission caught De Morgan's attention from the very beginning. Providing workers with books that would occupy their minds, teach them right reason, and preserve them from the kinds of excesses that erupted in the Swing Riots was the perfect occupation for an idealistic young man who did not have the patience to confront their "ignorance and folly" head on. In the years after he finished Cambridge, he wrote a number of essays—on acoustics, on statics, on projective geometry—but all of them were rejected as too technical.[26] De Morgan was far too deeply committed to the cause of popular education to be discouraged by these early failures, however, and so he adjusted. In 1830 the SDUK agreed to publish *On the Study and Difficulties of Mathematics,* in which De Morgan set out to share the mathematics of reason with the world beyond his London University classroom.[27] This time he succeeded in attracting an audience. *Study and Difficulties* was issued and reissued through the nineteenth and into the twentieth century.[28] For De Morgan, it marked the beginning of a decade and a half spent spreading his views of mathematics and reason by writing for the SDUK.

In January 1831 Knight launched the *Quarterly Journal of Education* in an effort to educate those people "on which depends the education of all the rest," in other words, teachers.[29] De Morgan was a major supporter of this journal from its inception. He contributed general articles with titles like "On Mathematical Instruction" and "On Teaching Arithmetic"; more focused ones with titles like "On the Method of Teaching the Elements of Geometry" and "On the Method of Teaching Fractional Arithmetic"; articles on educational institutions, including "Polytechnique School of Paris" and "Royal Naval School"; as well as numerous book reviews. The *Quarterly Journal* was not a commercial success, so Knight ceased publication after only four years, but during that time it was the perfect vehicle for De Morgan's thinking about the best ways for his countrymen to develop reason through mathematical education for people of all ages and social conditions.

The *Quarterly Journal* was just the first SDUK publication to welcome De Morgan's contributions. Equally receptive was the *Penny Magazine,* which Knight began publishing to compete with cheap pamphlets that flooded London after the passage of the Reform Bill of 1832. From the SDUK's middle-class point of view, the majority of these were "dangerous in principle and coarse in language." Knight's answer was a little magazine that contained no material to "inflame a vicious appetite; not a paragraph that could minister to prejudices and superstitions … no excitements for the lovers of the marvelous—no tattle or abuse for the gratification of a diseased taste for personality, and, above all, no party politics."[30] Instead, the *Penny Magazine* included travel reports, literary reviews, a series of articles on "Old English Ballads." De Morgan contributed tidbits on fractions and the impossibility of squaring the circle. Some members of the society at first found it "awkward" and "beneath the dignity of the Society" to be publishing a "penny weekly sheet," but Knight's little experiment was a great success; by the end of the year, England's readers were buying on the order of 200,000 weekly numbers of the SDUK's *Penny Magazine.*[31]

The *Penny Cyclopædia,* which first appeared in 1833, was born from the success of the *Penny Magazine.* Knight's hope had always been that providing a "ramble-scramble" of short articles in the *Penny Magazine* would serve to "enlarge the range of observation, to add to the store of facts, to awaken the reason," and "assist in the establishment of a sincere and ardent desire for information."[32] The *Penny Cyclopædia* was his answer to these newly created demands. His intention was to issue the *Penny Cyclopædia* in weekly penny pieces that could in the end be gathered into an eight-volume encyclopedia.

Knight's vision for the *Penny Cyclopædia* was as ambitious as Diderot's plan for the spectacular *Encyclopédie* that has long been hailed as the quintessential publication of the Enlightenment. Like the Frenchman, he had no interest in "an affair of scissors and paste"; instead, each article was to be an original contribution written by "the best that could be found" in that particular area. His determination to hire experts meant that Knight's project ran away with him, just as Diderot's had seventy-five years before. Within months, he began to see how difficult it was for scholars to "give us the very cream of [their] knowledge" and "pour out the fullest information" in articles that were "readable and perfectly intelligible to the popular mind." That he was also asking that these articles be written "in the most condensed form of words," while paying their authors by the page added impossible insult to injury. By the end of the first year the articles were longer, and the price rose to two pennies; by the end of the third year it was four pence.[33]

With each rise in price, the number of subscribers dropped until the *Penny Cyclopædia* was barely viable. Knight was nonetheless far too committed to let his project die. He and his writers persisted until the last of the *Penny Cyclopædia's* twenty-seven volumes appeared in 1843. These were followed by several volumes of supplements, the last of which appeared in 1858. In England, Knight's *Penny Cyclopædia* languished in a publishing backwater—too expensive for the working classes to afford and too cheaply produced to compete with more attractive middle-class publications. But the Enlightenment vision of universal education that drove its publication was very powerful outside of England, where it was embraced in countries as far away as Japan.

The writers who sheltered under Knight's financial umbrella were major beneficiaries of his dogged determination to spread knowledge throughout the world. These included the editor, George Long, who stayed with the project through its entire ten-year span, and the dedicated group of contributors who gathered around him. De Morgan was a central figure in this group. Like Diderot before him, Knight believed that in a reasoned encyclopedia "it was essential that one mind should have the almost undivided charge of Mathematics,"[34] and he placed De Morgan in charge of mathematics and astronomy. De Morgan embraced the challenge with gusto. In his astronomical articles, his goal was that "every discovery, or determination of fact, of any importance, was made as clear to the world as the subject allowed";[35] in his mathematical ones he meticulously laid out the mathematics of reason. These two tasks constituted a prodigious amount of work; by the time the last volume of the *Penny Cyclopædia* was completed in 1843, De Morgan had written almost one-sixth of it.[36]

The almost seven hundred articles De Morgan wrote for the *Penny Cyclopædia* over the course of more than ten years build on one another to create a remarkably coherent whole. This was not an accident. One of the disadvantages of covering a large subject in an encyclopedia is the way its alphabetical ordering undercuts the proper relation of subjects to one another. But as De Morgan wrote individual articles on topics from "arithmetic" to "trigonometry," with "logarithms" and "quantity" in between, he remained constantly aware of the relations among them, so that the whole series would be "complete and harmonious."[37] He succeeded so well that if someone gathered them into a single whole, they could have been used as a mathematics textbook.

In the case of the *Penny Cyclopædia,* the enormous task of article integration was further complicated by the fact that the volumes came out over
the course of more than ten years. This meant that all of the writers had to
find ways to respond to contemporaneous change. Knight was particularly
proud of the ways Long and his writers managed to follow the development of
British railways in their articles,[38] but he never mentioned the equally dramatic
change that took place in De Morgan's view of mathematics. Over the course
of the period in which he was writing the mathematical articles of the *Penny
Cyclopædia,* De Morgan was coming to a radically different view of the nature
of the subject. His articles for the *Penny Cyclopædia* place him at the cutting
edge of a new school of English algebra that developed in the 1830s and into
the 1840s. He never wavered in the conviction that mathematics was a model
for the practice of reason, which meant that as his mathematical ideas developed, his understanding of reason changed as well.

*Study and Difficulties* may be taken as the starting point of De Morgan's
journey to a new view of the mathematics of reason. In this overview, he presented the same tortured view of negative numbers that he had adopted in his
Bourdon translation three years earlier. By 1833, however, there are signs that
his position might be changing. In the article "Algebra," which appeared in
the first volume of the *Penny Cyclopædia,* he neatly postponed any discussion
of negative numbers by referring the reader forward to two not-yet published
articles, "Negative" and "Positive." This postponement may well have been a
response to the stringent length constrictions Knight imposed on all of his initial articles, but it also served as a way for De Morgan to keep the door open
while he considered the possibility of a different approach to the troublesome
negative numbers.

The person who in 1833 was leading De Morgan to his new approach was
his Cambridge tutor, George Peacock. Over the course of the 1820s, even as
he was encouraging De Morgan's mathematical studies, Peacock had been
carefully examining reports of different mathematical systems that English
explorers were encountering as they confronted various cultures around the
world. Some sense of the breadth of his thinking may be surmised from the
titles of the papers he presented to the Cambridge Philosophical Society,
which had in 1819 arisen from the ashes of the Analytical Society: "Greek
Arithmetical Notation"; "On the Origin of Arabic Numerals and the Date of
Their Introduction in Europe"; and "On the Numerals of the South American Languages."[39] Peacock's explorations of non-Western views of number

changed his understanding of mathematics as dramatically as Sophia's and Lady Byron's interactions with Rammohan Roy changed their understanding of religion. The variety of approaches to number that he encountered was so rich that he abandoned the historically static view in which arithmetic had been conclusively defined by the ancient Greeks. In an 1829 article titled "Arithmetic" in the *Encyclopædia Metropolitana,* he showed how all of the diverse number systems he had read about fit into an enormous pattern of development.

After spending a decade mapping the worldwide progress of arithmetical systems, Peacock was ready to turn his attention to algebra. In *A Treatise on Algebra,* published in 1830, he resolved the problem of negative numbers by embedding it in the same kind of progressive narrative he had so carefully mapped out in arithmetic. In Peacock's view, the history of algebra developed through two major stages. In the first, arithmetic equations like $4^2 - 3^2 = (4 - 3)(4 + 3)$ were generalized into the form $a^2 - b^2 = (a - b)(a + b)$ for $a > b$. In the second stage, such equations were further generalized by dropping the restriction $a > b$. Removing this restriction immediately opened the possibility that $b$ could be greater than $a$, and thereby allowed negative numbers. That these numbers had consistently yielded true results over the course of two hundred years supported his conclusion that negative numbers were valid.

Having laid out this historical narrative, Peacock enshrined his conviction that it reflected a universal truth in what he called the "Principle of Equivalent Forms." He worded this principle somewhat differently at different times, but the following may be taken as a basic statement of it: "Whatever equivalent form is discoverable . . . when the symbols are general in their form though specific in their value, [e.g., $a^2 - b^2 = (a - b)(a + b)$ for $a > b$] will continue to be an equivalent form when the symbols are general in their nature as well as in their form. [e.g., $a^2 - b^2 = (a - b)(a + b)$]."[40] Translated into everyday language, Peacock's Principle of Equivalent Forms asserted that any result of an algebraic equation that had been generalized from a legitimate arithmetic one was valid, even when it was not clear what that result meant.

Peacock's principle fundamentally changed the Lockean model of reasoning that Frend and De Morgan had been imposing on algebra. Accepting it entailed recognizing the validity of some algebraic results based on the processes through which their symbols were generated, rather than on the meanings of those symbols themselves. Accepting results of processes was just a temporary state, however. Even as Peacock opened the door to unclear mathematical entities that had been legitimately generated, he also insisted that in time their true meanings would appear. The truth of mathematics continued to rest on the meanings of its symbols, even in the cases where that truth was not yet clearly understood. In the end, his principle did not really change the view that mathematics was meaningful, which had for so long supported Locke's model of reasoning. What it did do was replace the circular view of history that had attended that approach throughout the eighteenth century with a nineteenth-century progressive one, in which human understandings of meaning expanded over time.

De Morgan was stunned when he first encountered Peacock's Principle of Equivalent Forms. His former teacher's proposal that mathematical truths could be found by following algebraic symbols beyond their clearly defined arithmetic roots left him deeply perplexed by a vision of "symbols bewitched, and running about the world in search of meaning."[41] But he respected the older man far too much to dismiss his ideas out of hand. Instead, over the course of several years, De Morgan foraged among his books for insights into the ways the relations between algebraic forms and their meanings had developed over time. As he delved into the mathematical past, he was increasingly impressed by the ways that arithmetically generated forms found ready interpretations beyond the realm of number where they had originated. Slowly, over the course of several years, he became convinced that Peacock was right: the historical record showed that, when carefully manipulated, mathematical symbols could generate true insights into whole new worlds much larger than those from which those symbols had originally been drawn.

De Morgan announced his change of heart in a positive review of Peacock's *Treatise on Algebra* published in the *Quarterly Review* in 1835. As always, the focus of his interest lay in "the *reasoning* of algebra, considered as a pure science," which meant considering algebra as the kind of "compendious language" he had described in the introductory lecture he delivered to the London University.[42] But having accepted Peacock's approach, he was beginning to view language in a newly dynamic way. He gave as an example the uses of the verb "to see." In its initial meaning, as a child first learns it, "to see" is a verb that "applies to perception by means of the action of light on the eyes." But over time, its meaning is expanded, and "to see" may include "any conclusion which the reason can draw," as when we say we can *"see"* that Rossini's music is different from Handel's. Whereas before, De Morgan would have condemned this kind of fluidity of meaning, after studying Peacock he came to understand that it was not only legitimate but essential; to insist that a word be used only in its original sense "would be to deprive the mind of several most important modes of invention."[43] Months spent mulling over Peacock's arguments led De Morgan to see that spoken languages develop progressively.

The same was true for the symbols of algebra. Those symbols had all originated as "representations of quantity," and the rules by which they were be manipulated had all initially been drawn from the essential definition of quantity.[44] When new results were generated by the rule-bound manipulation of those symbols, those results pointed to the need for an extension in the original definition of quantity. In mathematics strict rules governed the terms of these extensions. They had to be made in such a way that the extended meaning included all of the results that were true within the original, more restricted one.

De Morgan was supported in his conversion by the work of John Warren, who in 1828 had succeeded in developing a consistent interpretation of both negative and imaginary numbers on a Cartesian plane.[45] De Morgan hailed Warren's works, along with Peacock's, as "the most original which have appeared in England, in *pure* mathematics, since the 'Analytical Calculations' of Professor Woodhouse."[46] He still adhered to a view of reason in which the truth of mathematics rested on the interpretation of its symbols, but he now saw those symbols as the product of a process of discovery, in which progress guaranteed ultimate success. For De Morgan, as for Peacock, negative and impossible quantities came to epitomize this kind of change. He was never again to insist that his students explain the meanings of their symbols at every step of an algebraic process. He had come to recognize that those meanings were to be found at the beginning and the end of a symbolic process whose validity was guaranteed by a deep historical experience of mathematical progress.

Frend was appalled when De Morgan explained his new approach: "I am very much inclined to believe that your figment $\sqrt{-1}$ will keep its hold among Mathematicians not much longer than the Trinity does among theologians," he huffed. The older man defended his position with a return to the classical world in which geometry and algebra were separate studies, in a way that delegitimized all of De Morgan's interpretative examples: "Algebra requires not the aid of geometry though the two sciences sh'd be studied together," Frend insisted. He admitted that negative numbers could be useful in "science as in mechanicks," in ways that were proving to be "of great benefit to mankind." When mathematics was being approached as a model for reason, however, he insisted on "certainty not uncertainty, science not art."[47] Frend's objections fell on deaf ears. De Morgan was far too liberated by Peacock's insights to be slowed by the older man's classical objections.

As it turned out, accepting Peacock's principle did more than free De Morgan from Frend's crabbed eighteenth-century arguments about negative numbers. Embracing Peacock's providential view of mathematical development also bolstered his claims to be a gentleman by transforming him from mathematical mechanic into a cosmopolitan man of the world. De Morgan had been haunting the stalls of London booksellers ever since he left the libraries of Cambridge, but as long as he accepted Frend's static approach to mathematical truth, his collection was useful only for showing all the corruptions of meaning that had crept in over time. That changed when he adopted Peacock's progressive vision. Books became artifacts of a progressive history that lay at the very core of mathematical understanding, and De Morgan's library became essential to his mathematical work. Like a country gentlemen who pursued natural history on his estate, De Morgan became an urban gentleman who pursued the natural history of reason in his library.[48]

# CHAPTER 15

## Reasoning among the Stars

In the time Sophia was spending with Lady Byron, and Augustus with the SDUK, they were immersed in two separate worlds. The only men in Lady Byron's world were those like Frend or King whom she looked up to as mentors; the only women associated with the SDUK were potential readers. In these two different places, Sophia and Augustus were following parallel paths in their explorations of what Sophia would call "Truth" with a capital *T,* but their interests did not overlap. As Sophia began to engage with Eastern spirituality and Augustus pushed ahead with new ideas of algebra, they were following reason on two sides of a gendered divide.

Their paths continued to cross in London, however. The Frend house on Bedford Place was just blocks from De Morgan's bachelor apartment on Gower Street, and the young man soon became like one of the family. When Frend was not discussing mathematics with him, Sophia was consulting him about the books she was reading, Frances playing the piano to his flute, and Harriet regaling him with ghost stories.[1] Perched at the top of the chronological ladder of Frend's children, Augustus fit comfortably into the role of eldest son.

In the summers, De Morgan's dislike of the countryside made him the perfect house sitter for the Frends' London home. In the letters he and Frend exchanged, he joked that his most positive out-of-town activity was clipping hedges in an attempt to introduce some kind of regularity into the "viridity of extra-urban scenery."[2] He found the seaside, with the waves "slopping" and "fiddling at the sand"[3] equally nerve-wracking and even less susceptible to correction. Frend cheerily countered De Morgan's need for regularity with fulsome tales of family adventures: "Frances has trusted herself to the driving of Alfred & they came home with no other adventure than the wheels of their carriage being locked in by that of a cart";[4] Alicia was "quite delighted" with horseback riding;[5] and "Mrs. F is this moment come in with a parapluie over her head, there being about a drop to a square foot."[6] In letter after letter the friendship between the two men deepened, but Sophia only appeared as casually and haphazardly as any of her siblings. She and De Morgan were not coming together there.

They were, however, meeting in the context of one of De Morgan's primary "town delights." He had enjoyed the fellowship of the Astronomical Society ever since he came to London, and its fortunes were on the rise. In 1831 its members were finally given control of the *Nautical Almanac* and granted a royal charter that transformed them into the Royal Astronomical Society (RAS).[7] In the same year, De Morgan was made honorary secretary. It was an inspired appointment. The man who resisted leaving town happily called the meetings and wrote the minutes of a group whose members were often on the move. His eyesight was so limited he could neither view the stars nor read experimental instruments, but he was very well suited to recording the conversations and publishing the papers of those who could. Being honorary secretary placed De Morgan at the very center of the RAS.

Although not causally related, the Astronomical Society's elevation to RAS also marked the beginning of a new chapter within its leadership. In the earliest days of the Astronomical Society, Herschel and Babbage had constituted the essential Cambridge-educated core of the group. By the 1830s they were moving away, which left room for a younger trio to take their place. De Morgan was the youngest of this second wave of Cambridge graduates; the other two were Richard Sheepshanks and George Biddel Airy. All three had begun their careers at Trinity College. Sheepshanks had graduated as tenth wrangler in 1816, won a Trinity fellowship in 1817, and was called to the bar in 1825. He was, however, as unwilling to practice law as De Morgan was and instead lived off of an inheritance from his cloth manufacturer father. Airy finished as first wrangler in 1823, then earned his living first as Lucasian Professor of Mathematics, then as Plumian Professor of Astronomy, and finally, beginning in 1835, as the Astronomer Royal in the Greenwich Observatory. De Morgan, Sheepshanks, and Airy together formed what De Morgan called an "equi-tenacious triangle"[8] that throughout the 1830s came together for the meetings of the RAS and then retired to the Piazza Coffee House for joyous evenings as the "Astronomical Club."

Francis Baily was always present as well. As Babbage and Herschel were drifting away from the society they worked so hard to create, Baily's commitment was growing ever stronger, until he regarded "its welfare and interests as identical with his own."[9] Over the course of the 1830s, when Babbage was burying himself in calculating machines and Herschel observing at the Cape of Good Hope, Baily was securing rooms for the RAS in Somerset House, building up a considerable library, and nursing the society's finances to the point that it became "rich and independent."[10] He forged a close relationship between the RAS and the Royal Observatory at Greenwich and ensured that his protégé Airy was made Astronomer Royal in 1835. Technically the RAS elected a new president every two years, but in fact Baily ruled astronomy in England until his death in 1844.

On the surface, it would seem that Sophia could have no place in this part of De Morgan's world. The Astronomical Society did recognize Caroline Herschel's work with a gold medal in 1826 and the RAS named Mary Somerville an honorary member in 1835, but no woman was actually admitted to membership until 1919. Women were nonetheless very present in the early years of the RAS. Baily lived on the northern edge of London in a "pleasant house in a garden sheltered by sycamores," which was the perfect setting for large social gatherings. At home, he was the consummate host, "well fitted by his clear-headed steadiness of character, as well as by his excellent temper and geniality, to form the centre of a knot of friends sharing in the same pursuits."[11] Equally essential was his sister, whose presence meant that women as well as men could gather at the Bailys' house. As he opened his home for scientific meetings, Baily was following a tradition that dated at least to Robert Boyle in the early days of the Royal Society. But whereas other scientific hosts enforced a discipline that excluded women, Baily and his sister domesticated both the science and the practitioners of astronomy who came to their house in ways that included them all.[12]

Sophia was a major beneficiary of this inclusion. For years she had thrived among her father's friends, but she had trouble fitting in with her peers. Her upbringing had been essentially different from that of most young women her age, and in her early London years she "often shrank humiliated"[13] when she saw how irrelevant her knowledge was in their eyes. At the Bailys' house, however, she found herself embraced by a warm group of similarly educated women. Baily's sister was essential to her feeling welcome, but equally present were Airy's wife and her sister, Sheepshanks's sister Ann, as well the wife of George Long, the editor of the *Penny Cyclopædia*. Once or twice a month, this group gathered for parties at their various houses. Whereas De Morgan and his flute were stars of musical evenings at the Sheepshanks, Sophia stood out at Baily's house, where the entertainment was more literary and dramatic. Thirty years later she still remembered that "no house in London, I suppose, had held more happy parties than 37 Tavistock Place."[14] In the social world of the RAS, Sophia and Augustus found themselves together in a place where each of them felt they belonged.

Sophia and De Morgan both felt at home with the Bailys, but fitting into that world required each of them to move away from their individual worlds of reason. For Sophia, the adjustment was primarily social. With Lady Byron she was with a mentor who encouraged her efforts to follow the reason they had both learned from Frend. In the Baily household she was part of a group defined by their relation to the men of the RAS. Sophia's father had ensured that she could follow astronomical discussions, but she showed no signs of wanting to enter them. Instead, she happily joined the other women in negotiating the interface between their female world and the masculine one of the RAS.

The men of the RAS were quite content to let the women maintain the social interactions at 37 Tavistock Place so that they could focus on establishing the legitimacy of their work in the Victorian world that surrounded them. Herschel emerged early as the major spokesman for their gentlemanly ideal. In 1831 he published *A Preliminary Discourse on the Study of Natural Philosophy,* in which he presented the pursuit of science as the highest human endeavor. Herschel's scientific investigator was a natural philosopher whose vision was large enough to encompass everything in the universe. Never content with narrow explanations, he was always pushing himself to his limits. The new and unknown expanded ever larger before him, "refinement follows on refinement, wonder on wonder, till his faculties become bewildered in admiration, and his intellect falls back on itself in utter hopelessness of arriving at an end."[15] Being thus overwhelmed by the enormity of what he did not know, Herschel's natural philosopher moved seamlessly to the conviction that his intellectual existence would continue beyond death, and that in an afterlife "he shall drink deep at that fountain of beneficent wisdom for which the slightest taste obtained on earth has given him so keen a relish."[16] In the meantime, he found himself filled "as from an inward spring, with a sense of nobleness and power" that left him calm, refreshed, and more fully open to the good in the world around him.[17] For Herschel, the pursuit of science led its practitioners to a higher plane of insight above all of the tawdry pushes and pulls of the material world.

Herschel easily found an audience for his inspired view of the scientific enterprise. In the wake of the passage of the Supplemental Test Act and the Roman Catholic Relief Act, his descriptions of scientific practice resonated with England's Christian readership without prescribing particular religious affiliation. His *Preliminary Discourse* was a runaway best seller whose popularity only increased in 1833, when its author took off on a grand adventure to map the stars of the Southern Hemisphere from the Cape of Good Hope. From then on Herschel served as the epitome of a Victorian scientific gentleman.

The William Whewell who had dominated Trinity conversation when De Morgan was a student there agreed with Herschel about the uplifting qualities of scientific activity. In the decades after De Morgan fled Cambridge, Whewell became ever more firmly entrenched. From 1828 to 1832 he was professor of mineralogy; after 1838, he was professor of moral philosophy; and as he rose through the ranks of the university, he became a major voice for the Anglican intellectual establishment. His *Astronomy and General Physics Considered with Reference to Natural Theology,* published in 1833, focused largely on the ways astronomy works to strengthen the character of its practitioners.

Unlike Herschel, who was actively pursuing the science he was describing, Whewell relied on the example of past astronomers to demonstrate the character-building effects of scientific practice. Newton was one of Whewell's most powerful exemplars. From Whewell's point of view, the personal process of grappling with mathematics had given Newton a powerful glimpse into the mind of the God who had designed the universe. As a result, the man whose statue stood in the anteroom of the Trinity College Chapel was as saintly as he was brilliant. For Whewell, Newton's whole character stood as inspirational credit to the country and college that had supported him.

When it came to the essentially uplifting nature of mathematical reasoning, Whewell and De Morgan agreed completely. But for De Morgan and his colleagues in the RAS, the Cambridge man's exaltation of Newton was aggravating. The Royal Society had rested its claims of scientific preeminence on the character of Newton for more than a century, and in 1831 their position was strengthened with the publication of a hagiographic Newton biography by the Scottish philosopher David Brewster. For members of the RAS, Whewell's 1833 echo of Brewster's praise was the final straw. In 1835 Baily, who had been fighting against the Royal Society for years, retaliated with a carefully

researched historical reconstruction of a fierce argument that took place between Newton and the Astronomer Royal, John Flamsteed. For more than a
hundred years, Flamsteed had been assumed to be the unreasonable party in
this quarrel, but in an *Account of the Rev'd. John Flamsteed,* Baily argued that
Flamsteed was in the right. In Baily's rendering, Newton appeared as a con-
temptuous autocrat who used an obsequious Royal Society to deprive Flamsteed of his rights to material he had painstakingly gathered over the course
of decades. In Baily's view, it was Flamsteed, not Newton, who most exemplified the character of a gentleman of science because of the patient years he devoted to piling up data.[18] Whewell was appalled by Baily's *Flamsteed.* As one
of his allies put it, "If Newton's character is lowered, the character of England
is lowered and the cause of religion is injured."[19] Needless to say, De Morgan was unmoved by this argument, and in the concluding paragraph of the
otherwise highly laudatory article on "Newton" for the *Penny Cyclopædia,* he
carefully recognized that mathematical brilliance did not always guarantee
good character.

As De Morgan recognized in one of his doodles, the conflict between
Newton and Flamsteed was as much about the comparative values of theoretical and practical astronomy as it was about credit for the results (figure 7).[20]
Whereas the men of the Royal Society were telescopically exploring the stars,
the gentlemen of the RAS were using mathematics to catalogue them. Baily's
and Herschel's insistence on doing the calculations for their star catalogue
themselves was a way of modeling and defining an identity for a new breed
of mathematical astronomers. In the obituary he wrote after Baily's death,
Herschel carefully articulated what that identity involved: "far-sighted, clear-
judging, and active; true sterling, and equally unbiased by partiality and by
fear; upright, undeviating, and candid, ardently attached to truth, and deem-
ing no sacrifice too great for its attainment."[21] Such laudable qualities were
both grounded in and demonstrated by the meticulous care that Baily's astronomical work required.

By the end of the 1820s, the discipline of calculation was giving way to
the discipline of precision measurement as the shaper of character in the RAS.
One of the central, defining features of Baily's house on the edge of town was
that it was "well-enclosed on all sides, and from the nature of the neighbour-
hood, free from any material tremor from passing carriages."[22] This seclusion made 37 Tavistock Place perfect for exact experimentation, and over the
course of four years, from 1828 to 1832, Baily responded to the French creation
of the metric system with an effort to place the English standard of length

Figure 7: Augustus De Morgan. Caricature of Newton and Flamsteed.
RAS MSS De Morgan 3. By permission of the Royal Astronomical Society.

on an equally precise natural basis.[23] Pursuing this project entailed isolating
himself in a room for more than twelve hundred hours,[24] during which he
painstakingly measured the periods of more than eighty pendulums of different forms and materials in order to determine precisely the length of the seconds pendulum. Several years later the same room again witnessed him hard
at work, this time peering through a telescope and recording the marks of
an instrument that measured the specific gravity of the earth.[25] The push for
precision continued after Baily's death. In 1844, Sheepshanks spent months
furthering the pursuit of a precise standard of length by recording 89,500 micrometer observations in the stillness of the cellar under the RAS rooms in
Somerset House.[26] These kinds of precision measurement effectively distinguished the astronomy pursued in Baily's RAS from the observational world
of the Royal Society.

De Morgan was as incapable of pursuing precision experiments as he was
of observing the skies, but he fully embraced the RAS vision of astronomy.
Part of his job as honorary secretary was to educate the public, and through

him the SDUK became the mouthpiece of the RAS. The equi-tenacious triangle divided up the astronomical articles in the *Penny Cyclopædia,* with Airy
composing an overarching discussion of "Gravitation,"[27] while Sheepshanks
described the instruments that De Morgan couldn't operate. De Morgan's task
was to write the explanatory articles ensuring that "every discovery, or determination of fact, of any importance, was made as clear to the world as the subject
allowed."[28] He neatly laid out the history, theory, and results of Baily's efforts
to measure the specific gravity in the articles "Attraction," "Cavendish Experiment," and "Weight of the Earth."[29] And he opened up the night skies with
an elaborate series of star maps.[30] The SDUK began publishing a *British Almanac* along with a *Companion to the Almanac,* which aimed to present generally
useful subjects in such a way as to be "valuable to every class of readers."[31]
Every year for more than twenty years, De Morgan contributed a substantial
article on a topic of interest, from eclipses to decimal coinage to the history
of mathematics. He was an educator at heart, and writing for the *Companion*
was an annual joy.

———·———

   The return of Halley's Comet in 1835 was a golden opportunity for the
members of the RAS to demonstrate the value of their work. Accurately predicting the comet's position was a very public test of the power of the mathematical astronomy they were championing. De Morgan stood in the center
of these negotiations. As the honorary secretary, he was in charge of monitor-
ing and recording the event in the *Monthly Notices;* as the central astronomy
writer for the SDUK, he was in charge of communicating those events to the
larger public.
   De Morgan opened the discussion in the essay "Halley's Comet" that he
contributed to the 1835 *Companion to the Almanac.*[32] There, over the course
of ten pages, he carefully laid out the difficulties involved in being precise
about where and when the comet would reappear. In the eighteenth century,
he explained, the French astronomer Alexis Claude Clairaut and his helpmate
Mme Lepaute had devoted months to an attempt to construct an orbit that
took into account perturbations caused by Jupiter and Saturn. These efforts
led the two calculators to predict that the comet would arrive in 1759, a full
fifteen months after what others had expected. That the comet arrived within
a month of when they said it would convincingly demonstrated the value of
their precision. Seventy-six years later, De Morgan proclaimed that refine-
ments in calculating techniques meant that predictions would be even more
accurate, and he provided three feasible orbits for the comet. In the event,

however, even this range was not enough. When, after a month of fruitless searching, England's stargazers finally spotted the comet in mid-September, it was not in any of its predicted positions. The 1835 return of Halley's Comet posed an enormous challenge to all of the RAS's claims of precision.

Firmly entrenched in London's smoke, De Morgan wouldn't have seen the comet even if his eyesight had allowed it. The Frends, however, were vacationing in Hastings, where the skies were clear. Frend, who was approaching eighty, found that his "reverence for the aetherial traveler" was not enough to overcome his fears of encountering "the night air at a distance from the house."[33] But on September 21 he could see the comet from his back patio, rising above the neighbor's house with a vapor trail "as plain as a pike staff,"[34] and by October 8, it was bright enough that those with "younger eyes" could see it without the aid of a telescope.[35] Sarah Frend had always prided herself on having spotted an unpredicted comet in 1811;[36] now she joined her husband and daughters in feasting her eyes on this much brighter one.

Thrilled as the family was by the sight of the comet, Frend was acutely aware of the problems it was posing for the mathematical astronomers in the RAS. All three of their predicted paths were "sadly out," and when he began to trace its actual passage it threatened to travel beyond their map entirely.[37] Frend proposed that he and Augustus might collaborate on a map drawn to a larger scale that would locate the actual orbit among the three predicted ones. "I will bear the expence of the plate & the profits shall be divided between us or the loss will be mine not yours," he offered.[38] Frend clearly saw how important it was that the RAS create and disseminate an accurate description of the comet's trajectory, even though its astronomers had not been able to predict it.

Frend's particular plan did not come to fruition, but the RAS was equally concerned about the comet's path. By December, Stratford had calculated a new ephemeris, or table of positions, for the comet, "founded upon the revised elements, and embracing the period between August 1, 1835, and March 31, 1836."[39] Not satisfied with this preliminary effort, the editor of the *Nautical Almanac* turned to the public for help. He asked the "many observers who are either unaccustomed to, or have a distaste for" intricate calculations to "transmit their observations to me, with a full statement of all particulars necessary to an accurate estimate of their value" so that he could use them to figure out the comet's orbit.[40] The value of this concerted effort was not lost on De Morgan, who, in the society's annual report of February 1836, made a special point of commending Stratford on his success in bringing the comet under reasoned control.

Even as Frend was supporting RAS efforts to establish the comet's orbit, he was also experiencing it in other ways. In letters to Lady Byron he was at least as pleased by his success in teaching his manservant the astronomy behind the event as he was by seeing the comet itself.[41] He then used its appearance as a springboard for reflection. He found himself "lost in astonishment at the changes I have witness'd" and reveled in the evidence of religious tolerance growing around him. He recognized he would not be alive when the comet reappeared in seventy-five years, but he was ready to "leave the earth in fullest confidence" that the world was fast moving toward a better order of things.[42]

The comet provided Sophia with a different opportunity. In addition to observing it from the well-positioned gazebo in the family's back garden, helping her father to record observed positions on his map, and listening to his excitement about education, religion, and politics, she and the other women of the RAS were deep in writing a parlor play called *The Comet*. The dramatis personae of this piece included not only Baily, Sheepshanks, De Morgan, and Airy, but also Frend, Stratford, an otherwise unmentioned man named Rowbotham, and the Ghost of Halley. Over the course of the women's play, all of the reasoned pretensions of the men of the RAS vanish under the gaze of a powerful female Comet.[43]

The play opens with Sheepshanks as a lovelorn swain, gazing at the skies and pleading with Comet to reveal herself. Enticing her with a telescope and honeyed words does not have the desired effect, however, and his heavenly love remains silent and hidden. The second scene begins with Airy crisply directing his colleagues to

> Expect the Comet in Gemini,
> Either twenty degrees behind or before
> Eastward or Westward, I can't say more
> And 'twill pass through the square
> Of the greater bear—
> Between this time and Christmas 'tis sure to be here
> And with one or two tails it will surely appear.
> At first it will look rather misty & hairy
> I remain yours obediently, George Biddel Airy.

As his attempts at precision and reason descend into absurdity, ever-elusive Comet remains off stage.[44]

When in the third scene Comet finally does appear, she is far more interested in her freedom than in any of her astronomical pursuers. "Oh how I wish my tail were grown, that I might brush these tiresome fidgets off the sky," she grumbles; "'Tis teasing—here and there and everywhere/Are quizzing glasses mounted up to stare." She well remembers the joy of evening walks after "my dear Halley took me for his own," but now that "he has gone the dear departed!/A thousand lovers in his place have started," and she feels besieged.[45]

At least that is what Comet wants her audience to believe. Actually, she is rather a flirt, who can't resist taking a quick peek from behind a cloud to see "What Sheepshanks means by gazing thus at me." When Sheepshanks sees her, however, she again takes refuge behind her cloud. "Come my dear" Come along/They've all tried to find you, & all got it wrong," the Ghost of Halley soothes and the two of them exit "leaving Sheepshanks in a fit." As presented by the women of the RAS, the comet was wholly unimpressed by both the observational and the rational pretensions of the men of the society.

*The Comet* provides a glimpse into the dynamics of the group in which Sophia and Augustus were beginning to become aware of each other. The play is a distinctly amateur production—the characters ridiculous, the dialogue silly, and the action minimal—but it points to the ways the women negotiated their relationship with the men of the RAS. In public the women were completely supportive of the work from which they were excluded, but in private they could—and did—make fun of it.

───·───

This parlor play is just one of the remnants of RAS entertainment that has survived. De Morgan's "proof" that he was "<u>not</u> a gentleman" was undoubtedly written as a declamation to be performed at one of Baily's parties. Although De Morgan's interest in mathematics education was not common to that group, his concern about being a gentleman was. Maintaining Baily's vision of gentlemanly science required considerable discipline. Internally Baily's impact can be seen in the ways he tempered the behavior of De Morgan's equi-tenacious triangle. By De Morgan's own admission, his friends comprised a volatile group: "each for himself, deciding that he was a rational and practicable man, and that the other two, no doubt worthy and rational, were a couple of obstinate fellows."[46] However, Baily's firm insistence on the sociability to be expected among scientific men meant that there was "never a sharp word" among them.[47] Baily's presence transformed the equi-tenacious triangle into a triangular pyramid on top of which he stood as a firm and calming keystone.[48]

Sustaining the gentlemanly image of the RAS to those outside of the society also required considerable effort. Baily was quite willing to fight and fight hard to be sure that the RAS defined legitimate astronomy, but his aggressiveness did not fit the gentlemanly view of science that he wanted to uphold. When Herschel wrote Baily's obituary, he tied himself in knots trying to reconcile the man's belligerence with the disinterested gentleman he claimed to be. The best Herschel could come up with to justify Baily's pugnacity was to assert that "to enforce that exertion which is necessary to healthy life, there is always need of some degree of friendly violence, which, if administered without rudeness, and in a kindly spirit, leads at length the revived patient to bless the disturbing hand, however the urgency of its application might for a moment irritate."[49] Baily was equally uncomfortable with this part of his image. Perhaps alerted by the evidence of Newton's nefarious schemes that he had found ferreting through the historical record, he systematically purged the evidence of his own battles from his correspondence.[50]

Baily's protégés could be equally vicious. Their ferocity spilled into the public eye in a protracted battle against James South that engaged the men of the RAS throughout the 1830s. South was one of the earliest members of the Astronomical Society; he had teamed up with John Herschel in an extensive project of observing double stars that won the two of them a gold medal from the society in 1825. After this initial triumph, South moved his five-foot equatorial telescope to France, where he plunged into a study of multiple stars so impressive that the Royal Society awarded him the Copley Medal in 1826, and the French began a campaign to persuade him to remain permanently in France. The English responded by offering him a knighthood and a grant of £300 per annum to forward his work in astronomy. Pleased with this settlement, the now *Sir* James South brought himself and his instruments back to England, where he continued his pursuit of scientific power. Almost immediately upon arrival, he attacked the Royal Society with a pamphlet even more vitriolic than Babbage's,[51] while at the same time insisting that he be named president of the Astronomical Society.

South's demand put the membership in a bind. His quest after ever more expensive and flashy instruments posed a challenge to the meritocratic image of mathematical astronomy the Astronomical Society was trying to model, and his self-serving style did not fit with the image of respectable middle-class gentility they were working so hard to establish. Nonetheless, he was a founding member of their group, one who could be very useful in their efforts to gain the royal approval they needed to become the Royal Astronomical Society. After considerable deliberation, De Morgan's friend Stratford nominated South for the presidency, while at the same time making sure that

Sheepshanks—"a man guaranteed not to be intimidated by him"[52]—would be his secretary. The hope was that in this way the middle-class members of the Astronomical Society could use South's power for their purposes, even as they held his extravagances in check. In the event, that strategy did not work. By the end of South's two-year presidency, the group had been granted their royal charter, but South's sense of self-importance had so alienated everyone in the process that he was driven out of the RAS he had worked so hard to create.

South's departure was just an early skirmish in a war that Sheepshanks and South both pursued relentlessly to the end of their lives and beyond. The presenting issue for the next scene of this personal drama was the conflict between South and the well-known and respected instrument maker, Edward Troughton. These two first came together around a "magnificent achromatic object glass, the masterpiece of Cauchoix,"[53] which South purchased for the princely sum of £1,000. Having spirited this extraordinary lens out of France in 1829, he then hired Troughton to mount it for him. Troughton botched the job, however, and by 1834 the two men were fighting it out in court. Sheepshanks had no obvious interest in this conflict, but he followed his hatred of South into the middle of it. He pushed Troughton to take legal action against South's complaints, and then repeatedly counseled Troughton not to compromise. With Troughton as his excuse, Sheepshanks hounded South through the courts.

The same issues of class that had fueled Banks's fight against the mathematicians in the 1780s echoed through the struggle between the newly knighted James South and his lowly instrument maker. In the 1830s, however, the deck of class and reason was being reshuffled and the tables turned. The RAS was offering a new meritocratic view of science that challenged the aristocratic dominance of South and his expensive instruments. The members' criteria for knowledge, as well as for reason, were rooted in the mathematics they had learned at Cambridge. By this standard, however brilliant South might be as an astronomical observer, that he "does not know a sine from a cosine, and is not able to use a table of logarithms for the simplest computation"[54] meant not only that he was not a competent astronomer, but that he could not claim to be rational.

In 1838, the men of the equi-tenacious triangle took the conflict out of the courtroom and into the popular press. In 1838, Sheepshanks wrote a series of letters to *The Times,* challenging South to a "peacable duello" in which the two would submit to be examined on astronomical calculations in a room equipped only with "tables of logarithms, nautical almanacs, two policemen, and one strait waistcoat," that is, a straitjacket, for South.[55] Even as Sheep-shanks was thus publicly goading his opponent into a rhetorical duel, De Morgan and Airy each came close to engaging South in real ones. South easily matched Sheepshanks in verbal abuse, and the taunts he aimed at Airy and De Morgan led each of them independently to the point of naming seconds before they allowed the issue to die in exchanges of frosty letters.[56]

Behind all of this personal drama, the legal case between South and Troughton dragged on. Finally, in 1838, a full four years after the former first challenged the latter in court, a judge ruled in the instrument maker's favor.[57] In response, a furious South dismantled the telescope Troughton had built for his precious object glass and sold the pieces for scrap. Some blamed his be-havior on the men of the RAS for being "quarrelsome" and acting "in a con-spiracy" against South. They, however, took the sale as positive proof that their former president had gone mad; one hundred years after the event, the official historian of the RAS was still pointing to it as evidence of "the utter impos-sibility of getting South to listen to reason."[58] It is difficult to find anything rational in the battle between Sheepshanks and South, but those who gathered in Baily's house spent considerable effort casting it as confirmation that they were ladies and gentlemen of reason.

The conflict with South was perfect material for the parlor play writers of the RAS. Eight months after the success of *The Comet,* Sophia and her friends composed *The Knights of the Trinitie: An Anciente Ballade* about the struggle between South and the Trinity men of the RAS.[59] The setting of this second play is the annular eclipse of the sun, which was to be seen in a band through Scotland and the north of England on May 15, 1836. South traveled to Aln-wick Castle to observe with a party organized by the Duke of Northumber-land,[60] while Baily made his way to Scotland. Little remains of South's ob-servations, but Baily triumphantly identified and explained the small bright lights still known as "Baily's beads" that appear just before the eclipse becomes total. Barely eight months after Halley's comet had challenged RAS claims of mathematical control, the precision of Baily's observations bolstered its asser-tions of expertise.

*Knights* is a celebration of this RAS triumph. It takes place in Alnwick Castle, where the Duke of Northumberland has gathered together a group to observe "spots on y^e Sonne." South appears as a "Craven knight" who had no honorable right to be there because he had acquired his knightly accou-trements by marrying a "ladye of many lands" to gain "her golden store." The names of his ill-gotten gains reflected his astronomical pretensions—his spear was Transit, his shield Circle, his fiery steed Eclipse—and "his helmet was micrometer, because his brains were small." Ranged against this clattering pile of pretension were the "Knights of the Trinitie," an honorable group of gentlemen, whose power rested on the ease with which they could "measure earth and sky" rather than on ostentatious possessions. The scene was set for a showdown.[61]

The moving force behind the action of *Knights* is an underhanded effort by South to add Stratford's spurs to his crassly purchased collection of arms. Stratford, however, is a true gentleman and refuses to be bought. "He must gain his spurs, like me, in fight / For he'll never set foot in mine!" he cries. The Knights of the Trinity "all joined with sword and tongue / For all loved Stratford of Kensington / And vowed to right his wrong." Having thus declared their honorable position, Stratford and the Knights ride together to the lists. There as they wait, mounted and at the ready, South's horse "careered into the field / And downward fell the knight." Lying humiliated in the midst of his hardware, South vows "Eclipse no more to ride," while Sheepshanks crows, "Leave arms to those who love the right, / And swear by the Trinitie!" The *Anciente Ballade* ends at this point, when the gentlemanly virtues of mathematical astronomy have triumphed over South's crass materialism.[62]

Since De Morgan could not see the stars, he had little obvious role in plays devoted either to comets or to eclipses. However, he was becoming ever more interesting to Sophia as these parlor pieces were being composed. In *Knights,* he appears only as a figure who "studied volumes of eastern lore," that is, "Algebra. From Al.gebir—gibberish." He plays a somewhat larger role *in The Comet,* where the men of the RAS are concerned by what they see as the intellectual removal of their honorary secretary. When Baily complains that "you don't seem to care / Whether the Comet's here or there," De Morgan admits, "It's neither here nor there to me / I would not cross the street to see."[63] That he was able to gracefully accept this kind of teasing about his eyesight is striking testament to the level of De Morgan's comfort among the people who gathered in Baily's house.

Another character in the play was at least equally irritated by De Morgan's indifference. Even as Comet was brushing the other men aside, she was being forced to admit, "There's one that I should like to lead a dance / only to punish for his nonchalance." It is not particularly difficult to read Sophia into Comet's desire to entice De Morgan "to dance." For years Frend's oldest daughter flirted happily with her father's friends—decades later she remembered being beaten at chess by Baily, winning a dinner through a bet with

Charles Butler, traveling with Colonel Briggs to see Cromwell's skull—but none of these men were marriage prospects. A series of walks in Hastings with the poet Campbell were long and rambling enough to generate thoughts about age differences in marriage, but there is no evidence that either of them was serious about it.[64] Neither Frend nor Lady Byron seemed concerned, but as Sophia moved through her twenties, the women of the RAS were acutely aware that each passing year was moving her further from marriageable age.

In August 1835, Sophia began corresponding with Augustus through notes appended to her father's letters. The first of these was a playful reference to the kinds of things she and the other women of the RAS were writing into *The Comet*. Augustus was as capable of teasing as she was, and on hearing of her plans, he wailed that he could only prepare to die while leaving Stratford to "put you in the Nautical Almanac, as a new planet, always in opposition."[65] A full eight years after they first met in Stoke Newington, Sophia and Augustus were beginning to find a common ground in the convivial world of the RAS.

# CHAPTER 16

## *Expanding Consciousness*

Even as *The Comet* was providing entertainment, there were hints of impending trouble in the Frend household. In an August letter, Frend thought to mention to De Morgan that Harriet did well on the journey to Hastings;[1] in another De Morgan assured Sophia that he would not have guessed her sister had been under the weather.[2] But these were optimistic glosses on the condition of Sophia's "merry" younger sister, who by the New Year was dangerously ill. When healthy, Harriet had been so exuberant that De Morgan referred to her as "the crier of the court,"[3] but by the end of February, she was so weak that Frend couldn't hear her speak at all. Although her loving father held out hope to the very end, Harriet died in mid-March of 1836. A saddened Lady Byron wrote a deeply sympathetic letter, which also acknowledged "the consolations which I know that you, & and I trust that all your family possess."[4] She was right in recognizing her old friend would find comfort in knowing his daughter "left this world in the strongest confidence in the promises given to us in the gospel."[5] Nonetheless, Frend was never quite the same after Harriet died. He continued to write cheerful letters, but in them he began to ramble and repeat himself more than he had before.

Even as his daughter's death was shaking him to the core, Frend's views of the "promises given to us in the gospel" remained firmly rooted in the close readings of the Bible that had supported his Unitarian faith for decades. The understanding of death and resurrection he took from those readings can be found in a carefully preserved letter he had written to a completely healthy Harriet five years before her death. Upon returning from a funeral, the seventy-five-year-old had written to his then nineteen-year-old daughter, "You, of course, cannot think of death exactly as I do, nor is it fit you should." He felt, nonetheless, that it was good for both of them at times to look beyond their individual perspectives and "contemplate it in its true light."[6] For him, that "true light" was to be found in a careful interpretation of the fifteenth chapter of Paul's first letter to the Corinthians. Frend's beloved Handel had closed his *Messiah* with verses 51–53 of this chapter, where "in a moment, in a twinkling of an eye" all shall be changed, "the trumpet shall sound and the dead shall be raised incorruptible."[7] In his 1831 letter to Harriet, Frend focused on the buildup to this triumphant conclusion.

In 1 Corinthians 15:36, Paul points out "that which thou sowest is not quickened, except it die."[8] The same is true of the human soul the apostle goes on, "It is sown a natural body; it is raised a spiritual body."[9] Frend recommended Paul's simile to Harriet. "We are consign'd to the earth as seed to the ground," he explained. We do not know what happens next, any more than we know the processes that produce plants, but we do have "heavenly assurances that what was consign'd to moulder in the grave shall appear again in a new form."[10] Harriet's father was clear that we can have no idea about the nature of that new form nor can we understand what the world will be like when we next arrive. We do, however, know the all-important "fact" that in the world to which we are going, all religions, "nations, & languages shall be united together in the sole homage to the father of all & in brotherly love towards each other."[11] In Frend's view, all of the variety of human experience was just a means to the end of this ultimate unity.

After Harriet's death, her grieving father was completely clear that her life in this world had prepared her well for the unifying love Paul assured him would prevail in the world to come. She had grown up in a church that transcended all sectarian divisions by focusing solely on the worship of God, and at the time of her death, her understanding of Christian love was so expansive that one of her best friends was "a Jewess."[12] He took comfort in knowing that his daughter had died ready to enter the heaven that awaited her. In the meantime, he knew that as a Christian, he was "forbidden to grieve as those who have no hope." In his practice he tried to learn from an "eastern sage" who said "mourn for thy friend but then wash thy face & eat bread."[13] That Frend faithfully followed this advice may be seen in the fact that there are no further references to Harriet in any of his subsequent letters.

Frend's version of washing his face and eating bread was to travel. A scant three months after burying Harriet under a simple, flat gravestone in the dissenting section of the new Kensal Green cemetery, the Frends drove away from London.[14] They avoided Hastings, where they had so often vacationed with Harriet, and instead moved by stages to Exmouth. While the horses were being watered in Dorchester, they visited Lindsey's vicarage and church in Puddletown, and once arrived in Exmouth, Frend made a pilgrimage to his older brother George's grave.[15] Having thus honored their familial roots, the Frends moved on to the town of Clifton, near Bristol.[16]

De Morgan remained in the city, looking after the house, making sure the bills were paid, and keeping an eye on fifteen-year-old Alfred, who was fighting homesickness at school. He followed their travels but was not at all tempted to join them. Mr. Parsons's school had been in the nearby village of Redland, and he had no desire to return: "I learned to hate the country on Durham Down."[17] Clifton delighted Frend, however. He loved its industrial bustle, was buoyed by the flag fluttering on the site of the railroad being built to London,[18] and enthralled by the construction of the "exalted bridge over the Avon," which he could see through his telescope.[19] He searched out local astronomers and mathematicians and happily toured warehouses filled with immense quantities of Chinese tea.[20] Confident that his beloved daughter had been well prepared for heaven, Frend opened himself to all that was new and different in the world around him.

Sophia dutifully accompanied her family on their travels, but her thoughts were elsewhere. She had nursed Harriet night and day for months and was deeply shaken to find her beloved sister gone. She recognized that Paul's statements in Corinthians were the salient text for understanding what awaited Harriet, but on one critical point she could not accept the standard interpretation of that text. She was not prepared to agree that for the moment, Harriet was effectively dead and would remain so until the sounding of the trumpet at some far distant moment of resurrection. Even before her family left Stoke Newington, she had been hoping that the afterlife was "not just a dream."[21] Harriet's death added a new urgency to her need to "feel sure of another life, and that we might meet all we loved in it."[22] In the months after her sister died, Sophia set out on what would be a lifelong quest to understand the world of life after death.

From the beginning, Lady Byron was a major companion on this voyage. She understood Sophia's longing and encouraged her to see that insisting on Harriet's continued presence lay beyond the limits of reason that her father had placed on his own understanding. "Your feelings of grief form a veil between your mind's eye and her whom you have lost—but it will be withdrawn, like a mountain-mist, and you will then realize the existence of the immortal being," Lady Byron promised. Using the case of her mother as an example, she described a process in which she had moved from a sense that her mother "lived, but immeasurably remote from me" to the recognition that she was "near, and as perhaps more conscious of my feelings than I am myself." It would not happen immediately; "we must wait for all those changes thro' which our souls have to pass," she explained.[23] But with time and patience, Lady Byron assured her young friend, Sophia would directly experience Harriet's presence.

Sophia appreciated Lady Byron's assurances, but she was not about to settle a point as centrally important as the continued life of her sister by means of personal experience and feelings alone. All of her education had been directed toward developing her reason so that she could find the answers to her questions in the Bible. That Harriet still lived was a difficult position to pull out of Paul's text, however; not only her father, but also her great-grandfather, Francis Blackburne, had been absolutely clear that there was "no proof in the scriptures of an intermediate state of happiness or misery between death and the resurrection."[24] Sophia persisted nonetheless. The weeks she had spent by her dying sister's bedside gave her a kind of immediate knowledge that her father and great grandfather could not claim, and the confidence her father had instilled carried her smoothly past any disparities between her thinking and that of male theologians. In a series of letters to Lady Byron, Sophia began to explore the shape of life after death.

The theory of the afterlife that Sophia developed in the summer of 1836 survives only in the letters that Lady Byron wrote in response. Sophia's challenge was to somehow deal with all the time that stretched between Harriet's death and the blessed moment "when the trumpet shall sound and the dead shall be raised incorruptible." Lady Byron saw the problem to lie in Christian believers' insistence that the dead would all be raised on a single day. The "Inspired Writers" were writing figuratively when they wrote in those terms, she argued. The Day of Judgment was better understood as the particular moment that a "soul is awakened to the full consciousness of its past history and of the judgment of God upon that evidence."[25] Each person experienced this moment individually, which meant it would arrive at different times. With this reading, the trumpet could have sounded for Sophia's beloved sister immediately upon her death. It meant that Harriet could already have been "raised incorruptible" and was not lying dead in the ground.

As Lady Byron was reinterpreting divine time, Sophia was applying theories of psychological development to the mysterious transition from "natural body" to "spiritual body." As her ideas were reflected back in Lady Byron's words, "our permanent self"—which Sophia called "Consciousness"—was "perpetually employed in evolving itself from all the perishable stuff that gathers about it." Time and again this self cast off "its skin" as it outgrew its previous understandings. For the fully developed soul, death could be seen as a final molting, after which the "wholes of past existence will then be an offering to the Deity," and Consciousness will be free to rise above the matter below.[26]

Lady Byron interpreted Sophia's construction as a description of a movement toward "pure Faith," which she also found in the writings of David Hartley, Samuel Coleridge, and a friend of hers named Greaves.[27] What all of these theories had in common was the view that as they developed, individual humans followed a path within a larger spiritual progress, which could take place equally before and after death. All of Lady Bryon's sources recognized that the Day of Judgment might be far off for those who did not grow spiritually during their lives. But those, like Harriet, who "left this world, having brought a full offering to the altar" could "be regarded as now endowed with all the perceptions of the highest state of being."[28] For Lady Byron and Sophia, Blackburne's and Frend's interpretations were the products of a static eighteenth century in which this truth was hidden. But Lady Byron and Sophia were nineteenth-century women, who could see that history was progressive and human understandings were expanding. In their new century, it was becoming clear that death was just a step in a larger process toward participation in a broader Consciousness.

In summer 1836 De Morgan was as deeply involved in exploring the nature of transcendent consciousness as Frend, Lady Byron, and Sophia were. His subject matter was different, but the impulse toward transcendent understanding was the same. His texts were mathematical rather than biblical, filled with symbols rather than words, but he approached them with the same kind of close reading that Frend, Sophia, and Lady Byron were using. While the words of their texts were sanctified by their roots in a sacred past, the symbols of De Morgan's mathematics were sanctified by their generation through the progressive historical processes that Peacock had recognized in algebra. As Sophia and Lady Byron were following their reason to comprehensions of Consciousness, De Morgan was blazing a path through the philosophy of George Berkeley, the educational ideas of William Whewell, and the calculus of Augustin Cauchy to the very frontiers of mathematics, where whole new worlds of understanding beckoned.

The early eighteenth-century theologian George Berkeley was a major guide in De Morgan's explorations. He first encountered Berkeley at Cambridge, where the philosopher was best known for his wickedly witty description of Newton's fluents as "the ghosts of departed quantities."[29] One of the marks of Woodhouse's rebellion against Cambridge mathematics was to recommend Berkeley's work to any student of mathematics "for the purpose of habituating the mind to just reasoning."[30] But when De Morgan plunged into Berkeley's philosophy, he found himself caught up in an approach whose implications went far beyond the critique of Newtonian thinking. De Morgan was intrigued enough that he added an addendum to the biographical article "Berkeley" in the *Penny Cyclopaedia,* in which he described Berkeley as a radical idealist who denied the existence of the whole material world independent of sensations.[31] In the Anglican bishop's view, all experiences are actually just impressions, "communicated by the Creator without any intervening cause of communication."[32] Everything we experience as a sensation of an external world is illusory. For Berkeley, all that we can know are ideas beamed directly into our minds by God.

De Morgan could see the difficulties of denying the reality of the world that we experience. "To read Berkeley so as to give him a fair chance, some one else should turn the page over," he quipped, because "the touch of the paper periodically intervening is a snake in the grass." But he found the way Berkeley's idealism turned everything in the universe into an intellectual affair so intriguing that he resisted the temptation "to make philosophers of the fingers."[33] He recognized that he could never definitively establish the truth of Berkeley's position, but he liked to point out that it could also not be refuted.[34] De Morgan was not about to take a public stand on an issue that could not be settled through reasoned argument, but Berkelian idealism gave credence to the fundamental importance of the world of mathematics that so engaged him.

De Morgan had company in his idealistic approach. As he was writing about Berkeley's philosophy, Peacock was moving into a new life as dean of Ely Cathedral, and new ideas had started to slip through Frend's mind "like water through a sieve."[35] But William Whewell was beginning to take the place of these mentors in De Morgan's world. Whewell was a high Tory Anglican who fiercely defended the elite exclusivity of Cambridge, whereas De Morgan was a religious radical committed to educating all of England's people. Nonetheless, two fundamentally important points united Whewell and De Morgan's views of the world. The first was an idealism that held ultimate truth to be grounded in ideas; the second was a progressivism that saw the human understanding of those ideas developing over time. From the middle of the 1830s, Whewell and De Morgan's intellectual kinship blossomed in an epistolary relationship that lasted to the ends of their lives.

The foundations of their friendship were laid in 1835, when Whewell published *Thoughts on the Study of Mathematics as a Part of a Liberal Education,* in which he defended the study of mathematics in ways that could equally have been written by De Morgan himself. Many uneducated men are "full of conviction which they cannot justify by connected reasoning," Whewell there asserted.[36] The major function of the Cambridge liberal education was to rectify this problem by teaching students to recognize and develop the reasoned foundations of those beliefs. Getting to this point required practice, and mathematics provided that opportunity. Practicing the mathematics that lay at the center of the Cambridge liberal education was the best way to teach students the power of their minds.

When Whewell spoke of mathematics in this context, his model was always Euclidean geometry, in which reasoning was so direct and immediate that its results were "necessarily and inevitably true."[37] It might take weeks of study before a student truly understood a geometrical theorem, but attaining that understanding had all of the power of a conversion experience. In the instant that a student truly understood them, geometrical theorems were transformed from intellectual exercises into truths so immediate that it was impossible to conceive of a world in which they were not true. Frend called this kind of transcendent understanding "the truth." Sophia and Lady Byron called it "Truth" with a capital *T.* Whewell described it as "necessary truth," and it stood at the center of his educational project. "One of the most important lessons which we learn from our mathematical studies is a knowledge that there are such truths and a familiarity with their form and character," he explained.[38] Opening students to the experience of necessary truth by teaching them geometry would both connect them to the God who thought geometrically as he constructed the universe and teach them the kind of understanding they were looking for in all areas of their lives. De Morgan and Whewell may have disagreed politically, but the two men agreed completely about the central importance and power of the encounter with necessary truth.

There was an important flip side to the educational position that Whewell and De Morgan shared. Placing an essentially humanistic, reasoned mathematics at the center of the liberal education meant that students' ability to understand and work with mathematical ideas was an essential test for the viability of those ideas. In *Thoughts,* Whewell identified two areas of mathematics that were failing the acid test of student understanding. Within a year, De Morgan was addressing both of them in writings for the SDUK.

The first problem Whewell identified was the study of proportion in the fifth book of Euclid's *Elements,* which had left students frustrated and confused for centuries. De Morgan responded with *The Connexion of Number and Magnitude: An Attempt to Explain the Fifth Book of Euclid.*[39] Whewell's second problem was with calculus, which remained mired somewhere in the conceptual mud of Newton's *evanescent quantities,* the empty symbolism of Leibniz's *dy/dx*s, or some poorly defined "mixture of the two."[40] In *Thoughts,* Whewell declared himself ready to break out of this box by declaring that the "real fundamental principle" on which the calculus rested was "the conception of the *Limit.*"[41]

The language Whewell used to describe the limit—"the idea itself," the "fundamental idea,"—points to its place within a larger philosophy of knowledge that was taking shape in his mind. In the five years after he published *Thoughts,* Whewell came to see that the mark of necessary truth was that it was rooted in a "Fundamental Idea" that was intrinsic both to our minds and to external reality. In 1838 he published a three-volume *History of the Inductive Sciences,* in which the entire history of human knowledge became a narrative of the progressive identification and development of the fundamental ideas that structure the world. In 1840 he followed up with *The Philosophy of the Inductive Sciences Founded upon Their History.* The quintessential fundamental idea was Space, which had been recognized as the basis for all geometrical reasoning since the time of the Greeks. Recent history, however, was revealing more. From Whewell's perspective, Newton's triumph lay in having successfully identified the fundamental idea of Force that was the true foundation of cosmology. Since then others had begun to identify and explore fundamental ideas in subjects from chemistry to geology and beyond. When Whewell declared that "the conception of the *Limit*" was the "real fundamental principle" of calculus, he was bringing the subject into the fold. Recasting calculus in response to this proclamation engaged De Morgan for the next five years.

When Whewell pronounced the limit the "real fundamental principle" of calculus, he was following the lead of the French mathematician Augustin

Cauchy. In 1821 Cauchy published a textbook entitled *Cours d'analyse* (*Course of Analysis*), in which the limit served as the foundational center of calculus. Cauchy's understanding of what was involved in establishing mathematical foundations was very different from Whewell's, however. Whereas for Whewell truth rested on the conceptual immediacy of a fundamental idea, for Cauchy it lay in a set of precise definitions. For Cauchy *numbers* became "the absolute measures of magnitudes"; *quantities* became "*numbers* preceded by the signs + or –"; *variables* became *quantities* that could "take on successively many different values." In this context, the *limit* became the "fixed value" that is approached when "the values successively attributed to a particular *variable*" end up differing from it "by as little as we wish." The *limit* thus defined established the critically important definitions of the *infinitesimal* as a *variable* whose successive values fall below any given number, and the *infinite* as the *limit* obtained "when the successive numerical values of a given variable increase more and more in such a way as to rise above any given number."[42] Cauchy saw these definitions as putting to rest all of the questions that had plagued calculus since its inception.

Cauchy's definitions did not capture the richness of the ways the words had been used for centuries, but they had the advantage of being both individually clear and mutually dependent. His definition of a *variable* relied on his definition of *quantity,* his definition of the *limit* relied on his definition of *variable,* and his definitions of both the *infinitesmal* and the *infinite* relied on his definition of the *limit.* They were, in addition, briskly practical; so, for example, his definitions of the *infinite* provided a straightforward way to decide whether a variable was infinite or not. Tightly interwoven with one another, these clear, operational definitions formed a net strong enough to support virtually all of the analytic results of the previous century. And for Cauchy, that was enough. "In establishing precisely the meaning of the notation that I will be using, I will make all uncertainty disappear," he crowed.[43]

Others were distinctly less enthusiastic. They saw that embracing Cauchy's definitions entailed abandoning the view of reason Locke had modeled on the clear thinking of Euclid's geometry. Cauchy knew this, but he didn't care. He was a leader of the post-Napoleonic reaction that swept a generation of Enlightenment mathematicians from their posts at the École Polytechnique. In the introduction to his *Cours,* the counterrevolutionary dispensed with his predecessors' ideal of mathematics as reason in a sentence as brief as his definitions: "Let us cultivate with ardor the mathematical sciences, without wishing to extend them beyond their domain; and let us not imagine that we can attack history with formulas nor that we can offer the theorems of algebra or of integral calculus as moral training."[44] Thus relinquishing all of mathematics' claims to be an essentially humanistic subject did not trouble him. He was an ultraconservative Catholic who was unimpressed with Enlightenment claims for the power of reason. Truth, for Cauchy, was to be found through revelation and the authority of the church, not through some form of mathematical reasoning.

Cauchy's rigorous approach was as unacceptable to the practitioners of the subject that lay at the heart of liberal education at Cambridge as it was to the mathematical thinkers of the French Enlightenment. That his students stomped out when he first introduced it at the École Polytechnique certainly suggests that it failed Whewell's litmus test of student comprehension.[45] Nonetheless, within fifteen years, the efficacy of Cauchy's methods was so impressive that even Whewell was being forced to recognize their legitimacy. As he did so, however, the Cambridge don rejected the rigorous austerity of Cauchy's definitional approach. The *limit,* Whewell explained, may be approached through abstract definitions or intuitive axioms or some combination of the two, but "whatever course is taken, the foundation on which our conclusions rest is the idea itself,"[46] that is, the human conception that lies beyond mathematical definitions and symbols. When Whewell identified the limit as the "real fundamental principle" of calculus, he was claiming that it could be conceived as clearly as a geometrical theorem could.

De Morgan agreed, and from 1836 to 1841, he took up Whewell's challenge to translate the operational power of Cauchy's limit into a properly reasoned mathematics. De Morgan's *Differential and Integral Calculus* began as a series of twenty-five 32-page, one-shilling SDUK pamphlets, which were compiled into a single 785-page volume in 1842.[47] As he was writing this serialized text, he was also reiterating and clarifying his approach in a bevy of *Penny Cyclopædia* articles, including "Infinite," "Limit," "Magnitude," "Number," "Operation," "Negative and Impossible Quantities," and "Relation." Conspicuously absent from this list is an article on "Rigor." In its stead, De Morgan offered an article on "Interpretation," in which he explained, "It is interpretation which creates the distinction between algebra, as now known, and arithmetic with general symbols of number, or universal arithmetic." Whenever new and unexpected results emerge in mathematical work, he went on, "the rule always is, let the interpreted meaning of the new symbols be such as will make the whole of the process true by which they were obtained."[48] In this way, mathematics would remain always rooted in the human conceptual world of reason that constituted its truth.

The essential point that distinguished De Morgan's *Calculus* from Cauchy's *Cours* lay in definitions. For Cauchy, definitions were fixed points that defined the limits of mathematical legitimacy; for De Morgan, they were constantly negotiable way stations on the path to full understanding. De Morgan had learned from Peacock that mathematical understandings grew organically from murky darkness to blinding new insights. This meant that the true mathematician would be always on the lookout for ways to explore, develop, and expand the definitions of his subject to be sure they were large enough to include all possible scenarios. He insisted that allowing definitions to fix the boundaries of the legitimate and the illegitimate was a sure way to impede mathematical progress.

Adjusting the certainties of Cauchy's rigorous definitions to accommodate the richness of De Morgan's historical vision was a challenge. It took the Englishman five pages to cover the ideas of magnitude and number Cauchy covered in two sentences, more than ten pages to elucidate the definition of the limit that Cauchy had laid out in two neat lines, and another twelve pages to recognize the many intricately intertwined meanings of infinitesimal and infinity. As De Morgan twisted his way through one tortuous example after another, he explained that the convolutions "are not unimportant" because "it is of great consequence that the fundamental notions of mathematics" should be presented using "the rude and unrigorous form in which they are expressed in common life."[49] Even when he finally declared on his twenty-seventh page that "in future we shall use the theory of limits in all reasonings," he promised to bracket the paragraphs in which he did so in order to signal that this was just a temporary nod to convenience.[50] For De Morgan, the true mark of legitimate mathematics was the soil of human history that clung to its roots.

There were real mathematical consequences to Cauchy's and De Morgan's disagreement about the nature of definitions. The most glaring of these concerned divergent series, that is, infinite sums that did not converge to any clear limit. Divergent series had appeared in any number of mathematical investigations in the eighteenth century, but they could be strangely problematic. So, for example, the series $S = 1 + 2 + 4 + 8 + \ldots$ seems easily to meet Cauchy's definition of the infinite as the limit obtained "when successive numerical values of a given variable increase more and more in such a way as to rise above any given number." But simple algebra shows:

$S = 1 + 2 + 4 + 8 + 16 + \ldots$

$S = 1 + 2(1 + 2 + 4 + 8 + 16 + \ldots)$

$S = 1 + 2S$

$-S = 1$ or $S = -1$

So the sum of the series is not infinite, but rather negative one. This kind of unruliness represented enough of a challenge to Cauchy's program of rigor that he summarily ruled divergent series out of legitimate mathematics. He did admit that the decision to throw out whole areas of eighteenth-century work "may appear a bit rigid at first,"[51] but insisted that it was a small price to pay for certainty.

De Morgan vehemently disagreed. He saw no distinction between Cauchy's decision to get rid of divergent series and Frend's decision to eliminate negative numbers. He readily admitted that many divergent series are "perfectly incomprehensible in an arithmetical point of view" but "to say that what we cannot use no others ever can" was a refusal to learn from history. And so, he insisted on moving forward under a motto "contained in a word and a symbol—remember $\sqrt{-1}$."[52] For him, allowing all of the uncertainties of divergent series to remain in mathematics was a far better alternative than stopping mathematical progress.

Behind De Morgan's conviction that it would someday be possible to understand the true meaning of divergent series lay a faith in the conceptual reality grounding all mathematical results that was as rock-solid as the "strongest confidence" Frend placed in "the promises given to us in the gospels."[53] And in summer 1836, just as Sophia was expanding her father's views to incorporate her understandings of life after death, De Morgan was venturing into a mathematical world far more expansive than the one Frend had known. The younger man's reading of Berkeley was supporting his conviction that to do mathematics was to explore the ideas of a divine mind, and his reading of Peacock was pointing out a way to expand human understanding of that transcendent reality. In the mathematical books and articles that flowed from his London office, De Morgan was following his reason into an ever-expanding world of mathematical ideas.

It is not difficult to see similarities between the ultimate reality De Morgan was exploring in mathematics and the larger spiritual consciousness Sophia was investigating with Lady Byron. But there is little reason to believe that they were comparing notes as they left the static confines of Frend's Georgian world behind. The correspondence they carried on in the form of addenda to

Frend's letters was positively shrill in its giddy cheer. "Allow me to disthank you for writing polyglot letters to me about tan, & epi tan, which last I suppose means freckles," Sophia exclaimed.[54] Enjoy the world of "craniology, and mental-ology, and astrology, and folderololollogy," De Morgan replied.[55] But behind all the teasing, De Morgan and Sophia were beginning to recognize how much they had in common. As Sophia was reaching toward a "higher state of being" and De Morgan toward an understanding of divergent series, they were each, independently, laying the groundwork for a newly expansive vision of a transcendent world of mind that they would inhabit side by side.

# PART THREE

# Dividing Reason

# CHAPTER 17

# *Home on Gower Street*

Sophia's and Augustus's communications became somewhat more direct in September, when a "grand meeting" of the BAAS was held in Bristol. In the years since its first meeting, the group had been remarkably successful in its program of bringing science to the English people, and it was particularly proud of the special sessions opened to women. De Morgan, who characteristically did not attend, nonetheless told Sophia, "It would be very vulgar not to go."[1] She did not really need persuading and promised him a report "as distorted and exaggerated as ever my refracting mind" can make it.[2] But Sophia was disappointed in her hopes of learning from the "wisemen" of science. The scenes she described consisted of large crowds, addressed by speakers who "advanced science by telling the ladies to 'persuade their husbands and lovers that Science was as lovely as themselves.'"[3] She found little that qualified as science in the sessions open to "<u>the Ladies</u>"[4] and had too much self-respect to enjoy being patronized.

Soon after the close of the BAAS extravaganza, Sophia and her family returned to London. Just as she and De Morgan were adjusting to living blocks from one another, his life was dramatically changed by a boating accident that killed the man who had succeeded him as professor of mathematics at the London University. Desperate at the start of a new term, the school turned to De Morgan, who readily agreed to return as a temporary replacement. He was happy to be back in the classroom, but before he would accept a permanent position, he insisted that the concerns that had led to his resignation had been addressed. This meant that he could not be fired without cause, and that the board maintain the "exterior show" necessary to place its professors "in that advantageous position as to respectability which a gentleman requires."[5] After receiving legal assurance that these conditions had truly been met, De Morgan agreed to resume teaching on a permanent basis.

As De Morgan was working his way back into university teaching, Sophia was beginning to confront her feelings about the man she had known essentially as a sibling for so long. Her problems came to a head when De Morgan's beloved older sister Eliza died suddenly and unexpectedly in childbirth. Sophia rushed to support him. But when she offered the comfort she had so painstakingly drawn from Saint Paul, he picked her arguments apart so ruthlessly that she began to question whether he was a Christian at all. He was just displaying "that narrowness of mind, which has been spoken of as frequently attending the study of the exact sciences," Lady Byron soothed. All that really mattered was that he approached the world with "a religious feeling" illuminated "by the superior moral light concentrated in Jesus Christ."[6] Sophia's doubts nonetheless persisted. It was critically important to her that Augustus agreed on the essential tenets of her Unitarianism: that Jesus was "a man like ourselves" endowed with a special power of "receiving the spirit of God"; that in his lifetime he communicated that spirit to "erring man" through "words and miraculous works"; and that after his death he "was seen to rise to Heaven, from whence He sends the Spirit to those who are able to receive it."[7] Only after Sophia had secured Augustus's assurance on these essential points did she decide to marry him.

There is no comparable record of De Morgan's thinking before they became engaged, but there is evidence that he was pleased as his wedding approached. In a brief note he wrote to Babbage in April 1837 he joked that as the result of "living a quiet life and solving functional equations," he now had "a house to look out for, treat for, manage for, and prepare for the reception of a lady."[8] The same note of amused satisfaction is captured in the inscriptions preserved on the front page of his personal copy of *The Connexion of Number and Magnitude*.

In Augustus's handwriting, "W. Frend Esq., From the Author"
In Sophia's handwriting: "S. E. Frend, Took possession of this book, Oct 8, 1836."
In Augustus's handwriting: "And of the Author, Aug. 3, 1837."[9]

There is no evidence that Sophia got very far following De Morgan's lead through the fifth book of Euclid, but there is much to suggest that she captured his heart. That he captured hers is reflected in the pastel portrait Sophia drew of her fiancé with chubby cheeks and dimple on his chin (figure 8).

It was a happy couple that stood, surrounded by their families, in the Registrar's Office of St Pancras on August 3, 1837. Their religious discussions had resulted in an essentially Anglican wedding service, without Saint Paul's "exordium" that wives submit to their husbands.[10] Having thus taken care to preserve their individual autonomy, Augustus and Sophia took off for a honeymoon in Normandy. Upon their return, he got sick and she became frantic. As she tried to settle herself, her bedridden husband, and his "tolerably numerous"[11] books into their new house at 69 Gower Streeet,[12] she reached out to her father. "Panic does wonders on such occasions," he responded firmly, "and I cannot doubt that before we come to town you will be perfectly at your ease."[13] After almost three decades living under Frend's protective wing, Sophia had now to create a new home for herself and her husband.

However long it may have taken for the newlyweds to be "perfectly at ease" with each other, De Morgan was once again professor of mathematics, and once again called upon to give a lecture at the beginning of the new school year. Much had changed in the nine years since he had given his first introductory lecture. When he delivered his opening address in 1828, the London University did not have a royal charter and could therefore not grant degrees. That religion was the disqualifying issue became crystal clear in 1830, when the newly constituted Kings College, London, which was Anglican, had little difficulty being granted a charter. There was a real question of whether this kind of religious discrimination was legal after the repeal of the Test and Corporation Acts and the passage of the Catholic Emancipation Bill, but in 1834 Parliament allowed both Cambridge and Oxford to remain exclusively Anglican. The London University nonetheless continued to fight for its legitimacy until, finally, in 1836 a new agreement was reached. The London University would become University College London (UCL), and would join Kings College as part of a newly chartered, degree-granting University of London.[14] Under this arrangement, UCL remained secular, and Kings College remained Anglican, while all of their students were eligible to receive degrees from the University of London. In his lecture of 1837, later published as *Thoughts Suggested by the Establishment of the University of London,* De Morgan laid out his vision for the education that was suitable for the wide variety of young men who attended UCL.[15]

DE MORGAN, FROM A PASTEL BY HIS WIFE

Figure 8: Augustus De Morgan by Sophia De Morgan.
Frontispiece to Augustus De Morgan, *Formal Logic*
(London: Open Court Company, 1926).

The education De Morgan advocated for on October 17, 1837, mirrored the one he had received at Anglican Cambridge. He acknowledged that it might seem odd to draw from this example when addressing the fiercely secular UCL, but he insisted that the education pursued there should not be rejected out of hand. Except for their stubborn determination to "be the nursing mothers of those only who profess to hold one form of religious faith,"[16] De Morgan argued that the Oxbridge colleges offered a liberal education whose aim was to develop "the character and disposition of the individual" while cultivating a powerful and energetic mind.[17] This goal was certainly as important for the students at UCL as it was for those at the Anglican colleges.

At UCL as at Cambridge, De Morgan maintained that the best way to fully educate students was to take them through the process of learning at least one subject from the beginning through the middle to the end. He divided that process into three stages. In the first, the student was essentially dependent, forced to rely on others for help as he learned the difficult first principles. Much of this early learning might take place in lectures, where the teacher could guide beginners through every step. In the second stage of learning, the center of gravity shifted from lectures to private study. There were strong resonances with Christian prayer in De Morgan's insistence that students engage in "diligent study in the retirement of the closet."[18] A lecturer could point students in the direction of new ideas, but in mathematics as in religion, true understanding required that they explore its paths by themselves.

Even as De Morgan argued for the essential benefits of private study, he recognized that there were dangers. As the student began working "in the solitude of chambers," he was in the precarious position of discovering the power of the principles he had learned as a beginner with "nothing to humble the high notion which he will entertain of himself, his teachers, and his subject." De Morgan's student could not be allowed to rest in this self-satisfied state. The final stage of his education required that he be pushed to the limits of the subject where he "will begin to see that there is, if not a boundary, yet the commencement of a region which has not been tracked and surveyed, and in which not all the skill which he has acquired in voyaging by the chart will save him from losing his way."[19] Only when the student had reached this point could he "begin to form a true opinion of his own mind," and only then could he begin truly to understand what it meant to be a reasoning being in a universe that was much greater than he would ever be. "To make a subject teach the mind how to inquire,"[20] it had to be pursued beyond the point where knowledge was secure. It was only through this direct experience of the unknown that students could come to see both the strengths and the limitations of their minds—and recognizing both was essential to their success in the world.

Within days of delivering a lecture extolling the values of the liberal education he had received at Anglican Cambridge, De Morgan was fully engaged in imparting one at secular UCL. He was a consummate teacher, wholly committed to the development of the young men in his charge. His students were for the most part between fifteen and eighteen years old, and he tailored his teaching to their level. As they were trying to enter the subject in his beginning classes, he devoted most of his time to setting exercises and helping them when they had trouble. As they became more advanced, he spent more time introducing them to the kinds of new ideas and areas that the first principles opened up for them.

Whatever knowledge De Morgan's students might glean from his lectures remained essentially secondary to what could be attained through "diligent private study,"[21] however. He did allow that "the solitude of their chambers" might be mitigated by "the company of their books,"[22] and to this end he published a number of elementary textbooks on arithmetic, algebra, and trigonometry.[23] In addition, one of the most striking aspects of the De Morgan archive at UCL is the more than three hundred carefully handwritten notebooks he prepared to reinforce the material being covered in lectures.[24] De Morgan did all he could to support his students, but their true education lay in their personal engagement with the subject.

Approaching all of mathematics as an exercise in "sound reasoning and correct inference" required firm discipline,[25] and De Morgan was a firm disciplinarian. Students set their clocks by his passage to and from class, and he locked his lecture hall against those who were late.[26] Within his classroom, discipline was focused on practicing mathematics as reasoned subject. It required constant vigilance to keep the subject matter in focus as that subject matter was being explored through the manipulation of symbols that were so powerful. In his classroom students remembered the "bland 'hush!' with which he would suppress a suggestion which was simply stupid, and the almost grotesque surprise he would feign when one of them betrayed that he was following rules rather than thinking through the material on his own." His manner could be "exceedingly humorous, and gave a life to the classes beyond the mere scope of their intellectual interests,"[27] but there was always a goal behind his wit. De Morgan was determined to show his students the expanses of understanding that awaited those who reasoned their way through a subject until they found themselves on the very edges of comprehension.

De Morgan was totally devoted to his teaching at UCL, but the job did not pay very well. Professors' salaries continued to be tied to enrollments, which meant that they fluctuated. De Morgan's mathematics classes were relatively large, and before agreeing to return permanently he had clarified that no one else would be hired who could dilute their numbers, but he nonetheless never earned enough to fully support his family. As a result, he always had to supplement his professorial income by writing.

As Sophia was making a home for the two of them on Gower Street, De Morgan's writing was focused on probability theory. This was a relatively new subject for him; the 1830 SDUK treatise *On Probability* had been written by John Lubbock and John Drinkwater Bethune,[28] and De Morgan did not mention the subject at all in *Study and Difficulties of Mathematics,* published in the following year. There was a reason for this neglect. Although the study of probabilities was filled with equations, from De Morgan's point of view, they were not particularly interesting ones. Unlike astronomy, in which "the apparatus of mathematics which is required to establish results" was fascinating, in probability theory basic arithmetic was adequate to the task.[29] De Morgan had always recognized that the subject was useful for insurance and other business ventures, but mathematically he found it boring.

De Morgan's attitude changed in 1836, when he accepted a commission to write a book-length article, "Theory of Probabilities" for the *Encyclopædia Metropolitana.* The main source for this article was the 1814 *Essai philosophique sur les probabilités,* in which Pierre-Simon Laplace had essentially located all of epistemology within the theory of probabilities. Since most of our knowledge is only probable, Laplace wrote in his first paragraph, all of the most important problems we face turn on questions of probability. And so, he concluded, "The whole system of human knowledge is tied up with the theory set out in this essay."[30] The theory that he developed in the rest of the book was designed to provide a mathematical mechanism for making decisions in uncertain conditions. Laplacean probability required a significantly different approach to reason from the one that De Morgan had learned studying mathematics at Cambridge. Instead of locating reason in the human experience of learning, probability theory approached it as a regulated set of procedures.

All of the mathematics that Laplace laid out in his book was focused on calculating probabilities of ever more complex situations. His conviction was that as long as the probabilities of everyday events could be calculated, rational decisions could be made on those calculations. In practice, however, seemingly rational people often differ in their opinions. Laplace saw these individual differences of opinion as failures of reason, which could and should be eliminated by the determined application of an increasingly sophisticated probability theory. To reach the right conclusion requires "great precision of mind, a nice judgement, and wide experience in worldly affairs," he explained. "It is necessary to know how to guard oneself against prejudice, against illusions of fear and hope, and against those treacherous notions of success and happiness with which most men lull their *amour-propre*,"[31] that is, their excessive self-esteem. There was no place for personal judgement or opinion in Laplace's model of human rationality. When used properly, probability theory would generate a clear probabilistic answer to any legitimate question, and a rational person should base his decisions upon those probabilistic calculations.

De Morgan admired the probability theory that Laplace had expounded, but he would not follow the Frenchman in letting it define rational decision making. In the essay he wrote for the *Encyclopædia Metropolitana,* he carefully conscribed the conclusions that could properly be drawn from the power of Laplacean mathematical models. "It is wrong," De Morgan insisted, "to speak of any thing being probable or improbable in itself. The thing may be *really* probable to one person and improbable to another."[32] He elaborated on his meaning with the example of two people who have terminal diseases. To tell the first person, that a given treatment would raise the probability for survival from 60 percent to 70 percent would be good thing, but telling the second that a given medicine would raise the probability of survival from 90 percent to 100 percent would be immeasurably better. Mathematically, the difference between these situations is the same, 10 percent, but for anyone who is directly involved, they are not. Laplace's probability theory was a powerful tool, but it did not define reason. The essence of reason was its humanity, which meant that people could legitimately reason their ways to different responses in situations that were probabilistically the same.

De Morgan's insistence that "real probabilities may be different to different persons"[33] led him to divide probabilities into two categories. He used the phrase "moral probability" to describe situations that depend "upon the constitution of the individual, his knowledge of the circumstances, and the effect the event will produce." He used the phrase "mathematical probability" to designate the "moral probability in that case, and in that case only, in which the mind is disposed to consider" any outcome as of equal importance.[34] These definitions greatly limited the scope of Laplace's probabilistic ambitions by rendering the precision of the mathematical theory inapplicable to any situation in which one had a personal stake. It was, however, essential to De Morgan to maintain the distinction. Moral probability was human probability. By recognizing its legitimacy De Morgan was affirming the essential humanity of reason by defending people's freedom to reason their own way to life's important decisions.

De Morgan was deeply and personally invested in his defense of moral probability. His insistence that different people be allowed to draw their own conclusions was initially forged and repeatedly tried in his relationship with his insistently evangelical mother. As he moved ever further into Frend's Unitarian world, Elizabeth remained always concerned about his religious choices. Time and again she tried to bring him back into the fold. After the death of Eliza, the issue flared up yet again, and in a long letter Augustus pushed back against the piles of religious tracts she insisted on sending him. Religion is not a matter of feeling for me, he firmly explained; it is "to be tried by reason and evidence." But, he went on, "that is *by me,* for I do not object to anyone who thinks he can find truth by another method."[35] The distinction between moral and mathematical probabilities was essential to De Morgan's ability to recognize the legitimacy of both his and his mother's very different views of religious truth. He was willing to allow her to draw her own conclusions, even as he insisted that he be allowed to draw his.

De Morgan's plea for tolerance extended beyond his relationship with his mother. As he was drawing the line between moral and mathematical probabilities, he was living by choice in a secular world in which personal certainty did not guarantee assent, and in which one person could have absolutely no doubt about something with which others would disagree. His insistence on distinguishing between moral and mathematical probabilities recognized the implications of this diversity for probabilists as well as for all of the other kinds of ideologues and churchmen that filled his city. "The abomination called intolerance," arises from those who cannot see "that the real probabilities may be different to different persons," he explained.[36] Tolerance, particularly religious tolerance, was a fundamental tenet of De Morgan's rational world. He was not going to allow either probability theory or religious dogmatism to get in the way of the principle that all people must be allowed to follow their own reason to truth.

While Augustus was pursuing reason in his classroom and writings, Sophia was settling into her new life as his wife. In addition to establishing and maintaining a comfortably flowing household, she filled her hours with charitable activities. At the time of Sophia's marriage, Lady Byron was beginning to be drawn back into the emotional morass of Byron's surviving family, which left Sophia as an essential contact point for various charitable ventures. Sophia was somewhat apprehensive about introducing her new husband to her long-time patroness, but the two respected each other and quickly bonded over silly puns. More significantly, the introduction led Augustus to one of his most brilliant students, Lady Byron's daughter, Ada Lovelace. In the decade since Lady Byron first asked Frend's advice about tutors, Ada had sporadically pursued mathematics even as she married and became the mother of three children. She and Sophia had long known each other but the two did not get along very well. Sophia found Ada's flamboyance irritating, but Augustus saw her brilliance and for a year and a half tutored her in mathematics. Two years later Ada used what she had learned from him to explain some of the implications of Babbage's vision of calculating machines. De Morgan remained viscerally opposed to Babbage's efforts to develop a mindless mathematics, but he nonetheless took great satisfaction from his work with Ada.[37]

Augustus also brought a much-needed practical perspective into Sophia's charitable work. When Lady Byron asked him for his advice about a new "female benevolent society" being proposed by the Quaker reformer Elizabeth Fry, he assured them both that it had no chance of success. This particular effort may not have survived Augustus's critical assessment, but it did introduce the De Morgans to a woman who impressed them both as "one of the noblest of human beings."[38] Throughout the first spring of their marriage Sophia was busily working on an article on "Reform of Prisons" under Mrs. Fry's wing.[39]

On June 4, 1838, the De Morgans' life was totally transformed by the birth of their first child, Alice Elizabeth De Morgan. Lady Byron, who was traveling in Europe at the time, opened her house in Fordhook to the new parents. Sophia was very happy to be in the countryside with her baby, while Augustus was glad that Lady Byron's house was close enough to London that he could escape to his books or meetings of the RAS. His article for the *Encyclopædia Metropolitana* was finished, and during Alice's first summer, he was writing a more practically focused *Essay on Probabilities and Their Application to Life Contingencies*. This effort to capitalize on his probabilistic expertise almost backfired when his publishers accused him of plagiarizing himself. Although defending himself entailed "wasting a good deal of good grumbling,"[40] he succeeded in demonstrating the differences between article and book and in the end was able to earn some money to support his growing family.

After several weeks at Fordhook, the De Morgans went back home to Gower Street, where Augustus returned to teaching and Sophia to raising Alice. Sophia was completely entranced by her little girl, but Alice was not an easy baby. When the child was not yet three months old, Elizabeth Fry cemented her status as a saint simply because Alice fell asleep when she was reading aloud.[41] The rest of Sophia's friends were not so easily sympathetic, however, and Alice did not shine in Victorian parlors. One of Sophia's most vivid memories was of a visit to the home of the formidable Unitarian Miss Joanna Baillie, where one-year-old Alice refused to sit quietly on Sophia's lap. As the excited little girl raced around the over-decorated room, those gathered there let Sophia know that if Alice "had been properly managed from her birth, she would not have been so restless."[42] Looking back from a distance of forty years, Sophia could see that some children are "so mercurial that nothing can make them quiet, steady characters," and others "so heavy and dull that they cannot be roused into activity and interest," but she did not recognize this at the time.[43] Plenty of people offered advice about how best to "manage"[44] her little girl, but that advice was often contradictory, and Alice's reactions were unpredictable. The only thing that everyone seemed to agree on was that the responsibility for Alice's behavior lay with her mother.

Sophia responded with characteristic earnestness. On January 1, 1840, she opened a notebook with a title page printed for the use of students at University College London: "Lectures on *Education* Delivered in/ *near* University College London by ____" it begins.[45] By leaving the second underlined space on her title page uncompleted, Sophia left open the question of whether it was she or her children delivering these "Lectures on Education." The ambiguity was on purpose. In recording the events in her children's nursery, Sophia was following a path blazed in the eighteenth century by the radical Richard Lovell Edgeworth and his daughter Maria. In 1798, this father-daughter team published *Practical Education,* in which they brought together insights gleaned through years of observing "the conduct of a judicious mother, in the education of a large family."[46] Even as the Edgeworths compiled the results of years of child-rearing records, they recognized that their goal of making education "an experimental science" would require the "labours of many generations."[47] Sophia wrote her nursery journal in that spirit. She gave birth to her second child, William Frend De Morgan, on November 16, 1839, but on January 1, 1840, she found one-and-a-half year-old Alice far more interesting. Sophia opened her journal with high hopes of watching the development of reason in her beloved little girl, who was just beginning to talk.

Sophia recorded her phrenological reading of Alice's head on the second page of her nursery journal. What she found was that most of her daughter's "organs" were too unformed to be read, and those that she could read were "rather large," "large," or "very large."[48] Following this singularly uninformative analysis of her child's potential, Sophia was ready to begin recording her daughter's growth. Alice's behavior posed serious challenges to this project, however. Sophia opened her journal optimistically, with a little vignette showing Alice being gentle to William, but the sweetness was short-lived. Just two days later, when she asked Alice to be quiet so as not to wake her baby brother, the little girl screamed and walked out of the nursery "trotting back once or twice to set up her shoulders and look daggers." This act of defiance introduced a struggle between Sophia and her obstinate little girl that continued unabated throughout the two years of the nursery journal.

Sophia understood the dynamic in terms of compliance. Alice "has not yet a distinct idea of obedience," she wrote as her toddler was approaching her second birthday, and "this she must learn before she learns anything else."[49] Sophia's view of obedience reflected her early Victorian world. Isaac Watt, whom her father so admired, wrote of "the lovely character of a child obedient to reason and to his parents' will."[50] The more liberal Edgeworths questioned whether obedience was truly "the virtue of childhood" and recognized the challenge of expecting obedience "from children long before they can reason upon the justice of our commands."[51] Nonetheless, they too were clear that children had to do as they were told, albeit as a habit without understanding.

Most of the time, however, the Edgeworths recommended that parents avoid the issue entirely by creating a child-centered environment in which regularity reigned. This was difficult to achieve within the confines of the De Morgan household as the house at 69 Gower Street was not large, and its spaces needed always to be divided between Sophia's efforts to be a mother and homemaker and her husband's efforts to be a bread-winning educator and writer. Except for his nine o'clock and three o'clock lectures, Augustus worked in his home library. Every evening he had dinner with Sophia and spent a little time with his children, but otherwise Sophia, Alice, William, and the nursemaid Jane had always to fit themselves around his needs for peace and quiet.

Alice represented an enormous problem in this arrangement. Sophia began her journal with great hope, but her narrative quickly devolved into a series of efforts to control the little girl's behavior—by ignoring her, closing her into a different room, holding her hands to restrain them, or tying her to her chair. Each seemed to work for a while, but Alice kept upping the ante. On January

20, the child got so angry that "she screamed & fought Jane," yanked William's legs, and hit her mother.[52] This was just the opening volley of a week that devolved into "an almost incessant scene of crying, disobeying, holding hands & forgiving," until finally Alice was given "a grey powder & on Friday she was gentle & good with very few exceptions."[53] The "grey powder" that Sophia resorted to was undoubtedly one of the opiates that Victorians imbibed in startling amounts. Although it seems to have been effective in calming the little girl, it did have the side effect of inducing vivid dreams. Her mother did not make the connection between the medicine and her child's nightmares, but on some level it seems that Alice did, and "grey powder" joined the list of things she fought against.

The question of how best to respond to Alice's tantrums drew in everyone on both sides of the family. On one side stood the De Morgans, both mother and son, who wanted Sophia to spank her;[54] on the other stood Sophia's mother, sister, and Lady Byron, who insisted there were better ways to manage the child.[55] Sophia could sometimes see the little girl's "determined spirit of resistance"[56] as a good thing, because it signaled a strong will, but Augustus saw it only as an impediment to the requirement "that obedience must be instantaneous."[57] Sophia's mother-in-law maintained that "the certainty of a sharp slap instantly for disobedience" would solve the problem completely,[58] but her own mother was appalled when she spanked Alice. By July 1840 the conflicts had become so unbearable that Sophia gave up writing altogether. "The fact was," she reported when she again took up her pen in December, that "Alice became so very unmanageable & contradictory that I felt a foolish dislike of recording nothing but instances of contradiction, passion and perverseness."[59] For Sophia, who was always sensitive to the ways her own unusual upbringing marked her as an outsider, it was painful to watch her beloved firstborn refuse to conform to societal expectations.

Through it all, Alice continued to charge ahead. Her mother did recognize that when she began writing, her daughter did not really understand what was being asked of her. Sophia rejoiced when Alice was two and a half years old and could explain that being naughty meant being "conttadictus," that is, contradictious, when she did not do as she was told.[60] Sophia hailed this as a major turning point, but Alice remained obstreperous. Sophia consulted the work of Dr. John Conolly, who was known for introducing a "non-restraint system" into the treatment of the insane. If "Dr. Conolly manages 800 lunatics with no more violence than is required to lead them into their rooms when they are over excited," she declared, "I ought to control my poor little girl's outbreaks with as little difficulty."[61] But that was easier said than done. When Alice had just turned three, her mother had no trouble agreeing with a gentle neighbor who observed, after a meltdown, that "when the passion is once strongly excited it is really too powerful for the child herself to conquer."[62] Sophia sometimes formulated Alice's tantrums in religious terms, suggesting that "the little mind <u>must</u> be constantly occupied, or the devil gets possession, but at the same time over-excitement in any way, or fatigue, is very injurious."[63] To Alice, she spoke of a "spirit of contradiction" that built up inside and overcame her. Over the course of two difficult years, Sophia searched for ways to understand, while Alice grew steadily older and somewhat more manageable.

Sophia's entire program of producing obedience was really just a way to clear a space in which to encourage the growth of Alice's reason. From the very first day of her writing, Sophia was entranced by her little girl's mind. Much of her attention focused on recording, interpreting, and reveling in the many facets of her daughter's speech: "She used to say 'oh,' instead of 'Yes'—She has now learnt to say '<u>Ye</u>' and I heard her today say 'Oh dear!' & correct herself to 'Ye! Dear.'"[64] A year later, Alice could still be difficult to understand, but Sophia found it well "worth the trouble of puzzling it out."[65] One of Sophia's books encouraged her to record the "progress of a child's perceptions, or their gaining the ideas of form, place, distance, time &c."[66] that formed the bedrock of classical geometrical reason, but Sophia was much more interested in following the child's lead. She delighted in unexpected connections; when "Jane said something about <u>a jacket</u>," Alice began singing Jack and Jill; when Sophia told her Lady Byron's granddaughter's name was Annabella, Alice responded with "Umbilla—Keep yane off, Moma."[67] "Alice could frame a language" Sophia glowed when her three-year-old said "open a light" instead of "light a match."[68] She delighted in all of the unexpected connections and turns of phrase that provided glimpses into the world of her daughter's mind.

Alice's imagination was another source of fascination. From the age of about two and a half, she had an imaginary companion, "Marmee," whom she would often let stand in for herself as in "Mama, My Marmee will yore [roar] an wake up hi little brother dat <u>tiny</u> boy."[69] As her daughter grew older, Sophia began introducing other characters designed to carry messages about good behavior. When Alice dawdled getting out of her warm bed on chilly mornings, Sophia told her "a very interesting story" in which an imaginary "Louisa" had cured herself of the same behavior "by her own determination."[70] A year later, by which time Alice was three, Sophia had begun experimenting with more sophisticated stories that Maria Edgeworth reported having "read to some children for the purpose of eliciting their remarks."[71] Following Edgeworth's example, Sophia carefully adjusted her tellings to Alice's vocabulary and then recorded her daughter's responses.

Following the twists and turns of Alice's imagination could be fun, but telling stories could shade into lying, and lying could blossom into an even more serious problem than disobedience. Maria Edgeworth, whose child-rearing advice Sophia was following so closely, was also the author of the novel *Helena,* which had focused her and Lady Byron's attention on lying less than ten years before. As Alice became increasingly articulate and imaginative, her mother remained ever vigilant.[72] Sophia established her commitment to truth-telling in an incident she related when Alice was just three. The vignette opens with Alice telling her mother, "Don't look dis way" just after she had been told not to pick grapes. Having been thus alerted, Sophia saw that her daughter was reaching out the window to where the grapes were hanging. A back-and-forth ensued in which Alice at first tried to deny and then admitted she had intended to pick a grape. When, in the aftermath of this little drama, Alice asked to be forgiven for reaching toward the forbidden fruit, her mother clarified that the real issue was that she had denied her true intention. "I should not have been nearly so much grieved if you had picked the grapes." When Alice countered with, "But it's yong to pick the grapes," Sophia answered, "But it is much more wrong to say what is not true."[73] Sophia recognized that it was probably confusing to her three-year-old, but she was intent on teaching that lying was even worse than disobeying.

Storytelling constituted another danger zone. Sophia had little trouble distinguishing between legitimate storytelling and lying, but she was acutely aware of the slipperiness of the slope. One day when she went to pick up Alice from an afternoon with her Grandmother De Morgan, she found her mother-in-law "much amused by the earnest manner in which Alice insisted that she had <u>yote</u> down a <u>mettade</u> [wrote down a message] for her cousin Trevor." Sophia, in contrast, was worried. When she double-checked to be sure Alice realized that what she had produced was just an imaginary note, Alice re-sisted until "her eyes became so eaernest and her manner so eager that I did not know whether she believed what she said or not." For Sophia, this situa-tion was deeply problematic and required an immediate intervention. So she offered Alice a strawberry, but when the little girl opened her mouth to receive it gave her nothing. When Alice complained, Sophia explained that she was equally disturbed when Alice talked about her "make-believe" letters as if they were real. Sophia admitted at the end of writing this little story that she was not sure Alice actually understood what had happened, but she felt the issue was important enough that she had to make the effort.[74]

There was a strong social element to Sophia's concerns about telling lies, but she also had other grounds for her concern. Maintaining the clear connection between words and their proper meanings was absolutely essential to the Lockean program of reason. De Morgan was constantly being reminded of how hard it could be to maintain those connections as his students combined symbols in meaningless ways. He might be humorous in response, but he was relentless in his insistence that they maintain their focus on the proper meanings of the words and symbols they were using. The problem Sophia faced with Alice was in many ways more complicated. Even as she delighted in the poetry of her little girl's speech, she had always to be equally alert that Alice never lost sight of the proper connection between her words and their meanings.

At the same time that Sophia's daughter was beginning to talk, her father was losing his words. In a letter to Lady Byron announcing he was entering his eightieth year, Frend attributed the six weeks that transpired between his November birthday and the letter he sent in January to a mind that flitted from subject to subject "just as a child runs after butterflies."[75] Four months later, he was pleased by a visit to the Unitarian chapel on Newington Green, where he and his family had worshipped when they lived in Stoke Newington.[76] Both of these letters contained joyful affirmations of the lifetime he had devoted to bringing his vision of all-encompassing egalitarian reason into the public spaces of his world, but their rambling reflects the weakening of his mind. Frend's letter reporting the trip to Newington Green was his last.

In late 1841, a series of little strokes, or transient ischemic attacks, that were rendering Frend ever more weak, frail, and psychologically scattered culminated in "a stroke of paralysis," that led Sophia to rush to his side.[77] When he neither improved nor worsened, she returned to Gower Street until February of the following year, when his condition began noticeably to deteriorate. On the fourteenth of the month, Sophia went to be with her father, leaving her children under the care of the nursemaid. Day after day, she read the Bible by his bedside, while Augustus comforted his mother-deprived little girl with stories of King Arthur and his knights.[78] After several days of silent listening, Sophia's father surprised her by quavering his way through his favorite psalm: "As for man, his days are as grass; as a flower of the field so he flourisheth."[79] The next day, February 21, 1841, William Frend died.

Frend's death was a terrible loss for both of the De Morgans. In a loving obituary embedded in the "Report to the Twenty-Second Annual General Meeting" of the RAS, De Morgan rejoiced to be part of a group that recognized the value of Frend's open-armed embrace of the world. He marveled at the ways his father-in-law saw "every advance in art, learning or science,—every amelioration of social evils,—every improvement in the law,—every evidence, however slight, of disposition to act, think, or hope, for the better," as evidence of a God's loving plan for His beloved reasoning creatures.[80] Frend's ability to find the good in virtually anything "was a spring of comfort to his age which never ran dry" and remained "even after he was unable to speak or move," the younger man reported.[81] At the end of the long and complicated life Frend had devoted to reason, De Morgan hailed a man wholly at peace with his world.

Sophia's immediate responses to her father's death are lost in the mist of private conversation. For her, there was no RAS waiting for a eulogy, only a nursery journal in which she was determined to "name our loss" only "in connection with my children."[82] Talking to Alice about Frend's death faced Sophia with the challenge of conveying "as true an idea" as possible. She told her little girl that the doctor was trying to cure her grandfather, but that he would probably fail, and when that happened, he "will go away to a nice place where he will be made quite well."[83] This attempt to construct a child's-eye view of the theory of the afterlife that Sophia had developed after her sister Harriet's death seems to have made sense to Alice, who spent several days exploring the idea of this "nice place." When she asked her mother "whether the birds sang & the trees were pretty & had buds and fruit," ever-truthful Sophia admitted that she did not know because she had not been there. Thus freed of parental correction, Alice then pronounced her opinion that "he gathered the fruit from one of those trees, & eat it, & dat made him quite well." Sophia was thrilled. "What an extraordinary idea to enter a baby's head!" she exclaimed. Even though no one had told Alice anything about religion or the story of Adam and Eve, the little girl was talking of "eating the fruit of the tree of life."[84] Sophia was thrilled. All of her efforts to ensure that her daughter spoke the truth were rewarded by this glimpse into the world of pure consciousness still shimmering in the mind of her little girl.

With all of the time Augustus spent teaching and writing and Sophia spent entangled with her children in the nursery, the couple had only about an hour a day alone together. Sophia would talk lightly about her day, while her husband regaled her with "any interesting fact which he had come across in his investigations," but they did not engage in serious discussion. They instead talked about "matters of less importance, obscure derivations of words, and unsuspected translations, the origin of old customs, versions of nursery rhymes, and, above all, riddles, good and bad."[85] Augustus enjoyed reading aloud, and Sophia listened to the *Pickwick Papers* and *Nicholas Nickelby* from beginning to end. Secure in their shared commitment to each other and to leading lives of reason in their separate domains, at the end of each day the De Morgans found ease with each other.

# CHAPTER 18

## *Rearing Young Seedlings*

In January 1842 Sophia opened a new nursery journal. At this time, her phrenological analyses showed the head of two-year-old William to be "better balanced" than Alice's had been.[1] William was talking by this point, which means that he was somewhat more of a presence, but Alice, who was now three and a half, still dominated. William rarely ran afoul of his mother's strictures, perhaps because he was a boy and not being asked to sit still while his hair was being combed or to hem handkerchiefs with small neat stitches. Alice, who was required to do such things daily, continued to find herself in trouble because of "wiggling" or demanding to know the reasons why she was to do as she was told,[2] but overall, the conflicts between mother and daughter were becoming easier to negotiate.

Sophia continued to revel in the expansion of her little girl's mind. "Alice calls the feathery white clouds 'the juice of the sky' because I told her they were wet," her mother reported. "This morning she asked me if the dried African flowers under the glass case were dead, and on my telling her they were, she asked, 'Can flowers speak when they are alive?'"[3] When Alice learned that one of her mother's grandmothers was a Blackburne, she asked, "Was she black because she went after the coals, and burn because she went after the fire?"[4] As Sophia's daughter was finding her place in the larger world around her, the poetry of her conversation was becoming richer.

Religion provided a particularly interesting source of questions for Alice to navigate. Sophia was very clear that religious understanding was something that had to be developed, not taught, and she was determined not to teach religion to her daughter "till her own questioning shews that her mind is ready to exercise such impressions."[5] Unfortunately, Sophia's organic approach to Alice's religious instruction was unacceptable to her evangelical mother-in-law, who found not speaking about religion to be an offense against God. This conflict became heated enough that De Morgan wrote a firm letter to his mother insisting that it would be "fair on both sides" if "my wife, on a repetition of this subject shd find that the time of her visit is expired."[6] Clearly Alice was the daughter of committed Unitarians who were not going to allow anyone to confuse her as she reasoned her way to God.

By the time Alice was three and a half, Sophia decided that she was ready to be introduced to a loving God who lived in Heaven in the sky. Alice immediately began to exercise her reason on this construction and "henceforth she prattled glibly about 'my Good Father' in contradistinction to 'my real Papa.'" If she was the child of such a God, why did she find it so hard to be good, she asked. When Sophia explained that then "you would not have had the pleasure of conquering your naughtiness and of pleasing Him that way," her plucky little girl concluded, "Then it was right in Him. *He done it to give me a job!*"[7] When Sophia told her that God was not just her "Good Father" but had also created all things, Alice turned quickly to the question of where God had come from: "For you see, He could not make Himself because if He made Himself He must have had *arms* and if they were made He was made before He made Himself, and that could not be, you know."[8] All of these observations thrilled Sophia as she watched her daughter emerge from toddlerhood as a reasoning being.

Alice's family was growing almost as fast as her mind. At the end of the first nursery journal, on November 16, 1841, the De Morgans welcomed a second son, whom they named after Augustus's brother, George Campbell De Morgan. A year and a half later, in 1843, they gave their third son, Edward, the middle name of Lindsey, to carry on the legacy of the family of reason.[9] With four children under the age of six, the De Morgan family was bursting the walls of 69 Gower Street. The following summer, they moved to a larger house in the fresher air of Camden Town. The new house was close enough to UCL that Augustus could still walk to work every morning, but it was far enough that he did not come home between his morning and afternoon classes. This meant that from about 8:00 in the morning until 4:30 in the afternoon, 7 Camden Street was Sophia's domain.

Even as she bore the responsibility of child-rearing and negotiated the ongoing distractions of pregnancy, Sophia continued to pursue philanthropic work. The focus of her efforts changed dramatically in 1842 when startling accusations of white slavery resulted in the collapse of the Children's Fund. She was indignant at the accusations raised but was also relieved to be freed of what had become the rather exhausting job of raising money for England's largest charity.[10] She nonetheless remained involved in organizations for unwed mothers and continued to follow Elizabeth Fry's interest in the treatment of prisoners and the insane.

These philanthropic projects also provided material in support of Sophia's ongoing investigations into the power of mind. She was far from alone in her desire to investigate such powers scientifically. In 1838, John Elliotson, a professor of medicine at UCL began publishing articles in *The Lancet* that trumpeted his ability to create new states of mind in his patients by manipulating an essential force that flowed in and through humans. His particular triumphs came from the treatment of the O'Key sisters, two young women who suffered from some form of seizure disorder. Elliotson at first tried to bring the sisters through their seizures by interrupting and reorganizing unseen force fields around them with passes of his hands. Later he moved on to creating disordered mental states using the same methods. Over time, the sisters began to take charge of their sessions in ways that undercut Elliotson's credibility enough that he lost his job at UCL. The seed had nonetheless been sown, and Sophia joined her generation in a fascination with the medical potential of working with what were called "mesmeric" forces.[11]

Part of the impetus for Sophia's interest in mesmerism came from her experience of traditional medicine, which could be at once painfully invasive and frighteningly ineffective. The dynamic can be clearly seen in one of the rare moments that William emerged to dominate the narrative of her nursery diary. The trouble began when William was teething, which the early Victorians saw as an important developmental stage that was also mortally dangerous. On April 3, 1841, when he was a year and a half old, Sophia noted that only one of her son's "double teeth" was cut, and his gums were "sadly swelled."[12] Through the middle of the month, she casually reported that both William and Alice contracted and recovered from chicken pox, but then, on the twenty-fourth, William fell seriously ill. "Poor little darling, it came on suddenly with extreme difficulty of breath, weight & heat of the head," she reported.[13] Unable to do anything but cuddle her feverish child, she called the doctors.

The early Victorian physicians who came to William's bedside lived in a world where terms like "fever" or even "illness" covered a host of diseases we would now differentiate by causal agents. From their perspective, William's symptoms were manifestations of some kind of internal imbalance, but the only way to address them was through the use of external measures. Working on the body from the outside, they hoped to produce internal counter-effects that would return the body to its balanced state.[14]

The therapeutic regime William's doctors followed is a startling example of what this approach could entail. At first, they lanced his gums, "put leeches on his chest & gave him strong doses of mercury with Jalap"—a powerful purgatory. For a few days, this approach seemed successful in keeping the fever down, though it stubbornly returned for several hours every afternoon. Then, just when Sophia thought her baby was on the mend, the fever returned and "the paroxysms lasted longer than before & he seemed to have no intermission."[15] The doctors stepped up their efforts. They shaved off William's hair, put leeches on his temples, and steamed his head "constantly with spirits of wine." Again and again they lanced the little boy's gums, covered his stomach with hot flannels, and bathed "his poor little feet & legs" in mustard water. Alice was so stunned by what was happening "that for many days she did not speak much above a whisper." When she came upon her mother in tears, she did her best to offer comfort: "I don't tink Willy will go away. I will keep veddy kiet & den he will toon be 'well.'"[16] Sophia tried to take heart from her little girl's assurances, while following the doctors' instructions to the letter.

Fasting was another major part of the treatment. For three weeks the doctors allowed William to eat only "thin barley water & a very little milk," while dosing him with a large variety of cathartic medicines—calomel, James's powder, rhubarb, and ipecacuanha. When they finally began to allow him to eat a bit more, Sophia found it "very painful to hear his repeated cries for 'bed' (bread) 'mo bed'—when we dared not give it to him." William's first bath after five weeks of this treatment "was a grievous business." Sophia was appalled to see the "shrivelled skeleton," and the nurserymaid burst into tears when she found her emaciated little charge was unable to stand alone. Three weeks later he still "looked a perfect shadow" but was being allowed to eat, so Sophia could take comfort from the thought that "we had a fair prospect of seeing him gain flesh & strength."[17] A full year later Sophia was still noticing how pale he was,[18] and it is tempting to attribute the odd "high pitched drawl"[19] that characterized William's speech throughout his life to the dramatic treatment he received as a toddler.

Mesmerism offered a considerably more attractive approach to situations as unknown as a child's fever. Making passes over a child's body did not guarantee a cure—success required the action of invisible forces that were poorly understood—but the outcomes of leeching, blistering, lancing, and starving were not predictable either. By the middle of the 1840s, Sophia had begun to experiment on her own. She never claimed particular mesmeric prowess but reported that "many patients have spoken of *light* which they said they saw streaming from my fingers" when she made passes over their bodies. Even as she offered this credential, however, Sophia included herself among "those who had no power of vision," and therefore saw nothing. In one instance, when a neighbor brought her a ten-week-old baby whose legs seemed poorly aligned, however, Sophia succeeded in relaxing his legs with a series of passes "from the knees to the end of the little feet."[20] This single success was powerful enough to carry Sophia through years of exploration. Augustus left the experimentation to his wife, but he admitted being intrigued. At one point, he tried mesmerism "for the removal of ailments which required much medicine" and emerged completely convinced of the "curative powers of this agent."[21] The allure of contact with the powers of mind in the universe made mesmerism very attractive to both of the De Morgans.

While Sophia's household hummed around him, Augustus spent hours every evening in his library. He was certainly not alone among early Victorians in setting aside a room for his books. Just as he was emerging from Cambridge, English publishing was taking off into its own industrial revolution. Throughout the 1830s, 1840s, and beyond, an ever-increasing torrent of books poured from British presses and was absorbed into the private libraries of an ever more voracious reading public.[22] De Morgan may be seen as a leader among this group. His library was already large enough to daunt Sophia when they were first married; by the time they moved to Camden Street, it was significantly larger and growing apace. He began reviewing for the *Athenaeum,* which generated a stream of new books on topics from mathematics to astronomy to philosophy to theology and beyond,[23] while his ongoing rummaging expeditions through London's bookshops produced an equal number of old ones. By the end of his life, De Morgan had amassed and extraordinary collection of books.

De Morgan spent his library time writing at his cluttered desk or ferreting among the books and papers piled up on all of the surfaces of the room. It was a shabby place, but he took great pleasure in devising his own versions of "all the writing appliances and handsome bindings that ornament rich men's studies." Looking in from the outside, Sophia could only conclude that his "faculty of arrangement" manifested itself "more in classification than in tidiness,"[24] but he did work to maintain some kind of organizational system. "You have no idea what a provocative to order and decency is the having a number of little receptacles ready," he crowed after first encountering file folders,[25] and he thrilled at the discovery of paper clips. "I never could spell the word, but if *cowchoke* goes, I go too,"[26] he proclaimed of the India rubber that performed all of the functions that duct tape does today, and he devoted entire letters to explaining how to transform a half-penny steel pen into a creditable writing instrument by sticking it onto a swan's quill.[27] De Morgan reveled in all gadgets which promised to order the books and papers mounded in the library that was a physical extension of his mind.

Although few people actually entered the room, huge volumes of letters flowed in and out. Among De Morgan's most constant correspondents were Sir William Rowan Hamilton, John Herschel, George Biddel Airy, William Whewell, and George Boole. Hamilton was the most voluble of these. De Morgan seems to have prided himself on having spoken only once to the man he consulted on virtually everything from algebra to child-rearing, politics, theology, logic, and literary theory in a correspondence large enough to fill the third volume of the Irish mathematician's biography.[28] De Morgan knew others of his correspondents more directly. He greatly admired Herschel, and their friendly letters flowed back and forth for decades. Whewell served as the major sounding board for De Morgan's philosophical thinking and his early work in logic. He and Airy corresponded about the RAS; he and Boole about logic. De Morgan's correspondents were heterogeneous, both socially and politically. The central tie that bound them to the professor of mathematics at UCL was their common respect for the reasoned mathematics that called "into exercise some of the powers which most peculiarly distinguish man from the brute creation."[29] Day after day, in letter after letter, they joined De Morgan in his library.

Many of De Morgan's compatriots spent similar amounts of time reading and writing in their home libraries, but few saw their books as he did. His approach may be seen in the way he treated one cross section of them in his 1846 *Arithmetical Books*. He dedicated this bibliographical study to George Peacock, the man who had first opened his eyes to the mathematical significance of historical progress. As he brought this group of books together, he saw himself to be tracing the history of "the progress of mind."[30] He recognized that most history would focus on those pioneering works that became well known, but he nonetheless insisted on the value of "the minor works which people actually use, and from which the great mass of those who study take their habits and opinions."[31] No book was too insignificant to be part of this project. Even "the most worthless book of a bygone day" was worth preserving because, like a star so small it could only be seen through a telescope, it could serve "to determine the places of more important bodies."[32] De Morgan took care to acquire the books of great men like Copernicus or Newton, but he was no less intrigued by the efforts of their less exceptional coworkers.

The room that contained De Morgan's books was to him as much a natural history museum as it was a library. He did not include any book in *Arithmetical Books* unless he had actually held it, sniffed its binding, pondered its inscriptions and marginalia, and, of course, read it. Each book emerged from this process as an individual with a personality all its own. "For absurdity and dullness," a 1521 Parisian text on Boethius "ought to rank high among the mystic commentaries."[33] A 1687 work by D. Abercromby was "a real smatterer's book," which contained just enough names and ideas "to set up a man about town."[34] A 1696 book on arithmetic by Samuel Jeake was so weighty that "one would have thought arithmetic had been a branch of controversial divinity."[35] Despite this humor, each book was to him like a fossil, an individual artifact left over from a long process of progressive development.

In addition to being a catalogue, *Arithmetical Books* was De Morgan's fond farewell to the study of algebra. Already, by the time he moved to Camden Street, he was proudly declaring to Herschel that "there is now not a cloud in the heaven of algebra, except in the corner appropriated to divergent series." By this he meant that every symbol and every equation, no matter how obscure, "is not merely capable of interpretation but intelligible on the definitions of the symbols themselves."[36] Having spent more than a decade ferreting out and making sense of every algebraic expression he could find, he was beginning to see that discovering meaningful interpretations for algebraic forms was just a small part of a much larger project. "The best lessons for the future which a reflecting mind can have"[37] are to be found in the history of all language, he proclaimed. As he brought his algebra books together, he was getting ready to expand the methods forged on the equations we use when we calculate to include the sentences we use when we reason. De Morgan was getting ready to move into logic.[38]

De Morgan's shift from mathematics to logic was not an obvious one. The mathematics he learned at Cambridge was offered as an alternative to the Aristotelian logic that Locke had argued served "to perplex the signification of words, more than to discover the knowledge and truth of things."[39] Locke's charge was devastating enough that even in eighteenth-century Oxford, the study of logic withered on the vine. That it remained on the curriculum left an opening for its revival in the nineteenth century, however, and over the course of the 1820s, a group of Oxford dons began to consider the subject anew. In 1826, Richard Whately moved their discussion into the public sphere, first as an article in the Anglican *Encyclopædia Metropolitana* and then as a freestanding book entitled *Elements of Logic*. As he did so, he admitted that trying to revive "a study which had for a long time been regarded as an obsolete absurdity" could seem more difficult than just introducing a new one, but he had too much respect for the classical past to take that route. Although he recognized that the effort more resembled an "attempt to restore life to one of the antediluvian fossil plants, than the rearing of a young seedling into a tree,"[40] Whately set out to bring Aristotelian logic back to life.

The impetus behind Whately's work was the same problem that was gripping the embryonic London University at the time: how to define and establish a common ground on which different religious groups could carry out discussions. The founders of the London University ultimately ducked this issue when they decided there would be no religious discussion in their secular university, but Whately could not avoid the probem in that way. He was emphatically a member of the Anglican Church and of Oxford University. From these two positions, he saw the major goal to be preparing his fellow churchmen to defend their religion in the diverse world about to be opened up by the repeal of the Test and Corporation Acts and the passage of the Catholic Emancipation Bill. "Among the enemies of the Gospel now, are to be found men not only of learning and ingenuity, but of cultivated argumentative powers," he warned. Truth would prevail in the end, he asserted, but it could suffer real setbacks "if hasty assumptions, unsound arguments, and vague and empty declamations, occupy the train of close, accurate and luminous reasoning."[41] Learning to defend their faith through logical reasoning was essential to Anglican hopes of holding their ground against a rising tide of dissent.

Whately offered logic as the essential common ground on which to carry out these discussions. "If it were inquired what is to be regarded as the most appropriate intellectual occupation of MAN, as man, what would be the answer?" he asked. His response was "Reasoning."[42] Having thus defined the situation, he faced the challenge of finding the common denominators of the processes going on in so many different minds and reducing them to a system. For Whately, Aristotelian syllogistic logic constituted that system.

Whately's defense of logic was firmly rooted in the same linguistic tradition of reason that supported the mathematical curriculum at Cambridge. One of the first and most serious challenges to the Oxford man's efforts to revive the subject lay in the Lockean charge that the subject is essentially empty, that it takes place on the level of arbitrary words rather than true ideas. Whately countered by dissolving the distinction between words and ideas; some form of language is "an *indispensible instrument* of all Reasoning that properly deserves the name," he insisted.[43] He pointed to the example of Laura Bridgeman, an early nineteenth-century version of Helen Keller, who was often observed moving her fingers in signs even though there was no one there to read them. "Having once learnt the use of Signs," Whately explained, "she finds the necessity of them as an *Instrument of thought,* when thinking of anything beyond mere individual objects of sense."[44] For Whately words were never empty, and the logic that structured arguments produced with them formed the essential basis of reason.

Whately's *Elements of Logic* took England by storm, and it remained a centrally important text throughout the nineteenth and into the twentieth century.[45] De Morgan was one of the earliest converts. There is no trace of logic in the introductory lecture he delivered at the London University in 1828, but by the time he published *Study and Difficulties* in 1831, he was recommending that "all mathematical students" study logic before attempting geometry.[46] De Morgan's recommendation may stand as a testament to the power of Whately's defense of logic. Although Euclidean geometry and Aristotelian logic had each been held up as examples of reasoning since the time of the Greeks, the two were created and developed completely independent of one another. Over the centuries there had been some very occasional attempts to cast Euclidean proofs in syllogistic form, but they were massively complicated and routinely ignored as unnecessary. Modern logicians have since realized that Euclidean proofs actually cannot be reduced to syllogistic form, but neither De Morgan nor any of his contemporaries drew that conclusion. Inspired by Whately, De Morgan attempted a syllogistic proof of the Pythagorean theorem in *Study and Difficulties,* but the best he could come up with was "a specimen of a geometrical proposition reduced *nearly* to a syllogistic form."[47] Unfazed by his failure to complete the task he had set himself, he simply suggested that students work out the fiddly details on their own.

Even as De Morgan was embracing Whately's vision, the Scottish philosopher Sir William Hamilton (not to be confused with De Morgan's friend, the Irish mathematician William Roman Hamilton) was attacking it. In an

1833 review, the man who was widely recognized as the greatest logician in the United Kingdom dismissed the Oxford don as hopelessly parochial. No Englishman "seems to have studied the logical treatises of Aristotle; all are un-read in the Greek Commentators on the Organon, in the Scholastic, Ramist, Cartesian, Wolfian and Kantian Dialectic," he spluttered.[48] In his view, their ignorance of this rich continental tradition meant that they were fundamen-tally confused about logic's purview.

Hamilton's essential complaint lay in the way that Whately connected logic with language. From the Scot's perspective, Aristotle had presented the subject perfectly when he developed syllogistic logic. The only problem was that the Greek had failed clearly to specify the purview of the subject he had so definitively laid out. Hamilton credited Kant with having defined logic's proper boundaries once and for all when he made a clear distinction between analytic "laws of thought," which are properly logical, and synthetic "laws of things," which are not logical because they require the mind to move outside of itself into an external subject matter. For Hamilton, all of Whately's hopes of finding logic in essentially human processes of reasoning were based on a fundamental misunderstanding of this essential dichotomy. Therefore, all of the Oxford man's talk of locating logic in language was wrongheaded. The laws of thought were essentially fixed and immutable and captured in Aris-totle's syllogisms; any broader efforts to understand the nature of reasoning should be left to "other sciences, as psychology and metaphysic."[49] For Hamil-ton, logic was a pure and perfect subject that ought never be confused with the messiness of human reason.

Hamilton's quarrel with Whately was just a skirmish in a larger battle against the English tradition of Lockean reason. When Whewell published his 1835 *Thoughts on the Study of Mathematics,* Hamilton returned to the attack, this time against the mathematical view of reason that lay at the center of the Cambridge man's educational vision. Hamilton did concede that in geometry the laws of thought actually did correspond with the laws of things, that the structures of Euclidean space were actually the same as the structures of New-tonian space. But, he insisted, since geometry was the only subject that could legitimately claim this status, it was a poor focus for an education designed to prepare students for lives in the world.

Whewell understood Hamilton's position well: all of his talk of "neces-sary truth" was a way to recognize the special status the Scot was ascribing to geometry. He disagreed, however, about the uniqueness of geometry's claims to such truth. From Whewell's point of view, Newton's triumph lay in recognizing that the idea of "force" could be known with the same immediacy as the idea of space. For him, Newton's accomplishment lay at least as much in showing that necessary truths could be found outside of geometry as it did in accurately tracing planetary orbits.

Hamilton was infuriated by Whewell's claims that the kind of transcendent insight to be found in geometry could be attained in any other subject. Supported by what he described as a continental "cloud of witnesses,"[50] he insisted that geometrical understanding was an essentially limited and limiting model for reasoning. Encouraging students to take geometry as a model for true understanding would render them unable to engage with the huge variety of actual situations they would encounter as adults. The opposite would be the case for students who learned logic as the "form" of thought that transcended any particular "matter" of thought. These students, he claimed, would emerge prepared to recognize and respect the proper limits of human reason.

Thus, by the second half of the 1830s, the value of logic was being measured against the value of mathematics in a conversation about the nature and potential of human reason. Whately and Whewell agreed that the goal of a liberal education was to develop the power of reason in their students, but they disagreed about how best to do that. Whately, who saw learning the rules of reasoning to be worth the attention "of everyone who is desirous of possessing a cultivated mind,"[51] placed logic in the center of the liberal education. Whewell, who saw learning the rules of reasoning to be similar to learning horsemanship by book, preferred that students exercise their reason directly through the study of mathematics. Even as the two Englishmen disagreed about the value of logic, however, both of their arguments were constructed within the Lockean view of reason that had lain at the heart of the English liberal education for at least a hundred years.

Hamilton brought a powerful outside voice to this discussion. His separation of the "laws of thought" from the "laws of things" was a kind of analytic divisiveness that the English had confronted before. In both algebra and calculus they had turned to history in order to ward off continental efforts to separate symbols from their meanings. Hamilton's insistence that logic was a strictly formal subject, essentially separate from the complexities of human reason, constituted yet another challenge to the English Lockean view of the ultimate power of human reason.

De Morgan was as firmly embedded in the Lockean tradition of reason as Whately and Whewell were. But as professor of mathematics in UCL he was free to find his own way as an educator, a mathematical practitioner, and a student of logic. He saw no contradiction between practicing and teaching mathematics as an exercise of reason the way Whewell did while also exploring the rules of logic the way Whately did. In the article "Geometry," which appeared in the *Penny Cyclopædia* in 1838, he repeated his advice that students should study logic before venturing into this subject and referred readers to his forthcoming article "Logic" for further clarification of how to do so. As a writer for the SDUK, De Morgan could transcend the institutional differences between Cambridge and Oxford that were forcing Whately and Whewell to choose between the subjects.

As it turned out, De Morgan could not simply ignore Hamilton's objections to the Lockean view of reason, however. His articles for the *Penny Cyclopædia* were subject to editorial review, and by the time the article "Logic" was to be published, Hamilton's arguments had persuaded the general editor, George Long, that it was inappropriate to bring logic to bear on geometrical reasoning. So in place of De Morgan's promised article, Long directed readers forward to an article bearing the title of Aristotle's logical work, "Organon." Two volumes and several months later, when "Organon" appeared, Long admiringly laid out Hamilton's view that logic was a purely formal, classically defined study that hovered far above murky human processes of reasoning through language.

Never daunted, De Morgan proceeded to publish his ideas in a little volume entitled *First Notions of Logic (Preparatory to the Study of Geometry)*.[52] And then, two years and six *Penny Cyclopædia* volumes later, he pushed back against Long's Hamiltonian approach with a ringing endorsement of Whately's view in his article "Syllogism." "Every sentence in which different assertions are combined to produce another and a final assertion, is either a syllogism, a collection of syllogisms, or a mass of words without meaning. All that is called reasoning, and which cannot be made syllogistic, is not reasoning at all; and all which cannot easily be made syllogistic is obscure," he declared.[53] For De Morgan, as for Whately, logic formed the essential structure for all valid human reasoning.

De Morgan's "Syllogism" article was as much the articulation of a research program as it was a statement of support for Whately. Centuries of medieval scholarship had succeeded in codifying the syllogism into a very narrow and rigid form of argument. All Aristotelian syllogisms are formed from three propositions, each of which plays a somewhat different role. The first proposition is the *major premise;* the second is the *minor premise;* the final one, which is marked by "*Therefore,*" is the *conclusion.* All syllogisms can be cast in terms of these three forms of proposition (table 1).

TABLE 1

| Using Words: | Using Symbols: |
|---|---|
| All men are mortal. | All **Y** are **X**. |
| All men are rational beings. | All **Y** are **Z**. |
| Therefore, | Therefore, |
|    some rational beings are mortal. |    some **Z**s are **X**s. |

Propositions can themselves each be broken into four parts. In this example, "All men are mortal" is a proposition; "All men are rational beings" is a proposition; and "*Therefore,* some rational beings are mortal" is a proposition. The four parts are: the *quantifier,* the *subject,* the *copula,* and the *object* or *predicate* (table 2).

TABLE 2

| Quantifier | Subject | Copula | Object/Predicate |
|---|---|---|---|
| All | **Y** | are | **X** |
| All | **Y** | are | **X** |
| Therefore,<br>   some | **Z**s | are | **X**s |

In this construction the subject, and the predicate (or object) are nouns or noun phrases. The quantifier may be *universal*—in which case it is either "all" or "every"—or *particular*—in which case it is "some." Also possible is the negative quantifier "no" as in "No man is immortal." The copula is always a positive or negative form of the verb "to be." When De Morgan claimed that all reasoning could be made syllogistic, he was claiming that all reasoned arguments could be broken down into these terms.

De Morgan's position is far from obvious, but he insisted that as long as a sentence has a subject ("something spoken of") and a predicate ("something said of it") and "an affirmative connexion between them," it could be made a proposition. To illustrate the point he considered the sentence: "If he should come tomorrow he will probably stay until Monday." In this form, De Morgan's example did not fit the propositional form of subject, copula, and predicate, but it did after he translated it into a new format (table 3).

TABLE 3

| (Quantifier) | Subject | Copula | Object/Predicate |
|---|---|---|---|
| (Implied all) | the happening of his arrival tomorrow | is | an event from which it may be inferred as probable that he will stay until Monday |

This example suggests that given enough determination it may be possible to translate reasoned English sentences into logical propositions, but it also shows how awkward it can be to achieve that result.[54] But De Morgan had learned from Peacock that understandings of human reason could progress. Finding ways to expand the Aristotelian system until it was large enough to encompass "all that is called reasoning" was a challenge that was to engage him for decades.

De Morgan had little time to go beyond these beginnings in the several years after *First Notions* and "Syllogism" appeared. Until 1843, he was hard at work on the final volumes of the *Penny Cyclopædia* and until 1844, writing a series of four massive articles on algebra.[55] Then, in 1845, his fourth child, a second daughter, was born and named Anne Isabella after Lady Byron. By 1846, however, Sophia was between pregnancies, and De Morgan was ready to take the first substantial step toward expanding the syllogism so that it could fulfill its role as the defining form of legitimate reasoning. On November 9, 1846, "On the Structure of the Syllogism" was read to the Cambridge Philosophical Society. It was the first of five papers that De Morgan was to publish in the *Transactions of the Cambridge Philosophical Society* over the course of the next two decades. All of his original logical ideas appeared in these papers, which will hereafter be referred to as "On the Syllogism I–V."[56]

"On the Syllogism I" focused on the basic form of the syllogism. De Morgan's jumping-off point was a medieval refinement of the Aristotelian system, which organizes propositions according to a fourfold scheme in which each form is identified by a capitalized vowel (table 4).

TABLE 4

| Label | Form name | Quantifier | Subject | Copula | Predicate |
|---|---|---|---|---|---|
| **A** | universal affirmative | every | **X** | is | **Y** |
| **E** | universal negative | no | **X** | is | **Y** |
| **I** | particular affirmative | some | **X**s | are | **Y** |
| **O** | particular negative | some | **X**s | are not | **Y**s |

The relations among these propositions were presented in one of the earliest diagrams of the Western tradition, known as the Aristotelian "Square" or "Table of Opposition" (figure 9).[57] This configuration neatly shows the ways that the four forms of proposition relate to one another. The contraries and contradictories indicate pairs of propositions that cannot both be true; the subalterns and subcontraries indicate propositional pairs that clarify the relations among their terms. For centuries, the neat symmetry of this figure supported the conviction that the four Aristotelian propositions formed the true foundation of logic.

De Morgan accepted this basic configuration, but he questioned the traditional Aristotelian interpretation of the *particular negative* proposition, **O**. His

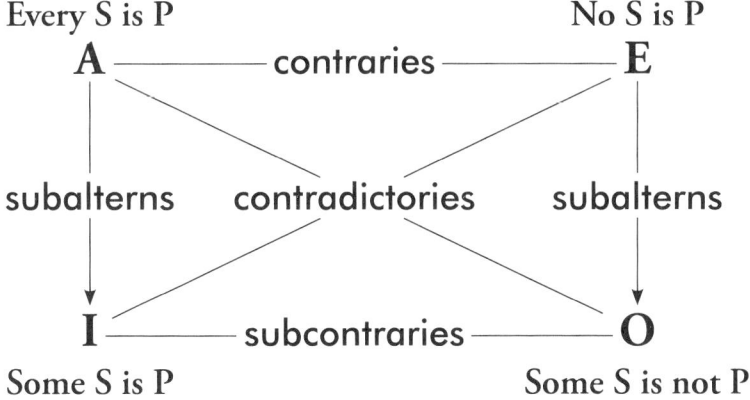

Figure 9: Aristotelian Square of Opposition, from "The Traditional Square of Opposition," in *Stanford Encyclopedia of Philosophy* (2012).

objection was that the copula should never be made negative. The predicate, he insisted, is defined as "that which *can* be said of" the subject, but in this case the predicate appeared as something which can *not* be said of the subject. Determined to keep the copula positive, De Morgan reconfigured the proposition (table 5).

TABLE 5

| Label | Quantifier | Subject | Copula | Predicate |
|-------|-----------|---------|--------|-----------|
| O | all (every) some | **X**s | are | not **Y**s |

At first glance this may seem a trivial adjustment, but no change in the Aristotelian system was trivial. In this case, De Morgan's decision to make the predicate "not **Y**" went against centuries of agreement that negative terms could not be considered as legitimate parts of logic. The problem was that given a positive term, like "man," the negative term, "not-man" was simply too huge to mean anything. De Morgan knew this well. "There can be little effective meaning, and no use, in a classification which, because they are not men, includes in one word, *not man,* a planet and a pin, a rock and a featherbed, bodies and ideas, wishes and things wished for," he grinned.[58] Any attempt to introduce negative terms into logic required dealing with the problem that they were meaningless.

However, De Morgan's determination to create a logic adequate to all of human reasoning required recognizing that rational people effectively use negatives all the time. In his first logic paper, he saw that what made this possible was that people place implicit limits on the subjects they are discussing; they are aware of what he called the "*universe* of a proposition."[59] Thus, to use one of his earliest and favorite examples, it would be possible for someone meaningfully to speak of a "*Briton*" and of a "*not-Briton*" or "*alien.*" What makes the use of such negatives possible is that the conversation is taking place within a particular propositional universe that is limited to people, or, to use De Morgan's words: "Let the universe in question be '*man*:' then *Briton* and *alien* are simple contraries."[60] De Morgan's use of the word "contraries" in this quotation, points to the fact that he started his investigation thinking in terms of negatives. But as he explored the idea, he began to see that the essential issue was not that one term is positive and the other negative, but rather that the two of them are "complements" that fill the entire "universe of their proposition." Thus, to use another of De Morgan's favorite examples, although "male" and "female" are both positive terms, they together fill the universe of "people."[61] De Morgan used lower case letters to designate the negatives he introduced, so that for "not **X**," he wrote **x**, and for "not **Y**" he wrote **y**. De Morgan's determination to add complements to logic added a new negative proposition for each positive one: "Every **X** is **Y**" was joined by "Every **x** is **y**," "Some **X**s are **Y**s" by "Some **x**s are **y**s," and so forth. This was a dramatic change that doubled the number of Aristotle's propositions.

De Morgan was very excited by the eight-propositional logic he had produced, and he eagerly sent a copy of his paper to the Scottish Hamilton. The reaction was as ferocious as it was unexpected. The logician had apparently been using an expanded system of eight propositional forms in his lectures for several years, and although De Morgan had no way of knowing that, the Scot accused him "both of an injurious breach of confidence towards me, and of false dealing towards the public." Completely stunned by this frontal attack on his honor, De Morgan responded in kind: "I beg to inform you that I shall wait till the 10th of next month [April 1847] for one of two things—your retraction—or the announcement of the specific time and manner in which you will publicly maintain the truth of the accusation which—excuse the expression—you have dared to bring against me."[62] Faced with this ultimatum, Hamilton quickly, though ungraciously, retracted the specific accusation of plagiarism. De Morgan stiffly accepted the apology, but the battle had been joined. De Morgan had the grace to recognize that the Hamilton who attacked him had been essentially incapacitated by a stroke, and he resolved not to engage in further controversy until the old man was dead. Nonetheless, for the rest of his life De Morgan's work in logic was spurred on by the challenge of responding to the logical work of Hamilton and his followers.

De Morgan took one powerful positive message from his initial confrontation with Hamilton. That they had both, independently, developed a logic with eight propositional forms struck him as "exceedingly remarkable," and it suggested that the eight-propositional logic he was exploring had "strong symbolic claims to being *something*."[63] Bolstered by the confirmation that the ideas he was exploring were valid, he set out to write a book entitled *Formal Logic,* in which he claimed the Scot's world for himself. Establishing this position required De Morgan to explain himself in ways he found deeply uncomfortable, however. Despite his wide reading and strong opinions, De Morgan routinely shied away from discussing the fundamental assumptions behind his work. He liked to say he "had no objection to Metaphysics, far from it, but if a man takes a candle to look down his own throat, he must take care not to set his head on fire."[64] But writing *Formal Logic* forced his hand. If he wanted to hold his own in a logical world filled with Hamiltonians, he was going to have to explain himself.

As De Morgan was raising the candle to his throat, he turned to Whewell for advice about how to negotiate the divisiveness that Hamilton's logic represented. The presenting issue for the correspondence was how to define the subjective and the objective. De Morgan had no trouble understanding the logical use of these terms, where the term "subjective" referred to the subject of a proposition like "All men are mortal" while the term "objective" referred to its object. In this format the subject "men" was the most concrete and essentially real part of the proposition, while the object, "mortal," existed only in relation to the subject. De Morgan was completely comfortable with these meanings, but as he was drawing together his thoughts for the introduction to *Formal Logic,* he had to contend with a completely different set.

De Morgan's challenge may be neatly located in the article "Subject, Subjective" that George Lewes wrote for the *Penny Cyclopædia.* In that article, Lewes declared his determination to introduce his countrymen to the proper modern uses of the terms "subjective" and "objective." "The Subject" Lewes explained, "is in philosophy invariably used to express the mind, soul, or personality of the thinker—the Ego. The Object is its correlative, and uniformly expresses anything or everything external to the mind; everything or anything distinct from it—the non-Ego."[65] In an effort to appear less foreign, Lewes claimed that these were the original Greek meanings of the words, which had then been muddied and confused by centuries of loose usage. Somewhat more credibly, he credited Kant with having developed these definitions in response to a century of continental adjustment to the rise of scientific thinking.

De Morgan was in no position to the question the validity of Lewes's interpretation. "I stop at Kant, whom I spell with a c and an apostrophe: I c'an't get through him," he quipped.[66] But he did have to come to grips with the meanings of "subjective" and "objective" that the younger man was insisting upon. The problem he faced was fundamental. The new meanings were not only logically strange, the way they divided the world was so alien as to be incomprehensible. "Is it likely that the ordinary antitheses of language should express an antithesis which people in general never think of?" he whined.[67] De Morgan found it so difficult to distinguish between ideas and experiences in this way that he was "always obliged to find out the meanings of these words afresh" every time he encountered them.[68] After several days of struggle, De Morgan decided that the "ideal and objective is the important distinction in practice."[69] Replacing the ego-limited term "subjective" with "ideal" restored the value of the experience of communion that marked moments of true understanding.

Whewell had no trouble understanding De Morgan's resistance to dividing knowledge into the subjective and objective. By the time De Morgan was asking him about the definitions of these terms, he had come to see that some form of this split was a precondition of the philosophy that was coming into England from abroad. In response to De Morgan's questions, Whewell had identified a whole list of oppositions—form vs. matter, theory vs. fact, necessary vs. contingent, subjective vs. objective—as "fundamental antitheses" that lay at the base of continental philosophy. All of these distinctions made Whewell as uncomfortable as they did De Morgan. Although continental philosophers considered "these elements of knowledge separately, they cannot really be separated," Whewell insisted.[70] He admitted, "Knowledge requires ideas. Reality requires things. Ideas and things coexist." Nonetheless, he continued, "Truth," which entails merging these oppositions, "*is* and is known." At a loss as to how to make sense of this contradictory situation, Whewell was forced to conclude that "the complete explanation of these points appears to be beyond our reach."[71] Thus, even as Whewell tried to help De Morgan understand the modern meanings of subjective and objective, the two men agreed that they were not helpful. Participating in the conversations of continental philosophers might require accepting these divisions, but in their English world, real understanding transcended the antitheses of form and matter, subjective and objective, ideas and things, words and meanings.

De Morgan devoted the introduction of *Formal Logic* to the explanation that Whewell feared was "beyond our reach." His starting point was the world described by Berkeley, in which all of our experience is beamed directly into our minds by God. This formulation rendered Hamilton's distinction between laws of thought and laws of matter meaningless. Doing so opened other questions, however. Particularly pressing was the question of how to distinguish between the real and the imaginary, which had been such a concern for Sophia and two-year-old Alice. In the case of his daughter, the solution seems to have come with maturity, but for his introduction he had to lay out a reasoned argument.

De Morgan began with the clear and indisputable recognition of his own existence. He then recognized the independent existence of other "sentients" who acted in ways that confirmed they were both similar to and independent from him. The existence of a reality separate from his own imaginings arose from communications with these other sentient beings. "Different minds re-

Figure 10: Augustus De Morgan. Picture letter. MS 913/A/3. By permission of the Senate House Library, University of London.

ceive impressions at the same time, which their power of communication enables them to know are similar," he explained. From these communications,
people are impelled to conclude that "there must be a *somewhat* independent
of those minds, which thus acts upon them all at once, and without any choice
of their own. This *somewhat* is what we call an external object."[72] De Morgan
thus located external reality itself in our communications with other people,
and as he did so he described a world in which "the true knowledge of words
will be the true knowledge of things."[73] Berkeleyan philosophy returned him
to the world of Lindsey and Frend, in which the essential reality of our existence is located in the language we use to talk about it. For De Morgan, then,
to study formal logic was to engage directly with the divine being who was
constantly engaged in communicating the world to his reasoning creatures.

De Morgan did not like to acknowledge that his move into Berkeleyan
idealism fundamentally changed the grounds of his understanding: "This *somewhat* is what we call an external object: and whether it arise in Berkeley's mode,
or in any other, matters nothing to us here," he insisted.[74] There is, nonetheless,
a significant difference between a Newtonian world in which external reality
is made up of objects in an absolute space and a Berkeleyan one in which that
reality is directly beamed to our minds from the mind of God. De Morgan
did acknowledge the reality of this difference in private: "I am a searcher after
things mental—not material," he proudly proclaimed, a man for whom "the interpretation of $\Sigma (x_m y_{m+1} - y_m x_{m+1})$ is practical—while the progress of electricity
through a wire is comparatively theoretical."[75] For De Morgan, real knowledge
was always ideal. Everything experienced was essentially secondhand.

As De Morgan tried to come to grips with the nature of logic and of
reason, he remained always fascinated by the children who were growing up
around him. In the Camden Street house, Sophia negotiated the ups and
downs of their everyday lives, but in the evenings Augustus focused on his
children's developing minds. He composed hieroglyphics to engage their interpretative skills (figure 10), and his letters to his friend, the Irish Hamilton
are filled with other ways to explore their developing minds. "Take a child and
say 'now we are going to draw a house,'" De Morgan directed, but then draw
one in which the chimney is hugely out of proportion. When the child points
out "that chimney is *too big*," he exulted, "the remark was dictated by the presence and action of the notion of relative magnitude."[76] Sophia no longer had
the energy to keep a nursery journal, but she too remained focused on cultivating her children's reasoning powers. Whately's call to rear young seedlings
of reason and logic reverberated through every room of the De Morgan home.

———◆◆◆———

# CHAPTER 19

# *Expanding Reason*

Through all these developments, the family continued to expand. In 1847, the De Morgans named their third daughter Helena Christiana after Augustus's grandmother, and in 1850, their final daughter, Mary Augusta, was born. Now seven children, all bound tightly into their parents' world by their names, filled every nook and cranny of the Camden Street house. A new generation of reason was rising.

Their parents were coming into their own as well. De Morgan continued to serve as honorary secretary of the RAS, which entailed faithfully attending meetings, publishing the society's *Notices,* and periodically trying to corral people to serve as officers. But Baily's death in 1844 marked the definitive end of the sparkling parties at Tavistock Place, and the merry social world in which the De Morgans first noticed each other was being replaced by a different one. At the new group's core lay De Morgan's fellow UCL professors and their wives. By the 1840s, slights against their "godless institution" were becoming more muted, but those who gathered in Camden Street were always proudly conscious of their status as outsiders in their still predominantly Anglican country.[1]

Challenging the pretensions of the Royal Society remained one of De Morgan's favorite pastimes. The society had become less overtly aristocratic since Banks's rule, but De Morgan always saw it as the established church of the scientific world, which worshiped Newton as its founding saint. In 1846

De Morgan composed a forty-page biography that showed Newton ruling the Royal Society as a king. "And never did monarch find more obsequious subjects,"[2] he scolded as he considered the ways Newton and his Royal Society henchmen treated Flamsteed. Even more reprehensible was the way its members had falsified records to ensure that Newton did not have to share the honor of having discovered calculus with Leibniz.[3] De Morgan was surprised that the Royal Society actually published two papers laying out his case—and basically relieved when they refused to publish the third so he could hold it up as evidence of their intransigence.[4] He was happiest as an outsider who championed reason in the face of a world that refused to honor it.

One evening, as De Morgan was thinking about the nature of mathematics, logic, and reason, he called Alice into his library to explore the ways she understood the world. His oldest daughter was by this time a fiery eight-year-old. Her father opened the conversation by asking her whether she could imagine stones defying the laws of gravity at the North Pole "where no one has been." She immediately countered with the information that people *had* been there since they had seen the natives "kissing with their noses." When he clarified that he meant at the actual North Pole, where no one had been, she countered that it didn't matter, since "there is something in the ground that draws the stones." When her father persisted, Alice was at first adamant: "No it's impossible. Perhaps the birds might take them up in their beaks, but even then they wouldn't go up of themselves." Under pressure she finally blurted out: "Oh yes, I can fancy three thousand of them going up if you like, and talking to each other too, but it's an impossible thing, I know." Now convinced that his daughter *could* imagine rocks defying gravity, Augustus switched to mathematical knowing, and asked her whether she could imagine seven and three making twelve at the pole. This time she paused for a moment before answering, "No, I don't think I can. No, it can't be; there aren't enough," and her father was finally satisfied.[5] Bursting with pride, he transcribed the conversation to show Whewell how clearly his daughter felt the essential difference between contingent and necessary truth.

Alice was still too young to go to school when Augustus was quizzing her about the physics of the North Pole. Her parents would send her brothers to the University College School, which was the high school counterpart to UCL, but girls could not attend. De Morgan had previously "held man-like and masterful views of women's powers and privileges,"[6] and he always liked pointing out that "when we overcome a difficulty we say we *master* it, but if we fail we say we *miss* it,"[7] but living with Alice changed him. As he watched his daughter grow into a reasoning being, he agreed with Sophia that she, as well as her younger sisters, deserved a school that would give them all the opportunities their brothers had to develop their reason to its fullest.

The De Morgans were not alone in wanting a school where girls could have "full scope and opportunity for the exercise of all of their faculties."[8] In 1848, the professor of literature at King's College, Frederick Denison Maurice, had succeeded in establishing Queen's College for girls. Alice was just ten at the time, but her parents did not want to send her there. Although the school made a point of admitting all girls, regardless of their religious affiliations, it was nonetheless an Anglican institution under the direction of men. The De Morgans threw themselves into creating an alternative, secular school in which women would hold the power.

The focusing force behind the formation of the Ladies' College in Bedford Square was a wealthy widow, Elizabeth Reid, whose parents had been among the original members of Lindsey's Essex Street Chapel. Reid liked to say that "a college for Women, or something like it, has been my dream from childhood,"[9] but it wasn't until Alice turned eleven that Sophia joined with a determined group of parents to make Mrs. Reid's dream a reality. Sophia and Augustus wrote out the prospectus and announcements for the new college, Lady Byron offered a list of patrons, and the De Morgans managed to rent a house in Bedford Square that was close enough to UCL that professors could walk to it. Caught out of town on family business, Mrs. Reid approved of their efforts from afar, writing the necessary checks while sending "love to my little friend Alice."[10] In October 1849, Alice was one of the first students at the Ladies' College in Bedford Square, which later became the Bedford College for Women (figure 11).[11]

One of the most striking characteristics of the new institution was the feminist nature of the group that created and supported it. Most were like Sophia in having very full lives as wives and mothers; their letters are filled with difficult pregnancies, whooping cough among their children, and, in the case of Sophia, an otherwise undesignated "serious illness."[12] They forged ahead nonetheless.

The insistence on female participation was played out not only among the founding lights, but also in the composition of the governing boards of the new school. In De Morgan's original prospectus, college governance was to be divided between a Committee of Ladies, who were in charge of arrangements, "which have reference to the conduct and convenience of the Pupils,"[13] and the professors, who were in charge of instruction. When the

Figure 11: Sophia and Alice De Morgan. RHC AR Bedford College Collection.
Archives, Royal Holloway, University of London.

school opened, these two groups were joined by a group of College Visitors, women who agreed to attend every class, since the girls could never be left alone with a male professor. A masculine board of trustees included Charles Darwin's brother Erasmus Darwin, his brother-in-law Hensleigh Wedgwood, and the progressive T. H. Farrer, but this group of men were primarily useful as names. As the subcommittee charged with appointing them put it: "Let the trustees be mere Treasurers, without discretion concerning the application of funds."[14] The administering of the Ladies' College in Bedford Square was determinedly female.

The faculty, however, was not. Despite some efforts to find women to teach in the school, the professors were overwhelming male, and at least in the first year, very impressively credentialed. In addition to De Morgan, who taught mathematics, the initial roster included A. J. Scott, professor of English literature at UCL, who taught English literature and moral philosophy; F. W. Newman, professor of Latin at UCL, who taught ancient history; and W. B. Carpenter, an examiner at the University of London, who taught natural science. Academic ambitions were high, and in the first year there were more students enrolled in moral philosophy, mathematics, and English literature than in the more traditional female subjects of French, Italian, and vocal music. The pupils, though, were a young and unformed group of girls who could not hope to rise to the level of their teachers. After a couple of terms teaching their subjects to twelve-year-olds, most of the faculty, including De Morgan, stopped teaching there, leaving the school struggling to find replacements.

Finding teachers was just one of the many challenges that faced the fledgling institution. None of the women founders had attended a school like the one they envisaged, and neither the men nor the women had any experience working across gender lines. Rickety though the organization often was, it worked well enough that after just two years there, Alice and her classmates had moved far enough beyond basic arithmetic to be struggling with logarithms.[15] There is no evidence that Alice's younger sisters enrolled in the school, but the connection to the family remained strong enough that Joan Antrobus, Augustus's and Sophia's great-granddaughter, proudly traveled from South Africa to attend for the year 1925–26. In 1985, the school merged with Royal Holloway College, where the De Morgans' vision of a school that gave girls the educational opportunities accorded to boys persists to this day.

Soon after Alice started school, her parents were drawn into the cause of abolition. Fighting against slavery was a long family tradition; one of the documents passed through the generations was a piece of anti-slavery legislation that Frend introduced into the Cambridge Senate in 1792. In 1852 the issue gripped the country again when Harriet Beecher Stowe's *Uncle Tom's Cabin,* exploded on the English scene. The book was a runaway best-seller in the United States, where it sold 300,000 copies in its first year; in England it sold more than three times that number.[16]

Sophia was so moved by Stowe's book that she vowed to do everything in her power to bring an end to the institution of slavery. In fall 1852, she drafted a letter to be signed by the people of England urging the people of America to give up their slaves. In her piece, she explicitly recognized that the English shared the blame for slavery since they were the ones who established the system at a time when "Americans were not under their own laws and legislature." Now, however, "uninfluenced by those personal interests which involve and obscure the question on its own soil," she wrote, the English had a clearer view of the pernicious effects of slavery than did those who were caught up in it. Having then laid out what she saw as the horrors of slavery, Sophia closed with the hope that God "will bring to your hearts a conviction of its enormity, & give you strength to abjure it."[17] In her attempt to address the problem through a combination of rational argument and theistic conviction, she showed herself to be her father's daughter, and like him her goal was political change. In this case, that meant bringing the letter to a larger audience.

Sophia's idea quickly caught fire among her friends. Within weeks of her penning the initial draft, Lady Byron was negotiating details of wording with the Boston abolitionist, Eliza Cabot Follen, who was then living in London.[18] At the same time, with fourteen-year-old Alice acting as amanuensis, Augustus and the Reverend J. S. Nicolay from King's College met to transform Sophia's discursive letter into a series of crisp bullet points.[19] They were not the only ones at work, however. One of the first people with whom Sophia had shared her idea was Rachel Chadwick, the wife of Edwin Chadwick, the English sanitation engineer who devoted his life to raising the standards of living of the poor. Edwin was very supportive, and in the week before De Morgan's and Nicolay's Sunday meeting he told the great reformer, Lord Shaftesbury, about the plan. This turned out to be a serious tactical error. Shaftesbury took the idea for himself, and on the same day that De Morgan and Nicolay were crafting their bullet-pointed statement, the lord was sending his version as a letter to *The Times*. When it appeared on Wednesday, November 9, a sheepishly apologetic Chadwick sent a copy to Sophia with an explanation of his part in the process.

Shaftesbury initially assumed that Sophia and Rachel would head the committee to collect signatures for his letter, but they would have none of it. It was not just that he had taken over their project. In doing so he had fundamentally changed it. Their letter was from the people of England to the people of America, whereas his was from the women of England to the women of America. Sophia and Rachel were incensed that he had deprived their project "of that dignity and weight which it would have as coming from 'Englishmen and Englishwomen.'"[20] Nicolay agreed that Shaftesbury's all-female version would not "succor the cause nor America,"[21] but he also knew that the lord was far too "vain" to listen to their objections.[22] In this, Nicolay was right. When Sophia and Rachel refused to cooperate, Shaftsebury simply turned the project over to the Duchess of Sutherland, and England's female glitterati flocked to the cause.

The result of their effort was extraordinary. A veritable army of women collected signatures block by block in all of England's major cities and town by town in the countryside. Single sheets began with the polished signatures of ladies of the manor and moved through the cruder writing of their personal servants to conclude with the scrawled Xs of scullery maids. Signatures flooded in from Australia, Canada, New Zealand, and Palestine. By the end of six months, the duchess's committee had collected enough signed petitions to fill twenty-six folio volumes, and it arranged to bringing Stowe to England in order to receive them.[23] The celebrated author crossed the Atlantic in spring 1853 and was rapturously received by the English public. After a couple of weeks traveling through cheering crowds in Scotland and northern England, she made her way to London, where she was feted at a dinner which included the city's mayor and Charles Dickens. Sophia's abolitionist cause had swept over England like a wave.

Sophia and Rachel were emphatically absent from the crowd of dukes and duchesses, lords, ladies, and archbishops who gathered in the Duchess of Sutherland's palatial residence, Stafford House, on May 7, 1853, to turn the bound volumes of signatures over the Stowe. But their predictions of the petitions' failure proved all too true. Stowe was totally unprepared to wield institutional power, and nothing came of her hopes to organize an American group to spread the abolitionist message. Sophia judged that other than pleasing some abolitionists and irritating some slavery proponents, the letter she had initiated "had no effect at all."[24] Subsequent historians have been equally unable to decide on a response, but it remains difficult not to be affected by the outpouring of concern reflected in the signatures of thousands upon thousands of women who had no other way to express their political commitments.[25]

Even as Sophia regretted the fate of her petition, she remained deeply committed to the cause of abolition. She met Stowe several times at Mrs. Reid's house, but at those gatherings she found herself even more interested in Mary Webb, a woman of mixed racial parentage—in Sophia's words "her mother was half-blood, and her father a Spanish general of good family"[26]— whom Stowe had met on the boat coming to England. Webb was well known for her dramatic readings and so impressed Stowe that she wrote a version of

*Uncle Tom's Cabin* for the talented woman to perform. Sophia found talking to Webb to be conclusive evidence against those who believed in the impossibility of "all social intermixture of the whites and blacks."[27] For her, all of the drama that surrounded Stowe's English visit was eclipsed by the experience of actually meeting a black woman.

Even as she supported abolition, Sophia's investigations of mind were gathering steam. Over the course of the 1840s, her interest in mesmerism expanded beyond experimentation at home and in the neighborhood to possible applications in her work with the insane.[28] Augustus remained cautious. When Hamilton's daughter received chloroform for an operation on her eye, De Morgan recognized that when it "can be ventured upon it is more speedy than mesmerism, which requires some previous trials."[29] When it came to making decisions about himself, though, his Berkelean conviction that mental phenomena were more real than material ones supported a willingness to accept mesmeric approaches. While admitting that mesmerism might not be as reliably effective as were other forms of medicine, both De Morgans continued to embrace its healing potential.

As Sophia continued her experiments, she began to see that the powers she had first observed in a medical context might extend to other spheres. In 1849, when she had induced a mesmeric trance in an effort to treat "fits" in a "young and ignorant girl," she found herself a startled witness to "the state of clairvoyance." While under the mesmeric influence, the girl talked Sophia through the streets of London to a house where she observed with minute detail the room in which Augustus was visiting with one of his friends. Although she never left her chair, the girl's descriptions of the house, the room, and the conversation within it were so complete, detailed, and accurate that both Sophia and Augustus were completely convinced that she had made an actual "mental" journey to the place she described.[30] That this unschooled girl had the power to see reality while traveling in thought was powerful support for the existence of a transcendent world of mind that lay behind the material one.

The De Morgans were far from alone in their continued interest in the powers of mind. Michael Faraday's experimental researches into electricity, light, and magnetism were feeding hopes that people could somehow interact with a previously unrecognized world of invisible forces. In 1851, fascination with the effects of hitherto unrecognized physical phenomena exploded in what contemporaries called an "epidemic" of table-turning. Convinced that bodies carried electrical forces, England's middle classes began joining hands on the tops of tables in hopes that by focusing their forces they could actually move the tables. Opinions were divided on the reality of the effects they described, but such disagreements served only to make the experience more exciting. By the summer of 1853, the English interest in table-turning had reached such a pitch that Faraday pushed back with letters to *The Times* and the *Athenaeum* in which he described a mechanism that would allow table-turners to check their experiences against an external reality.[31] His intent was to put an end to the whole discussion, but that was not the result. Instead of discouraging people, he succeeded in legitimizing table-turning as something to be investigated.[32]

All of the De Morgans' fury at having been patronized by the Duke of Sutherland erupted in response to Faraday's efforts to dictate the terms of investigations into the world of invisible forces. De Morgan sent off a "strong critique" to the *Athenaeum,* while Lady Byron huffed that even if Faraday was right, and the force at work when people observed tables turning is "a Self-deceptive Wish," that wish was itself worth investigating.[33] Many others also resisted Faraday's interference. The editors of the *Athenaeum* were so overwhelmed by "the flood of communication" Faraday's letter unleashed that they printed none of the replies, concluding that the whole question of table-turning "will soon need a journal to itself."[34] The gentlemen of science's efforts to invite the public into their work was showing its downside. Participation was a way to garner support, but controlling the conclusions of independent investigators like Sophia proved to be very difficult.

The problem was not confined to table-turning. In 1853, an American spiritualist, Mrs. Hayden, burst onto the London scene. This formidable woman was not content to watch passively as tables turned or tipped. She was interested in moving past impersonal forces to a world of spirits that could use those forces to communicate with people. This project was very attractive to Sophia, who had for decades been convinced that the dead, including particularly Harriet and her father, still lived in some otherworldly realm. Lady Byron was unimpressed by her first visit to Mrs. Hayden, and Sophia wanted to believe herself skeptical, but her defenses began to weaken when, within the first hour, Mrs. Hayden delivered the message that Harriet was "happy." When, on her second visit, her father tapped out *"Why do you doubt the holy attributes of God, when this is in perfect accordance with His teaching?"* Sophia was so impressed that she invited Mrs. Hayden to Camden Street so that Augustus could meet her.[35] He began with a test in which he asked questions "mentally," that is, *"in my mind,* without speaking," and was deeply impressed when rapping spirits gave him the right answers. After that, Mrs. Hayden made frequent visits to the De Morgans' house, where many others had similar experiences. Sophia's interactions with Mrs. Hayden led to an investigation of the spirit world that would occupy her for the rest of her life.

---

De Morgan declared himself satisfied with "the reality of the phenomenon,"[36] but his primary interest remained logic. When *Formal Logic* was published, he had sent a copy to the Scottish Hamilton ostensibly as a courtesy but probably also as a challenge. When the philosopher returned the package unopened, De Morgan just grinned and kept the book as a memento and filing cabinet for all of his thoughts and correspondence on logic. When the book overflowed completely with these interleavings, he tore it apart and transformed what had begun as a cheaply bound book into a handsome two-volume work, bound in red leather. In this form, De Morgan's personal copy of *Formal Logic,* now in the archives of the University of London, stands as a record of decades of his thinking about the subject.

De Morgan had company in his pursuit of logic. Whewell remained an ally who shepherded his articles into the *Transactions of the Cambridge Philosophical Society,* but De Morgan's most immediate intellectual companion was the young schoolteacher George Boole. The son of a Lincoln shoemaker, Boole began teaching at the age of sixteen to support his family. For years he climbed through the ranks, from monitor to head of school, while learning mathematics from the books in the local Mechanics' Institute. The enterprising young man emerged from his self-education deeply interested in algebra. He published his first mathematical paper in the *Cambridge Mathematical Journal* in 1841, and in 1844 was awarded the gold medal of the Royal Society for a different paper published in the *Transactions of the Royal Society.* Despite these triumphs, Boole's lack of traditional education meant that he was always an outsider to the world of Cambridge mathematics. In 1842, when he was the headmaster of a school in Lincoln, he introduced himself to De Morgan with a letter praising his *The Differential and Integral Calculus.* This effort at contact succeeded and their correspondence continued for decades.

Most of Boole's and De Morgan's interactions were in the form of letters, but in January 1847 Boole actually visited De Morgan in his Camden Street library. At the time, both of them were shifting their foci away from algebra per se in an effort to bring the symbolic power of algebra into logic. It would seem the perfect moment for collaboration, but De Morgan's experience with

Hamilton made him wary. As soon as he realized that he and Boole were working on the same challenge, he temporarily cut off their correspondence. "I would much rather not see your investigations till my own are quite finished," he cautioned in May 1847.[37]

De Morgan's skittishness meant that after May the two men worked independently until Boole's *Mathematical Analysis of Logic* and De Morgan's *Formal Logic* were published within weeks of each other that fall. After reading what Boole had published, De Morgan declared himself glad that they had not corresponded, because "some of our ideas run so near together that proof of the physical impossibility of either of us seeing the other's work" could well be important to protecting any claims of priority.[38] From a distance of more than 150 years, however, his concerns seem unfounded. In fact, the differences between De Morgan and Boole were so fundamental that one frustrated modern logician has labeled the Boole-De Morgan letters a "correspondence without communication."[39] For years the two sent their ideas to each other, but it was a rather solipsistic exercise in which neither ever really responded to what the other was doing.

The chasm between De Morgan's and Boole's approaches reflected the difference in their educational experiences. The mathematics De Morgan had learned at Cambridge, which was reinforced by Peacock's progressive historical view of algebra, was a model of meaningful reason. Boole, who had picked up algebra on his own, saw the subject differently. From his point of view, Peacock's insight was that algebraic symbols were in and of themselves devoid of meaning, that all of their value derived from their place in a larger consistent system. From this perspective, bringing the power of algebra to logic entailed embedding logic into the algebra by assigning logical meanings to algebraic symbols. So, for example, Boole used "$x + y$" to mean "$x$ or $y$"; "$xy$" to signify "$x$ and $y$"; and "$1 - x$" to indicate the complement of $x$.[40] In this way, he poised himself to use the relations of algebra to open up logic. De Morgan had no problem understanding what Boole was doing, but it did not fit into his program. From De Morgan's perspective, Peacock's essential insight lay in attributing the power of algebraic symbols to the long history that gave them meaning. In the case of logic, that it had a different history than algebra, meant that it needed different symbols.[41]

De Morgan took the first step toward developing a properly logical set of symbols in an 1850 article entitled "On the Symbols of Logic, the Theory of the Syllogism and in Particular of the Copula" ("On the Syllogism II"). He had already in his first logic paper and in *Formal Logic* introduced various shorthand symbols, "without which I should hardly have had patience for the many hundreds of cases which this paper has required."[42] However, each of the symbols he there used was as specific to its situation as arithmetic symbols were to theirs. In "On the Syllogism II" he recognized that if logic were to progress, it needed a single consistent set of symbols that could open it up to expansion the way algebraic symbols had opened arithmetic. But not just any set of general symbols had the power to push a subject forward. The key to progressive symbols was that they be functional, that is, they could be manipulated according to fixed rules to give valid results.

Creating a functional set of symbols required considerable effort. De Morgan's first step was to make a basic adjustment to the standard Aristotelian propositional form of subject and predicate, an adjustment known as "quantifying the predicate." In logic, "quantifying the predicate" means just what it says, that is, expanding the basic form of the Aristotelian proposition by adding a quantifier to the predicate (table 6).

TABLE 6

| Quantifier | Subject | Copula | Quantifier | Predicate |
|---|---|---|---|---|
| all (every) | | | all (every) | |
| some | **X** | is | some | **Y** |

Translating this logical language into everyday English can be clumsy, but it carries significant logical advantages. First, quantifying the predicate removes some real ambiguities in the Aristotelian system. So, for example, quantifying the predicate of the proposition "All men are rational beings," forces a decision between "All men are some rational beings," a construction that recognizes the existence of other rational beings, or "All men are all rational beings," which asserts that men constitute all rational beings. Awkward though such constructions may be, they introduce a higher level of precision into the traditional Aristotelian proposition. Second, and at least equally important to De Morgan's project, quantifying the predicate makes it possible to switch the order of the subject and predicate without changing the meaning of the proposition: "every **Y** is every **X**" means the same thing as "every **X** is every **Y**"; "every **Y** is some **X**" means the same thing as "some **X** are every **Y**." That propositions were "commutative" in this way was essential to his efforts to create a functional symbolic system.[43] Quantifying the predicate thus promised him both the precision and the flexibility he needed to create a functional symbolic system.

De Morgan's efforts to create a symbolic logic began by using parentheses to denote quantifiers. An enclosing parenthesis, as in **(X**, meant "every **X**"; an excluding parentheses, as in **)Y**, meant "some **Y**." At the level of propositions, he pulled the parentheses to the middle, so that **X)(Y** meant "every **X** is every **Y**"; **X))Y** meant "every **X** is some **Y**." But after weeks writing propositions this way, he saw a way to use the commutative properties of the quantified predicate to avoid the labor of writing **X** and **Y** all the time. Since it is always possible to change the order of variables in a proposition, De Morgan saw that he could arrange the **X**s, **Y**s, and **Z**s so that they would proceed in alphabetical order through the syllogism. For the writing-weary De Morgan, this meant that as long as he followed the convention of keeping the letters in alphabetical order, he could just leave them out. The result is a system in which **)(** meant "every **X** is every **Y**," and **))** meant "every **X** is some **Y**." On the level of syllogisms, "**(( ))** therefore **( )**" meant "some **X** are all **Y**; all **Y** are [some] **Z**; therefore, some **X** are [some] **Z**." De Morgan's Hamiltonian detractors scoffed at his clouds of parentheses, which they called "spicular," from the Latin *spica*, "ear of grain,"[44] but De Morgan just grinned and adopted the term himself. The parentheses that they saw as grains of wheat were for him a carefully crafted set of symbols that revealed the basic flow of the syllogism.

The power of De Morgan's symbols may be seen in their portrayal of what in Aristotelian logic is known as the "middle term." In any syllogism, like the one above, the first two propositions state the relationship between one term—**Y**—and two other terms—**X** and **Z**. Then, in the conclusion, the middle term, **Y**, disappears, leaving a new relationship between **X** and **Z**. De Morgan's spicular notation neatly captures this dynamic. In the example above, getting from the first two propositions—represented as **(( ))** to the conclusion—represented as **( )**—involved simply dropping out the middle two parentheses, which represented the quantifiers of the middle term, **Y**. This procedure, which De Morgan dubbed the "erasure rule," is precisely the kind of functional manipulation that he was reaching for as he developed his symbology.[45]

By the end of 1849, De Morgan was so proficient with his spicular notation that for thorny questions he found it easier to begin "with the symbol in my head" than to think the problem through.[46] Six months later, he was confident enough to go public. In March 1850 he sent "On the Symbols of Logic" to Whewell for publication in the *Transactions of the Cambridge Philosophical*

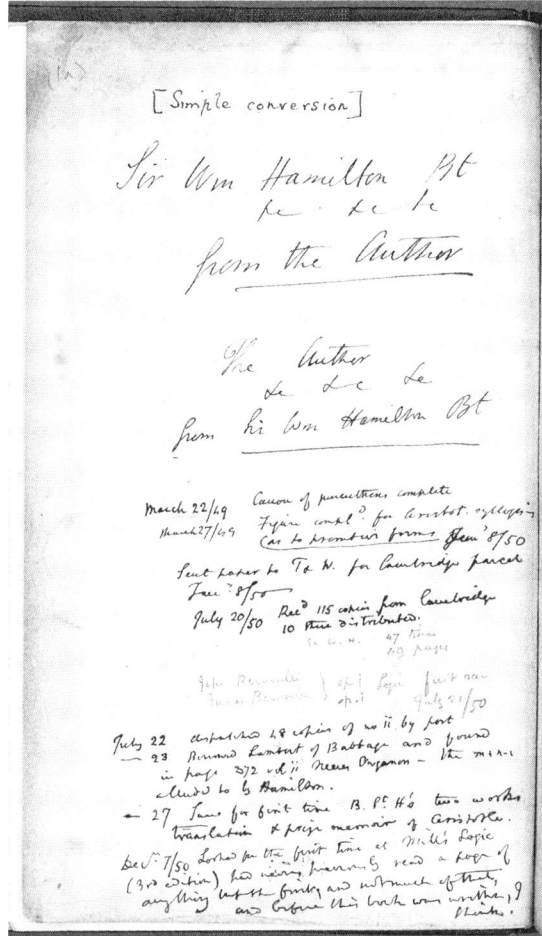

Figure 12: Flyleaf of Augustus De Morgan's personal copy
of *Formal Logic, or, The Calculus of Inference, Necessary and
Probable.* MS 776/1–2. By permission of the Senate House
Library, University of London.

*Society.* Then, on the flyleaf of his personal copy of *Formal Logic,* he carefully
traced its progress toward publication (figure 12).

Jun' 8/50 Sent paper to T & W[hewell] for Cambridge parcel
July 20/50 Rec^d 115 copies from Cambridge
10 then distributed
July 22 Dispatched 48 copies … by post.

From De Morgan's Peacockean point of view, successfully creating a functional logical symbology was a tremendously important accomplishment.

De Morgan's experience with the Scottish Hamilton had rendered him ever cautious about plagiarism. Therefore, as he was breathlessly moving toward publication, he carefully sent Boole each of the proof sheets to be sure he had not unwittingly borrowed one of Boole's ideas. Boole's only response was a letter asking whether the London man had heard of "any situation in England that would be likely to suit me."[47] Boole's obtuseness in the face of De Morgan's obvious excitement is an excellent example of the ways the two men for years passed each other like ships in the night.

As De Morgan and Boole were corresponding without communicating, he and Hamilton's followers were communicating without corresponding. Hamilton remained ill, but the Oxford theologian Henry Longueville Mansel took up the cause, and within a year of the publication of "On the Syllogism II" wrote a highly critical review of De Morgan's logical work. It would be impossible to overlook the "striking fact of a considerable amount of revived interest in the study" of logic, Mansel began, before turning his attention to "the important question, how that interest may best be controlled and directed."[48] More than two decades after Hamilton's first attack on Whately, Mansel was continuing his crusade to prevent logic from entering areas where he claimed it didn't belong.

For Mansel, De Morgan was the epitome of someone who was brazenly allowing logic to run rampant. For him, De Morgan's determination to expand syllogistic logic until it covered "all that is called reasoning"[49] represented a dangerous attempt to allow human rationality to infringe on areas beyond its rightful purview. Mansel's primary concern was that the sanctity of theology be protected from the dangerous inroads of reason. The very concept of revealed religion required that there be limits to human reason, and he saw De Morgan's logic as a fundamental threat to sacred epistemological boundaries. Mansel freely acknowledged that by limiting logic he was limiting the human rational capacity, but he was happy to defend this position. "It may be humiliating," he noted, "to know that man's powers are thus restricted; but the restriction is one which his Maker has thought fit to impose on him, and, regret it as he may, he cannot escape from it."[50] For Mansel the function of logic was to stand firm at the boundary of reason in order to show "clearly the nature of the pure laws of understanding, and the exact limits within which they are operative."[51] Beyond these limits lay great fields of meaning, including not only revealed theology but mathematics and physics as well, which logic could not properly touch.

Mansel could not really object to De Morgan's decision to quantify the predicate, because Hamilton had done so as well. He did, however, object to virtually every other one of the Londoner's innovations. De Morgan's "whole theory of a material universe, with its positive contraries is extralogical," Mansel insisted; "it is not by logic that we learn that real and personal fill up the universe of property,"[52] or that *Briton* and *alien* fill up the universe of nationalities. Mansel thus refused to accept the logical legitimacy of the relationship of complementarity that De Morgan illustrated by these examples. He was adamant that a truly formal logic could not be based on anything so clearly material.

Mansel also challenged another aspect of De Morgan's work. In traditional Aristotelian logic, quantifiers included terms like "some," "every," "all," or "many," but these are not the only quantifiers to be found in English. Hamilton had recognized "most" as a legitimate logical quantifier, which opens up a whole new set of possibilities: "Most **X**s are some **Y**s. Most **Y**s are some **Z**s. Therefore some **X**s are both **Y**s and **Z**s." De Morgan went further to consider syllogisms like "All **X**s are three-quarters of the **Y**s. Half of the **Y**s are all of **Z**s. Therefore, at least one-quarter **X** is some **Z**." Mansel was appalled by what he saw as an illegitimate effort to sneak mathematics into logic. He warned, once he concedes "that as a logician he is bound to know that two and two make four," the floodgates would be open, and "there is no art or science, knowledge or device that he would not be required to include in his system."[53] Mansel was willing to allow the quantifier "most" into logic because Hamilton had done so, but he drew the line at "one-third" or "three-quarters." For him, De Morgan's explorations of numerical quantifiers constituted a travesty of logic.

Behind all of these objections lay Mansel's indignation at the ways De Morgan's search for a logic adequate to human reason violated the fundamental distinction between form and matter, thought and things. He used Hegel as a straw man to describe someone for whom "Thought and Being become one and the same: the reasoning process is a continual creation of the universe; and Logic, the science of pure thinking is, at the same time a revelation of the whole mystery of existence."[54] De Morgan was emphatically not a follower of any German philosopher, but Mansel's depiction was actually a rather insightful description of De Morgan's Berkelean point of view. Boole may have been De Morgan's logical ally, but classically educated Mansel and De Morgan understood each other in ways that the autodidact never really did.

De Morgan's vow not to engage with Hamiltonians until after the old man died meant that he did not publicly answer Mansel. He could respond in private, however, and over the course of several days in February 1853, he returned to his youthful practice of "drawing mathematics."[55] This time his focus was logic, and he carefully crafted a series of diagrams which revealed the powerful symmetrical structures that supported his logical work. He never published these images, but they were essential expressions of the logical vision he was pursuing, and he carefully kept them, bound into his personal copy of *Formal Logic.*

The diagrams that De Morgan drew by and for himself grew out of the medieval tradition of the Aristotlean square. He called the first "Table of Relations of Propositions." Like its medieval forbearers the image is a square, with propositional forms, represented by De Morgan's parentheses, on its corners. However, since his system has twice as many propositions as Aristotle's, propositions also appear at the midpoints of each of the square's sides joined by a wealth of connecting lines—some solid, some dotted, some dotted and crossed—to indicate the relations among them. De Morgan carefully adjusted the angles in these lines to create a star-like image (figure 13).

De Morgan was not satisfied with his star, so he tried again, this time with an octagonal diagram. He was much happier with the way this eight-sided figure represented the configuration of propositions in the logic he was developing, and having found it, he was ready to expand his vision. He divided the eight propositions of his "Octagon of Relations" into two groups of four. Each proposition is embedded in a basket of nodal lines that connect to the syllogisms that would generate them. The four propositions that run horizontally through the center of the diagram are the conclusions of six syllogisms; those located above and below the central horizontal axes are generated by only two syllogisms (figure 14). This difference reflects the classical distinction between "universal propositions," like "All **X**s are **Y**s," and "particular propositions," like "Some **X**s are **Y**s." Their position in these diagrams may be seen as an example of De Morgan's program in action: what was before a distinction based on the meanings of the propositional terms becomes one based on their position in an overarching structure.

Four syllogisms lie at the right and left edges of each of the diagrammatic baskets at the bottom of figure 14. De Morgan called these eight syllogisms "strengthened syllogisms" and saw them as his most fundamental contribution to logic. These strengthened syllogisms were formed by flipping one or both of the parentheses in the middle term of a syllogism. In De Morgan's symbolic

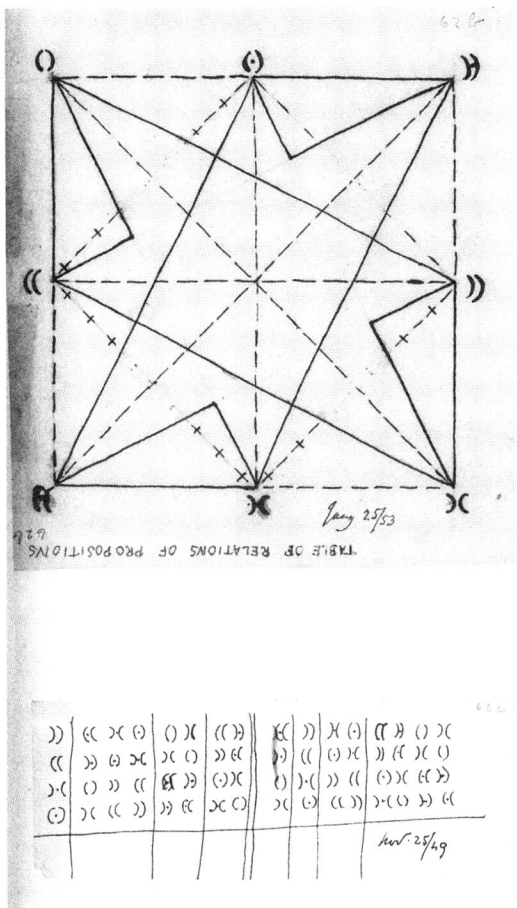

Figure 13: Table of Relations of Propositions from
Augustus De Morgan's personal copy of *Formal Logic,
or, The Calculus of Inference, Necessary and Probable.* MS
776/1–2. By permission of the Senate House Library,
University of London.

system, the validity of the erasure rule would require that this kind of flipping
make no difference in the conclusion. However, flipping parentheses in this
way can make a significant difference. So, for example, the syllogism "Every
Englishman is a tea drinker. Every tea drinker is a milk drinker. Therefore
every Englishman is a milk drinker." is clearly valid, but "Every Englishman is
a tea drinker. Some tea drinkers are milk drinkers. Therefore every Englishman
is a milk drinker." is clearly not. As this example shows, changing the quanti-

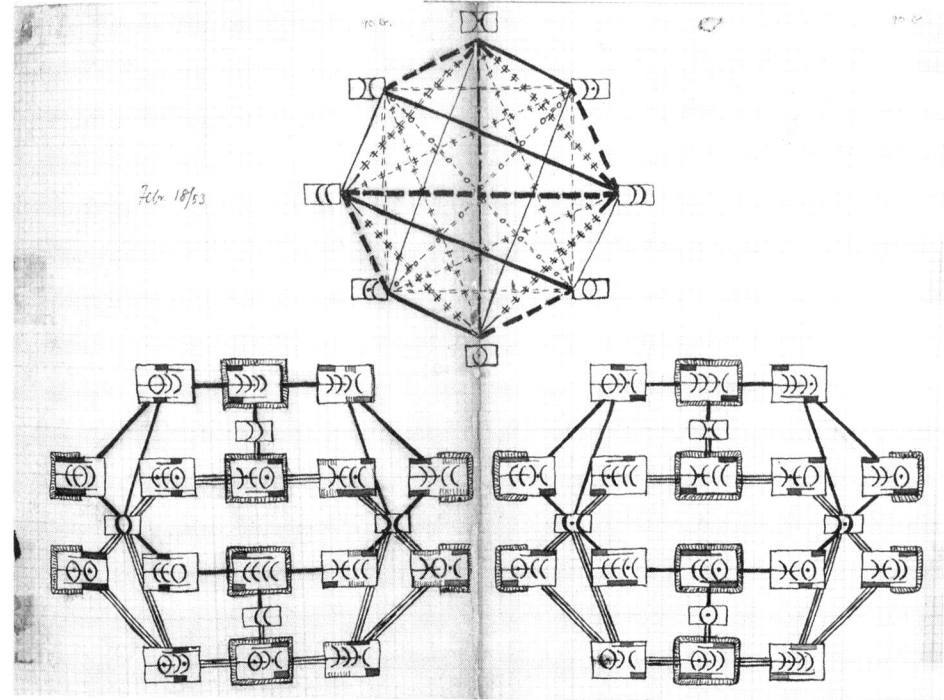

Figure 14: Logic diagram (2) from Augustus De Morgan's personal copy of
*Formal Logic, or, The Calculus of Inference, Necessary and Probable.* MS 776/1–2.
By permission of the Senate House Library, University of London.

fier of the middle term can make a real difference in the validity of a syllogistic
argument. For De Morgan this problem signaled the fundamental importance
of strengthened syllogisms. Like the negative numbers in algebra or divergent
series in analysis, they were legitimately created by the functionality of his system, but they were apparently meaningless or even false. He knew that they
were true because they were a natural product of a formal system; that they
did not appear to be so was the clear indication of a point that needed further
work. As had been the case with negative numbers and divergent series, the
challenge was to figure out how to interpret them.

   In the case of logic, De Morgan attributed the difficulties of this project
to the ways the development of the English language itself had been hampered by strictures imposed by the artificial limitations of Aristotelian logic.
"If technical forms could produce such an effect as either to supplant more

natural ones, or to prevent the natural growth of the most natural ones, the Aristotelian forms have had every advantage given them, and may have done it," he insisted.[56] De Morgan found this insight wonderfully liberating because it allowed him to accept the most convoluted English sentences as perfectly legitimate ways to interpret the results his symbols generated. Many of his interpretations stretched the syllogism to its breaking point, but he did not hesitate to explore them. Nothing could be allowed to stand in the way of his goal of finding an interpretation for the symbolic system he was lifting from the syllogism.

There are no obvious asymmetries in the diagrams at the bottom of figure 14, but De Morgan was still not satisfied. For his next effort he carefully folded a large piece of paper into a triptych. In this configuration, which can be seen as a folded image (figure 15) or an unfolded image (figure 16), the eight strengthened syllogisms are transformed into a gate that opens to a complex tapestry of thirty-two legitimate syllogisms with the eight propositional forms neatly and firmly woven into their midst. The various shadings and line constructions have specific meanings that explain the particular relations between the various syllogistic forms, while the symmetrical harmonies among them proclaim the validity of the whole.[57]

Having thus captured in a single image the entire syllogistic universe that was opened by his strengthened syllogisms, it might seem that De Morgan would have rested. But he did not. Instead, he constructed a diagram that is neither as elaborate nor as functional as were those that preceded it (figure 17). In it, his octagon of relations floats in the midst of a larger configuration of syllogisms and propositions. The initial effect of the enclosing structure may appear to be that of a three-dimensional cube, but the connecting lines undercut this impression. The oddness of this diagram may be a distortion attributable to De Morgan's monocular vision,[58] but it may also signal that the image is of a logical, not a geometrical, space. Its peculiarities might then be reflections of the significant difference that exists between a world in which there is an external reality made up of objects in an absolute space and one in which that reality is directly beamed to our minds from the mind of God.

This interpretation is supported by De Morgan's final diagram (figure 18). Here the structure of syllogisms and propositions of figure 17 is grounded in the heart-shaped face of a calmly alert being whose features are described by logical parentheses. Then, in the space above the octagon supported by this strange, but basically benign, visage, De Morgan sketched the kind of cha-

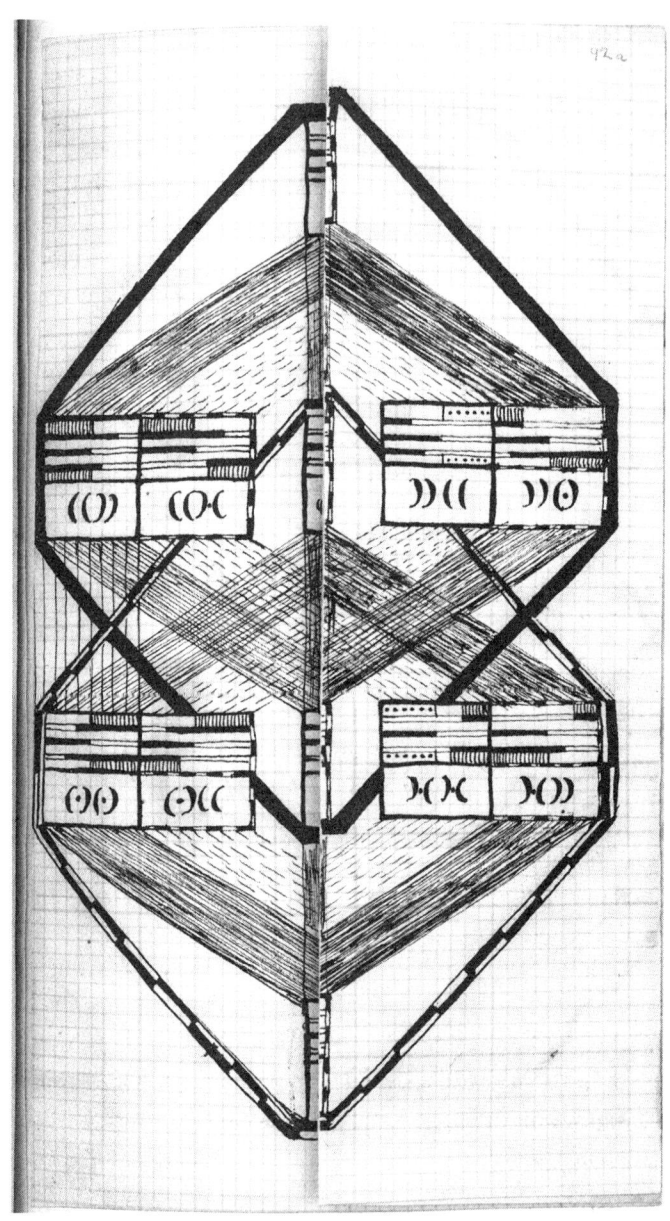

Figure 15: Folded triptych from Augustus De Morgan's personal copy of *Formal Logic, or, The Calculus of Inference, Necessary and Probable.* MS 776/1–2. By permission of the Senate House Library, University of London.

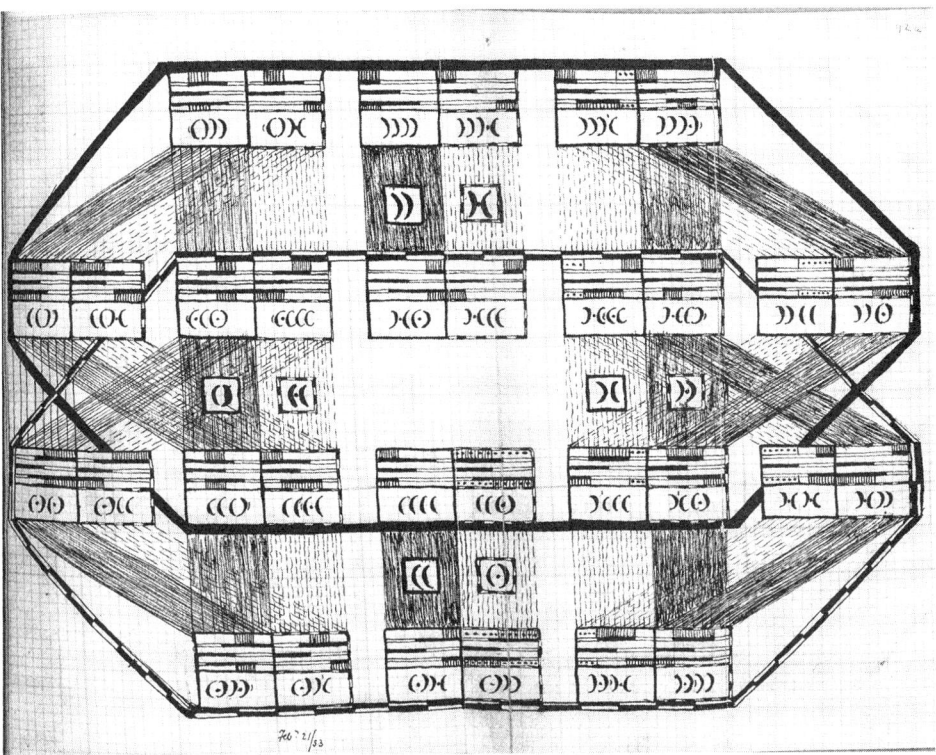

Figure 16: Open triptych from Augustus De Morgan's personal copy of
*Formal Logic, or, The Calculus of Inference, Necessary and Probable.* MS 776/1–2.
By permission of the Senate House Library, University of London.

otically human scene a father of seven might be confronting daily. Because of
these crude sketches, figure 18 might seem to be the least serious of the drawings De Morgan bound into his book (figure 18). For him, however, this image
of the human and the divine meeting in logical space was more than a passing doodle. It was an expression of his aspiration to find a divinely saturated,
logical alternative to the geometrical space of Newtonian cosmology, a map
of reason that encompassed both the human and the divine mind. He labeled
it "Table of Propositions and Syllogisms in the system which admits contrary
terms, February 22, 1853" and signed it with a flourish.

Thus the De Morgans devoted their lives to advancing the vision of reason
they had inherited from their forebears. In their community, they followed
the family tradition of engagement with the world of the seven children who

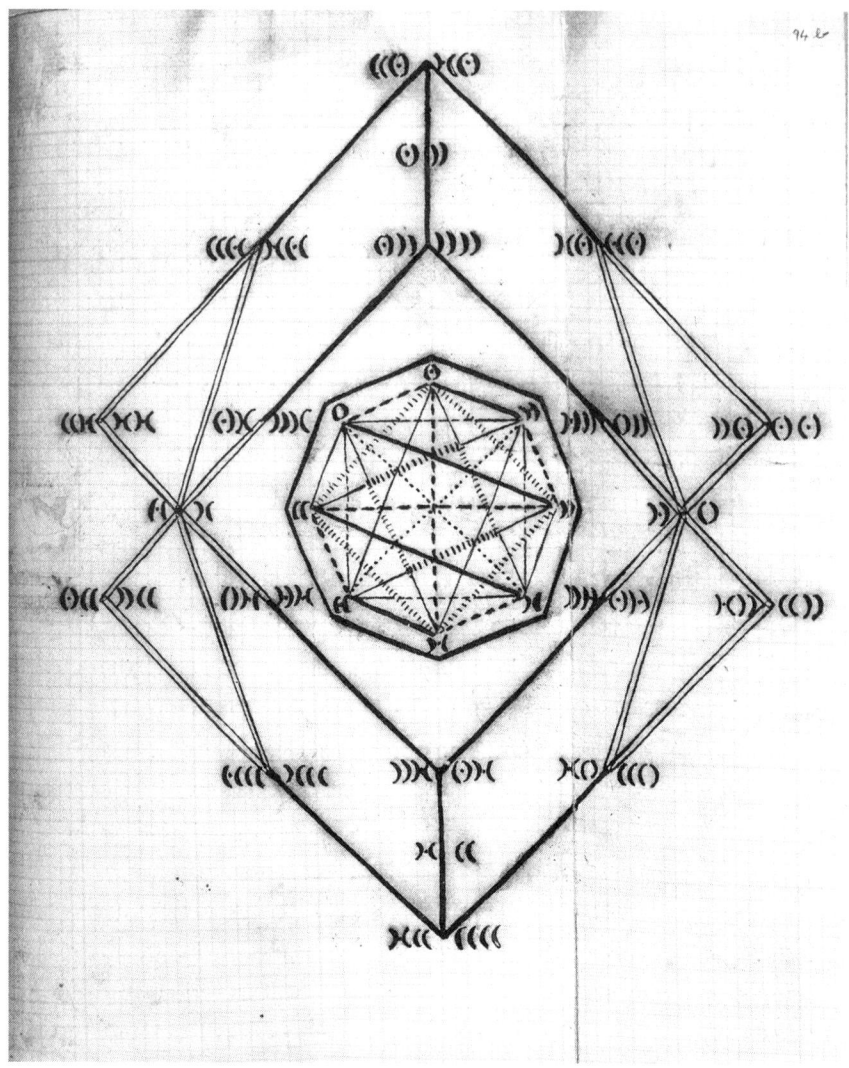

Figure 17: Floating diagram from Augustus De Morgan's personal copy of
*Formal Logic, or, The Calculus of Inference, Necessary and Probable.* MS 776/1–2.
By permission of the Senate House Library, University of London.

were flourishing around them. At home they were both deeply engaged in
what Augustus described as "that revival of serious thought upon mental subjects in which we now live, and which is far from having attained its full development."[59] Embedded as they were in a deeply gendered society, Augustus and Sophia took very different approaches to this effort. Sophia pursued

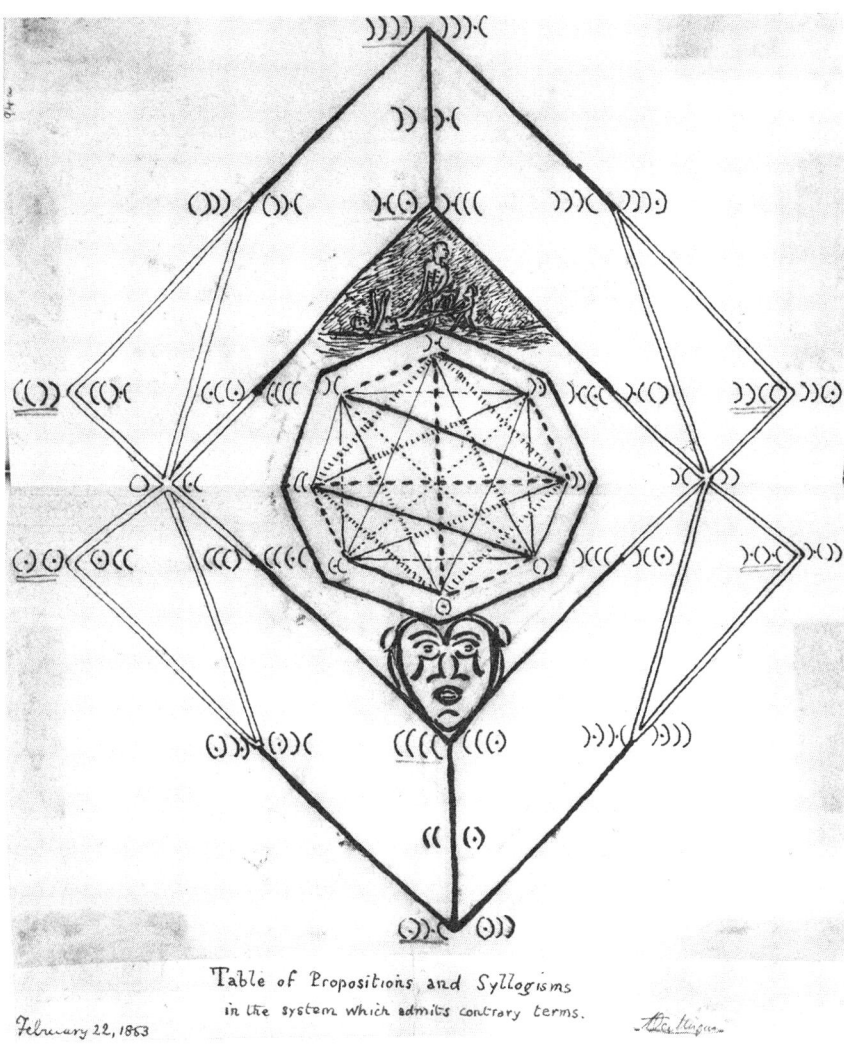

Figure 18: Floating diagram with doodles from Augustus De Morgan's personal copy of *Formal Logic, or, The Calculus of Inference, Necessary and Probable.* MS 776/1–2. By permission of the Senate House Library, University of London.

experiments with friends in their home, while Augustus closed himself into his library to investigate the deepest structures of thought. Despite these differences, the De Morgans were fundamentally united in their commitment to shine the light of reason on every aspect of the Victorian world in which they lived.

*—◆—*

# CHAPTER 20

# *Beyond Matter*

Augustus felt so empowered by his logical successes that he actually agreed to accompany his family for the first few days of their annual vacation in August 1853. Sophia organized excursions, while Augustus cracked Dad jokes—"What is the most genteel name for a street? Street Street, because that is a *square!*"[1]—and entertained the family with a stream of puzzles, word games, and mathematical conundra.

The most intriguing of the challenges Augustus posed for his home-grown audience grew from an observation one of his students, Frederick Guthrie, had made sometime in the fall of 1852: only four colors are required to construct any map in which all countries that share a border would be different colors.[2] When he first heard this claim, De Morgan found it strange, even unlikely. But after an evening spent drawing little pictures in hopes of proving it wrong, he instead emerged absolutely convinced of its truth.[3] He trumpeted the result to his friend, the Irish Hamilton,[4] who was not particularly interested.[5] De Morgan was not discouraged, though, and continued posing Guthrie's fact as a challenge to his friends and neighbors. As one by one his friends and family came to see the truth of the assertion, De Morgan understood himself to be watching the progress of reason in action; something that had not been known before was becoming self-evident before his very eyes.

In December 1853, De Morgan moved to expand his experimental pool by posing the question to William Whewell. At the time, the Cambridge man's concept of fundamental ideas was under fire from Henry Mansel, who was forcing him to concede that historically emergent ideas were not as immediate as was the classical idea of Space. In response, Whewell began to call necessarily true ideas that had emerged over time "axioms" in order to distinguish them from those like Space that had been known since antiquity. Whewellian axioms could be known as certainly as could the classic ideas, but grasping them clearly could require considerable time and education.[6]

De Morgan saw Guthrie's observation that four colors suffice to color a map as the perfect example of a historically emergent axiom. "By an axiom I mean a proposition which cannot be made dependent upon obviously more simple ones," he wrote in a December letter. The axiomatic character "consists in seeing the proposition" so clearly and immediately that no further proof is necessary. De Morgan then devoted two pages to reproducing the kinds of little diagrams he had used as he moved from a sense that it was incredible to the certainty that it had to be true. Having thus presented his own experience as exemplary, he tried to enlist Whewell's aid in confirming that it was universal. He urged his friend to try people with it, "but you must speak doubtingly yourself—as if it were not settled in your own mind."[7] De Morgan had no doubt that the experience of drawing maps would convince anyone who tried it that four colors were all that was needed. He was absolutely thrilled at having identified such an unambiguous example of progressive reason in action.

De Morgan never followed through, however. Less than a week after mailing his letter to Whewell, he sent a very different one to his Irish friend, Hamilton. His eldest daughter, Alice, had been "attacked with a gradually growing fever which now places her in great danger,"[8] he reported. Over the course of the next eight days, the De Morgans and their medical advisors tried desperately to bring the sixteen-year-old's fever down, but they failed. On December 23, 1853, Alice De Morgan succumbed to "a complication of nervous fever, inflammation of the lungs and very rapid development of tubercles."[9] The De Morgans were shattered. Sophia had apparently been trying to combat the teenager's "weakness and delicacy" for some time, but Augustus "did not realize the degree of illness till the end was near, and the blow fell heavily upon him."[10] Twenty-five years later, Sophia was still unable to write of these events, and Augustus never tried. The clearest description of the family's experience is to be found in the semi-autobiographical novel Alice's brother William wrote at the end of his life. Almost seventy years later, his memories of helplessness remained so vivid that he wrote, "I *am* that boy, the growing panic of that moment is on me still, and the gloom."[11] All of the De

Morgans had to reconstruct themselves to accommodate the world in which Alice had died.

When Augustus returned to his office after Alice's burial, he felt as if he "had been suddenly carried off, all round the world, and set down again at his desk."[12] In mid-February he was still too frozen to write any "notes at all beyond obligatory ones."[13] The only work that penetrated his grief was an anonymously published book, *Of the Plurality of Worlds,* which speculated about whether there was life anywhere else in the universe.[14] The issue had exercised modern thinkers since at least the end of the seventeenth century, when Bernard Le Bovier de Fontenelle cast his presentation of Copernican astronomy as a series of *Conversations on the Plurality of Worlds.* In the first years of *Evening Amusements,* Frend had happily proclaimed that the universe was full of inhabited worlds in which God's goodness could express itself. Fifty years later, *Of the Plurality of Worlds* came to the opposite conclusion, asserting that the potential of human reason was so enormous that the existence of the whole universe could be justified as an exercise of its power. All of the infinities of space that followed God's immutable laws existed in support of the intellectual harmony that was unfolding between God and the reasoning creatures God had created on planet earth, its author claimed.

De Morgan's first response to this anonymous work was to "throw it by for better days,"[15] but when he learned that Whewell was its author, he picked it up out of the grief-strewn chaos and found himself intrigued by the possibilities of intelligences far out in space. That humans were created at a particular time and in a particular place became for him a mark of a much larger dynamic. He saw that "all the stars and planets may be in their several progresses from $-\infty$ to $+\infty$, and every one at a different part of it," and that "this progression is infinitely varied in space."[16] Beyond its apparent emptiness, he began to consider a universe that teemed with progressive possibility. By the time Whewell sent him the second edition of *Plurality,* De Morgan was becoming convinced that there were "inhabitants" of other planets, though he could not say "whether conscious or unconscious, intelligent or unintelligent, &c., &c." What he did see was that these otherworldly inhabitants "have *uses* independent of us," which he strongly suspected were "also *trusts,* and therefore suppose *responsibilities.*"[17] De Morgan was clear that with her death Alice had passed out of the material realm, but in the universe he posited, she might still be an "inhabitant" with "*uses,*" "*trusts,*" and "*responsibilities.*" In answer to Whewell's expanses of emptiness, De Morgan offered a universe teeming with intelligences in which his daughter's life would not be wasted.

While Augustus was finding solace in infinite possibility, Sophia was searching for ways to hold her daughter near. Her image of the universe that Alice had entered was much more familiar and domestic than were her husband's thoughts about *uses, trusts,* and *responsibilities.* She never claimed to be a "sensitive," which meant that her personal glimpses into Alice's world were rare and fleeting, but she energetically engaged with those who claimed that capacity. In the fall of 1857, she began a diary in which she followed her daughter through a visionary world that included a rich cast of characters from Cupid to glimpses of God and Christ.[18] It is difficult to make narrative sense of Sophia's descriptions of the place where she found Alice to be living. What is clear is that Alice's mother was doing everything she could not to lose touch with her beloved child.

For the next several years, De Morgan found casual human interaction so painful that he walked home between classes. Nevertheless, teaching remained a mainstay. Sometime in the fall of 1854 he was recruited to the Decimal Association, whose mission was to replace the arithmetically chaotic English monetary system with a straightforward decimal one. De Morgan had devoted his 1848 essay in the *Companion to the Almanac* to the topic, and he threw himself into the cause with a will. All of his efforts came to naught—the English did not adopt a decimal coinage system until 1971—but in the years after Alice died, giving lectures recommending the change provided a comfortingly finite focus for De Morgan's thinking.

Sophia also slowly began to move out of her home. The neighborhood around their Camden Street house was a poor one, and toward the end of the decade she turned her attention to creating a Playground Society that would transform vacant lots into areas where children could play. She was a skilled fundraiser, but in this instance her efforts were undercut by her credulity, and the group dissolved after the man hired as an executor embezzled the funds. The idea was nevertheless a good one, and in the decades to come a similar group succeeded in bringing Sophia's vision to fruition.[19] Although neither of their efforts was immediately successful, the Decimal Association and the Playground Society marked Augustus's and Sophia's return to lives of reasoned activism.

By 1858, De Morgan's library was beginning to feel familiar again and his work was coming back into focus. Alice had been dead for five years, the Scottish Hamilton for two, and he was ready to defend his reason-based approach to logic. In February of that year, "On the Syllogism III" was read to the Cambridge Philosophical Society, and two years later that body heard "On the

Syllogism IV." Also in 1860, he pulled his thinking together in two synthetic pieces: a commissioned article, "Logic," for the *English Cyclopædia,* which was a successor to the *Penny Cyclopædia,* and a separately published *Syllabus of a Proposed System of Logic.* He published one more logic article in 1863, but "On the Syllogism V" was essentially a series of afterthoughts. The bulk of De Morgan's logical thinking was complete in 1860.

"On the Syllogism III" was a full-throated defense of De Morgan's commitment to creating a logic adequate to all of the complexities of human reason. He saw no excuse for limiting formal logic to the syllogisms Aristotle recognized, and he brushed aside attempts to use formal purity to protect the sanctity of the Aristotelian syllogism. In response to Mansel's charge that he had been sullying formal logic with the material confusions of reasoned thought, De Morgan asserted: "Logic is to consider the whole form of thought: *your* logic either contains the form of this thought, or it does not. If it contains the form of this thought, shew it: if it does not, introduce it."[20] Nothing in the twenty-five years since De Morgan published "Syllogism" in the *Penny Cycopædia* had changed his determination to expand logic to cover all of reasoned thinking. For him, every thought, no matter how material, had a formal structure. It was the job of the logician to identify and analyze it.

Defining and defending himself as a mathematician doing logic was central to De Morgan's program. In 1851, Mansel had claimed that De Morgan's mathematical predilections disqualified him from understanding the proper parameters of logic. In 1858, De Morgan argued that mathematicians were actually the best people for the job. Philosophical thinkers talked about form and matter, but they had no experience actually working with their categories; mathematicians engaged with the distinction all of time, although in most cases they did not realize they were doing so.[21] De Morgan supported his position with a view of history which turned Aristotle into a geometrician who pursued logic in order to understand the form of his thinking, transformed ancient Greek and Hindu mathematics into a logical tradition devoted to investigating the form of thought "for its own sake,"[22] and made Kant a mathematician who only turned to logic later in life. These men, he claimed, had actually negotiated the boundary between the matter and form of knowledge. Logic simply made explicit a process that they already knew through the practice of mathematics.

De Morgan recast the algebraic work he did in the 1830s and '40s to fit the narrative structure of matter and form that he was developing in the 1850s. From his perspective, the material subject of classical arithmetic was number, and the formal part comprised equations like $3^2 - 2^2 = (3 + 2)(3 - 2)$. Algebra

was created when the formal skeleton of arithmetic—that is, the equations—
was abstracted or pulled out of the numerical material in which it was embedded and allowed to stand alone as the material for a new study. The ultimate legitimacy of algebra's non-numerical equations, like $a^2 - b^2 = (a + b)$
$(a - b)$, resided in their genesis as the formal structure of arithmetic. But once
abstracted and cleansed of the material from which they came, these equations
became the material basis of a new subject that was dedicated to investigating
their formal properties.

In the late 1830s, De Morgan spent considerable time teasing out the formal
properties of algebraic equations—he is now recognized for having laid out
the basic formal structure of field theory—but he downplayed the importance
of that work. Approaching algebra through its forms alone was analogous to
putting a puzzle together with the image facing down, he explained. On the
other hand, those who look at the fronts and convert their "general knowledge
of the countries painted on them into one of a more particular kind by help
of the forms of the pieces, more resembles the investigator and the mathematician."[23] Twenty-five years later De Morgan had not changed his position. In
Mansel's insistence that logic be strictly formal De Morgan saw a call to put
the puzzle together without looking at the picture, and he proudly responded
as a mathematician who was able to comprehend both the material of the picture and the forms of the pieces. Developing this ability to move between form
and matter was what made learning mathematics useful "as a discipline of the
mind."[24] De Morgan's pursuit of a mathematics that recognized both form and
matter marked him as a true heir to the tradition of reason that linked language
and meaning which had guided Lindsey and Frend through their lives.

De Morgan cast logic as a subject with the potential to make the same
progressive move from form to matter that he had identified in mathematics.
Logic, he explained, was "the *form* of thought, the law of action of its ma-
chinery."[25] Just as an engineer could "watch the machine in operation without
attending to the matter operated on,"[26] so too could a logician focus on the
instrumental operations that constituted the form of thought without being
distracted by its matter. The process of identifying those pure forms was a
laborious one, but just as generations of engineers had succeeded in pooling
their individual experiences to come to a true understanding of the mechanics of machines, so too could generations of logicians come to understand the
pure forms of thought. De Morgan did not claim to have arrived at that goal,
but he was proud to be part of the process that would over time lead to understanding the essence of reason.

In July 1858 De Morgan hinted at his next step in a joke he sent to his friend, the Irish Hamilton. A nun, he there grinned, was so often visited by a young man that her mother superior became concerned. The nun tried to assuage suspicion by explaining that her visitor was a near relation, but the dogged mother superior pressed for more details. Finally, the nun blurted out that her visitor was "very near, indeed; for his mother was my mother's only child!" In De Morgan's telling, this answer left the mother superior satisfied that the nun and her visitor were indeed close relatives, and "she did not trouble herself to disentangle it."[27] He, however, was not so easily satisfied. As he was sending his joke off to Hamilton, he was beginning to turn his attention to the logic of relationships.

De Morgan had been dealing with logical relations from the very beginning of his career, but like the mother superior, he had at first not troubled himself to disentangle them. His first use of what might be labeled a logical relation occurred in his failed attempt at a syllogistic proof of the Pythagorean theorem in his 1830 *Study and Difficulties of Mathematics.* In the course of that exercise, he unselfconsciously cast the simple mathematical inference—$X$ is equal to $Y$, $Y$ is equal to $Z$, therefore $X$ is equal to $Z$—into syllogistic form by treating the relation "is equal to" as the copula. Nowhere in that piece did he explain or consider what it was that allowed him to substitute the relation "is equal to" for the copula "is" in this way.

By the time De Morgan wrote *First Notions* in 1840, he had begun to recognize that there were real problems involved in substituting a different relationship for the standard copula "is." In this work, then, he took a more conservative position, which left the copula as a form of the verb "to be" (table 7).

TABLE 7

| Subject | Copula | Object |
|---------|--------|--------|
| $X$ | is | equal to $Y$ |
| $Y$ | is | equal to $Z$ |
| $X$ | is | equal to $Z$ |

This approach had the advantage of keeping the copula pure, but as it did so it created a serious problem with the middle term that was so important to the functionality of De Morgan's system. In this construction, the object of the first proposition, "equal to $Y$," became something different, that is, "$Y$," as it moved into its position as the subject of the second proposition.

In *First Notions,* trying to explain this discrepancy left De Morgan flailing. In *Formal Logic* he returned to his first position, in which "is equal to" is the copula (table 8).

TABLE 8

| Subject | Copula | Object |
| --- | --- | --- |
| X | is equal to | Y |
| Y | is equal to | Z |
| X | is equal to | Z |

As De Morgan proposed this configuration for a second time, he was beginning to see that it contained the seeds of a change even larger than the one he initiated by introducing negatives into Aristotelian logic. He saw that "is equal to" is not the only relation that could work as a copula in this way; "is tied to" or "is a sibling of" would work equally well. De Morgan noted that these relations were all "convertible"—that is, propositions containing them would be equally true if reversed—but that was not really the issue. The relation "talks with" is also convertible, but that "**X** talks with **Y**" and "**Y** talks with **Z**," does not mean that "**X** talks with **Z**." What is more, relations like "is greater than," "is older than," or even "is a brother of" can be cast in this form but are not convertible. De Morgan called the essential characteristic of these relations "transitivity," but in *Formal Logic* abstracting the essential form of that relation remained a distant ideal.

By 1850, De Morgan was recognizing that transitivity could be expanded yet further. In "On the Syllogism II" he began looking at *bicopular syllogisms* in which two different copulas in the major and minor premises could lead to a third copula in the conclusion (table 9).

TABLE 9

| Subject | Copula | Predicate/Object |
| --- | --- | --- |
| X | is a brother of | Y |
| Y | is a parent of | Z |
| X | is an uncle of | Z |

In this case, the copulas "is a brother of" and "is a parent of" push forward into a final "composite copula," "is an uncle of." The relation between copulas was itself a critical factor in the outcome. As these kinds of examples multiplied around him, De Morgan saw the need to identify the essential structures that underlay the whole. He set out to develop "a formal mode of joining two terms which carries no meaning, and obeys no law except such as is barely necessary to make the forms of inference follow."[28] Doing this would support a theory of abstract copulas that rested not on their meanings but rather on the way they acted.

Mansel responded immediately. In 1851, he charged that all of De Morgan's considerations of relation fell outside of the proper bounds of formal logic because all of them rested on the specific meanings of the various copulas he was exploring.[29] De Morgan was not concerned. In his second logic paper, he had already argued that the standard copula "is" was as filled with meaning as any other relation. "Historically speaking, the copula has been material to this day," he claimed.[30] By thus categorizing Aristotle's copula as material, De Morgan turned the tables on Mansel. Now the philosopher could not claim to be doing formal logic either. As long as the validity of a syllogism depended on the meaning of the copula, including the Aristotelian copula "is," a truly formal logic was out of reach.

And there the issue rested for almost fifteen years, but De Morgan had not forgotten it. In "On the Syllogism III" he offered the analogy of a nutcracker to sharpen his image of the way formal theory could be derived from a mechanical process. Take "two levers on a common hinge," he began, and put a nut between the levers to crack it. Faced with this scenario, an engineer might consider his nutcracker to be the "<u>form</u> of <u>nut-cracking</u>" or even the "<u>pure form</u> of nutcracking." However, when presented with the full range of nutcracker variations, like "the screw, the hammer, the teeth, &c.," the discerning engineer would begin to see that his previous thinking was materially constrained and become ready to recognize that the essence of nutcracking is "a strong pressure applied to opposite sides of the nut."[31] There are no references to levers or screws or hammers or any other material object in this final statement. It represents the pure form of a nutcracker.

For De Morgan, this nutcracker analogy captured the situation he faced with the logic of relations. Clinging to the special status of the Aristotelian copula "is" when other copulas were equally transitive was like insisting that the leveraged nutcracker was the pure form of nutcracking. De Morgan judged all of the different copulas he was contemplating to be like "the screw, the hammer, the teeth" of nutcracking. And just as recognizing that there are many kinds of nutcracker forced the engineer to reconsider his position, recognizing that there are many kinds of copulas required logicians to reconsider theirs. It was no use for a logician to "hug" the transitivity of "is" and insist on it as the defense of the purity of Aristotelian logic. Aristotle's copula is just one of any number of transitive copulas, like "is equal to" or "is tied to." The differences among all of these various alternatives—including the Aristotelian "is"—are material. The legitimate response to their materiality was to take the next step and abstract the formal structure of logic from the materiality of the Aristotelian copular syllogism.

As De Morgan was joking about nuns, he was beginning the effort to abstract the pure form that underlay all of the materialities of particular copulas, in order to move logic to a place in which all that was left of the Aristotelian syllogism was the "*pure form* of the proposition, divested of all matter."[32] By 1860, he was confident enough to present a purely formal logic of relations which effectively reduced all of Aristotelian logic to a rather small and very simple subset of a much larger field of reasoned possibility. In "On the Syllogism IV" he developed a totally new set of symbols that allowed him to rise above the particularities of the Aristotelian syllogism in order to enter the larger world of relations. And with that De Morgan had finally countered Mansel's 1851 attack on his logical program. In "On the Syllogism III" he had responded to Mansel's needling by laying out the terms for a truly formal logic; in "On the Syllogism IV" he showed that he could meet those terms. He was vindicated.

The logic of relations laid out in "On the Syllogism IV" is widely seen as De Morgan's most important contribution to modern logic, but he did not see it that way. He loved dissecting "logical questions into their very atoms"[33] but was put off by the sterility of formal systems. Developing the logic of relations entailed approaching the compendious language of reason with the puzzle pieces face down. As was the case with algebra, or calculus, De Morgan could certainly do it, but ultimately he saw it as an empty exercise.

For De Morgan, the real intrigue of logic lay always in the complexities of concrete situations like the one described in his joke about the nun. While he was preparing his fourth logic paper for publication, he was deeply involved in the more immediate logics of "consanguity and affinity"[34] that underlay human relations. Elizabeth De Morgan died in 1856, and in response her son opened a notebook in which he carefully recorded the names, marriages, and children of her sixteen siblings. In another, he painstakingly recorded all that he could find about the lives of his father and his father's relations. He devoted considerable effort to his paternal relatives, tracing the stories that swirled around the death of his grandfather Augustus, recording the story of the next generation's early lives as foundlings in London, and trying to sort through the rights and wrongs of his father's struggles with the East India Company.

As De Morgan was thus documenting his family, he was also musing about his place within it. Uncomfortable with his role in a family of soldiers, he developed a coat of arms in which logical symbols took the place of the "birds, beasts, or fishes" of traditional heraldry, and "the letters ADM in geometrical cipher, double and invertible" formed the crest (figure 19). "This shield cannot be dishonoured," he carefully explained. "Reverse it and the same story is told in a different order."[35] Even as De Morgan chafed against the militarism of his family, he fully embraced its code of honor.

Below this heraldic image De Morgan placed another: "The Zodiac of the Syllogism." In his *Syllabus of Logic* he developed this Zodiac as a mechanical calculator, which could be rotated in such a way that it could produce all of the syllogisms he had laid out so beautifully in his earlier diagrams.[36] In print, he could only describe his vision in words, but alone in his study he was not so constrained. There the Zodiac wrapped around his crest, and the symmetries of his initials were reflected in the symmetries of his calculator. De Morgan gave no explanation of this design, but there was really no need to explain. As the aging man contemplated his life and his work, he was comforted by the image of himself bounded by a circle of his logical symbols, a deeply meaningful ring that at once protected him and served as a gateway into the "somewhat" that constituted the great universe of reason.

While De Morgan was privately musing about his relations to his family and the universe, Sophia was caught up in a world of rapping mediums, sensitive servants, and dreaming children. Her husband was not a regular participant, but he completely supported her efforts to explore the powers of mind in the spirit world. His willingness to stand behind Sophia's ongoing investigations put him at odds with the gentlemen of science who were resisting spiritualist intrusions into what they saw as their territory.

The most central figure on the scientific side of this conflict was Michael Faraday. In spring 1854, he had responded to the failure of his first attempt to discredit table-turners by organizing a series of lectures on mental education. For several weeks, large audiences gathered to hear England's scientific luminaries explain and defend the sanctity of science against the enthusiasms of mesmerists, table-turners, and spiritualists.

As he led the charge, Faraday sounded rather like Mansel. He opened his lecture by making an "absolute distinction between religious and ordinary be-

The above shews the bearings which A. De M. prefers to drawings of birds, beasts, or fishes. There is not a herald in the country who could blazon them, or set them forth in technical terms. So far as their comprehension goes, they would describe the coat as <u>Argent</u>, the forms of syllogism, the universals <u>sable</u>, the particulars <u>gules</u>. Crest, the letters ADM in geometrical cipher, double and invertible. This shield cannot be dishonoured. Reverse it, and the same story is told in another order. By turning two leaves will be seen the insignia preferred by the others of the name: in which there is just exactly as much mark of the beast as mark of the cross.

Figure 19: Logical coat of arms with the Zodiac of the Syllogism. MS Add. 7 (De Morgan Family Papers), University College London Library Services, Special Collections.

lief."[37] The spritualists' efforts to investigate the spirit world were rooted in a fundamental error, he explained; the ways of God were *revealed* and could not be discovered by reason. The best way to cut through their delusory efforts was to educate people's "judgment" so that they would not be misled. The most important lesson of all was *"teaching the mind to resist its desires and inclinations until they are proved to be right."*[38] He asserted that proper humility in the face of evidence was necessary for true scientific work, and he was particularly insistent that before launching any investigation, seekers needed to maintain clarity about the implications of the subjects they were investigating. "Before we proceed to consider any question involving principles, we should set out with *clear ideas* of the naturally possible and impossible," he explained.[39] Certainly all of Newton's physics, with its clear theory of gravity, should not be thrown out of the window by a dinner party of people who thought their table moved.

At the time Faraday made these comments, Sophia and Augustus were both too overwhelmed by Alice's death to pay much attention, but as the years went by, Faraday's determination to confine reason to the everyday material world began to rankle. In March 1857, De Morgan struck back in the *Athenaeum*. In an unusually long review of a lecture entitled "On the Conservation of Force" that Faraday had delivered the previous month, De Morgan attacked the eminent man's 1854 claims that all physical investigations should begin "with a clear idea of the naturally possible and impossible." Surely, De Morgan protested, "mature minds" recognized that perceived "limits of possibility and impossibility" were only a "mirage which constantly recedes as we approach."[40] If taken seriously, he insisted, Faraday's attempt to shut down the spiritualists would actually prevent scientific progress.

From De Morgan's point of view, Faraday made this mistake because he was an experimental investigator who had devoted his life to investigating the material world. This focus had led him to wonderful discoveries, but it did not confer the right to limit the possibilities and powers of reason. It took a mathematician to be truly open to the powers of thought that structured Faraday's material world. He himself "once knew all about what was possible and what was impossible," but his experiences with clairvoyance had changed him.[41] He was now beginning to "suspect that the *maintenance* of the universe is an energy of the same character as its *creation:* for aught we know, perpetual creation of something may be an actual condition of that maintenance."[42] All of these possibilities were closed to Faraday because of his narrow focus, but for De Morgan the only way to come to a true understanding of the material world was to remain open to realms of possibility far beyond what anyone could imagine.

Both of the De Morgans were actively involved in exploring the dynamic world of possibility that Augustus defended in his review, and six years later the two of them presented their visions in *From Matter to Spirit: The Result of Ten Years' Experience in Spirit Manifestations, Intended as a Guide to Enquirers.* The title page of the book stated simply that it was "by C.D. with preface by A. B.," but those pretensions of anonymity were fleeting. Within days, everyone knew that the book was written by Sophia, the preface by Augustus.

*From Matter to Spirit* may be seen as the epitome of the kind of work that Faraday and his allies had railed against at the Royal Institution in 1854. In his preface Augustus had fun describing the events that so concerned the self-proclaimed protectors of science. Clearly, something was happening when "among a people who have the reputation of a dry, practical, unimaginative temperament," an epidemic broke out and suddenly the land was "spotted with mediums before the wise and prudent had had time to lodge the first half dozen in a madhouse."[43] Even as he saw the humor of the sudden fascination with spiritual phenomena, he refused to dismiss it as a fad. Instead he held open the possibility that it was the harbinger of a great step forward in human understanding. In support of this interpretation, he proffered a theory of intellectual development in which progress was like the development of a crab or lobster, which is usually bound within a hard shell. Periodically, however, such animals throw off their armor to "take a good growth before the new shell is hardened." When it came to spiritualist thought, De Morgan suggested, "Our ghost shell has certainly cracked; but what and how much of growth the notion will get before the new integument is stiff enough for a philosopher, is still to be seen."[44] In the meantime, the spiritualists should be applauded for their willingness to follow in the open-ended experimental footsteps of the early days of the Royal Society, and their detractors should be recognized as similar to those who dismissed early navigational astronomers as "*star-shooters*" or pendulum clocks as "*swing-swangs.*"[45] De Morgan declared himself a neutral party, but when faced with members of "the higher class of obstructives who, without jest or sarcasm, bring up principles, possibilities, and the nature of things,"[46] his loyalties lay with the spiritualists who were bravely following the path of progress.

Sophia never displayed Augustus's sense of humor, and the book that followed his wittily defiant preface was more sober in tone. In it she followed the model of reason she had learned from her father into the wider nineteenth-century world in which she lived. One of the changes that had occurred in the twenty-five years since she tried to make sense of the death of her sister Harriet lay in the variety of sources she found relevant. She devoted nine pages to a close reading of the discussion of death in 1 Corinthians 15:35–57 that had supported her earliest conviction that Harriet still lived, but those pages are embedded in a fifty-five-page chapter that included sources from Plato to Swedenbourg. The Christianity she learned from her father remained Sophia's touchstone, but he had also taught her to be open to other perspectives. Her book was the product of a nineteenth-century Victorian who gave credence to many sources beyond the Bible.

It was not easy for Sophia to carry her father's and husband's views of reason beyond the sheltered classrooms of Cambridge and UCL, into her woman's world of deathbed scenes, near-death experiences, and ghost stories. A number of her sources were men, but they were talking to her in ways they did not with other men. Her task was to bring reason to bear on material that was usually relegated to the margins of respectability.

Sophia's primary challenge was to establish and describe a spirit world that was making itself known through a bewildering array of blowing curtains, turning tables, spirit writings, and trance descriptions. She fully recognized that she was not a medium, which meant that she had to collect information from others. She was like her father in being willing to listen seriously to those around her. But his strict standards of linguistic rationality did not work in her world of six-year-olds, nursemaids, neighbors, and mediums. She could, and did, routinely screen her informants for truth-telling, but she could not guarantee that they either spoke or wrote precisely and properly. The power of reason was thus transmogrified as it moved from the masculine world of its usual defenders into the predominantly female world that existed by its side.

Sophia was nonetheless determined to use reason to identify the basic structures that underlay all of her disparate forms of evidence. She organized spiritual experiences on a hierarchical scale of materiality. The least exalted experiences were those like "table-turning," which occurred on a material level that even she could observe. Somewhat higher up the scale was "spirit writing," in which someone holding a pen was guided by spirit power. In this case the act of writing could be observed by many, but the force behind it was experienced only by one (or sometimes two) people.[47] Highest of all were visions, dreams, and voices perceived directly in the mind, because these had no intersubjective material manifestations at all.

An elaborate theory of human development underlay this hierarchy of experiences. All people, Sophia explained, are made of a material body, an animating spirit, and an ever-developing soul. At the moment of death, the soul "passes away" from the material realm "and, animated by the spirit, becomes the body of the next life."[48] In this new form, the process of development continues; the spirits move ever closer to God and farther from the material world. This developmental model explained and justified Sophia's ordering of earthly experiences. Spirits who communicated through material manifestations like table-turning were at the lowest level of spirit development, whereas those who communicated without such manifestations were higher on the developmental scale.

Sophia distilled these hierarchies of experience and development from her conversations, readings, and experiences, but she needed a rational ground for her spiritual theorizing. She rested hers on what she called "the Principle of Correspondence," defined as "the law by which the external of one state agrees with the internal of that below it."[49] She acknowledged that this principle might at first glance appear "mystical and imaginary," but she insisted that it was "intelligible enough" to render any conclusions drawn from it "as certain as any branch of knowledge which can be deduced by well-marked steps from indisputable principles."[50] Sophia's Principle of Correspondence supported her spiritual theorizing in the same way that Peacock's Principle of Equivalent Forms had supported his algebraic investigations; it provided a stable platform from which to evaluate a set of otherwise confusing and untethered phenomena. Peacock drew his principle from the world history of algebra; Sophia drew hers from her spiritual readings and conversations. In both cases, as they delved into them, the material tenets that were at first obscure became ever clearer until they became incontrovertible. This self-evidence marked them as principles that could support their work in the same way that self-evident postulates supported geometry.

Once established, the Principle of Correspondence served a variety of important functions. Early on Sophia used it to resolve one of the more immediate problems that dogged her investigations of the spirit world. From the very beginning of her interactions with mediums, it had been a major problem for her to imagine Frend saying things like "*we long to clasp you in our arms in this bright world of glory*"[51] and simply impossible that he would make spelling errors that turned "'Beautiful' into '*butiful*,' 'writing' into '*riting*'&c."[52] But with the Principle of Correspondence, situations that before had raised questions "either of the honesty of the medium or the orthography of the spirits" became understandable epiphenomena of a communication process in which "the material brain of the medium seems to be the apparatus used for the transmission of thought."[53] What this meant was that "in the case of communication from spirit to mortal, the *imagery* or *ideas* contained in the medium's brain takes the place of words."[54] This interpretation allowed Sophia to make sense of the language and spelling of the messages she was receiving through poorly educated mediums. Their utterances severely challenged the exact world of reason, in which clear words corresponded to simple ideas. Although she had learned of that world from her father, she was quite willing to let it go in order to commune with him again.

More centrally, the Principle of Correspondence provided a concise description of the essential relationship between the material and spiritual worlds. Spirits that were internal in the human world became external in the spirit world. These two manifestations of spirit fit each other like a statue and its mold. The Principle of Correspondence allowed Sophia to expand this relation from the human microcosm to the ultimate macrocosm. "Extending the principle from individuals to the mass" led her to see that "the inner state of the material world forms the outer or phenomenal form of the spiritual sphere."[55] In death, the spirit that had before been lodged in a material body was released into the spirit world unencumbered by the matter that before had enveloped it "like a husk or shell." In its new state, the whole of what before had been the inner spiritual world corresponded "in all its details" to its new external state, "as soul to body and as spirit to soul."[56] According to this explanation, the experience of death that was so jarring in the human world became just a moment of passage from one state of existence to another.

De Morgan had devoted years to convincing himself of the truth of the Principle of Equivalent Forms for himself, but he was not willing to do the same for Sophia's Principle of Correspondence. Nonetheless, he fiercely defended her right to develop it. He began his preface to his wife's book with the declaration that although he had both "seen and heard" phenomena that were "*called* spiritual," he was unable to accept "any explanation which has yet been suggested."[57] Nonetheless, the confluence of his and Sophia's ways of thinking is manifest throughout her book. Her portrayals of the relations between matter and spirit mirror the relations between matter and form laid out in her husband's logical papers, and her descriptions of the movement from the material to the spiritual world reflect the movement from matter to form that constituted progress in his logical world. Augustus recognized this complementarity when he closed his preface with the observation that "between us we have, in a certain way, cleared the dish; like that celebrated couple of whom one could eat no fat and the other no lean."[58] Augustus had no desire to cross the boundary that separated him from Sophia's female world, but he recognized and respected its legitimacy.

Sophia was no more interested in following the twists and turns of her husband's logical thinking than he was in her spiritual efforts. She did, however, recognize that his work was an expression of his Berkeleyan conviction that we depend on the sustaining power of a divine being for our "absolute existence" as well "as for support and guidance through life."[59] In her spiritualist investigations, Sophia was following the expressions of that celestial mind in the physical world, while with his elaborate images and detailed writings, Augustus was exploring the webs of connection that supported that world. In their different ways, both of the De Morgans were pushing beyond the constraints of the material into an expansive world of divine reason.

——— ▸◂◆▸◂ ———

# CHAPTER 21

# *Beyond Reason*

In 1859, the De Morgans left Camden Street. It was hard to break away from the house in which three of their children were born and the family had thrived for almost two decades. Alice died on Camden Street, but it was also the place where she returned in the household's dreams and visions. Nonetheless, the neighborhood was going downhill, and Sophia attributed her periodic illnesses to the increasingly bad air of the urban environment. It was time for the De Morgans to move.

The De Morgans' new house on Adelaide Road was well suited to their growing brood.[1] In the immediate aftermath of Alice's death, Augustus admitted that he could "*understand*" but not "*feel* that six left made any set-off against one gone,"[2] but now both he and Sophia were again enjoying their children's company. At the time of the move, William was twenty, George eighteen, and Edward sixteen. A contemporary portrait captures the three of them on a stoop, with William playing the role of rakish art student, George a stolid scholar, and Edward a somewhat impish younger brother. Not pictured are their sisters — Annie fourteen, Chrissy twelve, and Mary nine — who were a rather precocious group that was apparently home-schooled. The De Morgan children formed a tight group. They enjoyed musical evenings in which everyone sang while Annie played the piano, Edward the violin, and Augustus his flute.[3] They played elaborate games in which one would draw a series of pictures and challenge the rest to write the stories to accompany them. They shared an interest in anagrams; "Great gun, do us a sum!"[4] is just one of of two hundred created from Augustus's name which Sophia carefully preserved. As time went by and the boys began to venture out on their own, they still returned all but daily for time with the family on Adelaide Road.

De Morgan was beginning to draw back from commitments that had structured his life for years. In 1858, a quarrel with the editor of the *Companion to the Almanac* brought more than twenty-five years of annual contributions to an abrupt end. His ties to the RAS were loosening as well. Sheepshanks's death in 1855 had definitively destroyed the equi-tenacious triangle, and after the move to Adelaide Road he used its more distant location as an excuse to retire from the "Club" that gathered for dinner after regular meetings. This decision was collegial, but when Airy lost the presidency to a Dr. Lee, De Morgan determined the election to have been rigged and resigned in a huff. He had been at the center of the RAS for decades, but with this election he could see that it was passing him by.

What did not change was De Morgan's teaching. At fifty he was just as committed as he had been when he taught his first class at twenty-two. A portrait sketched by one of his students shows him as a portly man with dark hair bushing behind a receding hairline. In front of him stands a desk with notes and a pointer he could use to pick out particular parts of the equation-strewn blackboard behind him (figure 20). There are other pictures of De Morgan at this age, but this is the best image of the man who admitted to his friend Hamilton that "all I do arises directly out of teaching, and has in some way or other reference to what can be brought before a class, and especially with a view to mathematics as a discipline of the mind."[5] For decades he covered all of mathematics—from the first book of Euclid and the most elementary forms of algebra, to the furthest reaches of calculus—in four different classes that met three times a week. And year after year, as he brought the subject down "from the altitudes of calculation and mechanical work to plain and simple considerations,"[6] he found new and exciting directions for research opening before him. The process of guiding students to the full use of their reason through mathematics was for him an endless source of excitement and new ideas.

After days spent talking mathematics with students, De Morgan came home for dinner with his family, and then retired for four or five hours with his books. One of the first projects that engaged him after the move from Camden Street took him back to his origins. His birth and long family history in India led him always to see himself as "a Briton unattached," at least as much Indian as English.[7] His interest in the long traditions of mathematics in

Figure 20: De Morgan teaching, sketched by a student. MS Add. 7 (De Morgan Family
Papers), University College London Library Services, Special Collections.

India dates at least to conversations with his Uncle Briggs in the Frend household in Stoke Newington. In the article "Algebra" that appeared in the first
volume of the *Penny Cyclopædia,* he located the roots of the subject in India
and declared that the Indian mathematicians had pursued it further than did
the classical Greek, Diophantus. He spent the next decade gathering all that
he could find about Hindu astronomy and mathematics, which he published
in an article on the principal Hindu algebraic text, the "Viga Ganita."[8] In his
newly arranged library on Adelaide Road, De Morgan returned to the subject.

The focus of De Morgan's attention in 1859 was a textbook written by
a mathematics teacher in Delhi named Yesudas Ramchandra.[9] De Morgan
saw great promise in Ramchandra's *Treatise on the Problems of Maxima and
Minima,* and he made sure that the Indian would receive an award for it. In
addition, he shepherded an English edition through the press with a care-

fully written preface in which he revisited the mathematics of the land of his forefathers. Whereas most Englishmen thought they were best able to teach mathematics to Indians, De Morgan's primary goal was to set the "Hindoo mind" free to "work out its own problem." The "Indian mind" excelled at algebra, he explained, and so he encouraged "young Hindoos" to enrich their mathematical tradition with the cultivation of geometry.[10] His hope was that branching out in this way would allow the Indian people to develop their reason and mathematics in their own distinctive ways.

Preparing an edition of Ramchandra brought De Morgan out of himself, but his other work was becoming more inward looking. After his burst of logical activity in 1860, his interests became more diffuse and miscellaneous. His final logic article, "On the Syllogism V, and on Various Points in the Onymatic System," was essentially a renewal of his arguments with the Hamiltonians that contained little that was new. In a historical article he traced the earliest uses of the signs + and − to the commercial world of the fifteenth century; in another he turned his attention to the harmonics reflected in the music he was beginning to share with his children.[11] And questions about the nature of the infinite that became so pressing in the period after Alice's death continued to spark his interest. By the early 1860s his focus had shifted back into mathematics. As a student, he explained to Hamilton, he had vowed not to "settle definitely and irrevocably the true foundation of the Diff[erential] Calc[ulus]" until he was "on the wrong side of fifty." Now thirty-eight years later he had passed that milestone and come to the conclusion that "infinity is a subjective reality," a reality that he had come to accept, though he was willing to "leave other minds to stand on their own bases."[12] When it came to the bases of calculus he had become a neo-Leibnizian who saw no way of "getting out of the concept—concept without image—of an infinitesimal" in understanding the calculus.[13] "We should constantly flounder and get wrong in using limits, if we had not the light of infinity to guide us," he declared.[14] De Morgan was very careful not to try to impose his conclusions on anyone else, but his experiences contemplating the infinite had assured him of the marvelous possibilities of the universe he inhabited.

In 1863, De Morgan ventured in a different direction with a signed column in the *Athenaeum* called "A Budget of Paradoxes." In the first of these he explained that what he meant by "paradox" was "something which is apart from general opinion, either in subject matter, method or conclusion."[15] According to this somewhat idiosyncratic definition, anyone—from circle-squarers to Copernicus to the author of *From Matter to Spirit*—who stood behind a position unorthodox for his time was a paradoxer. In the *Athenaeum* columns, which appeared sporadically over the course of the next three years, De Morgan introduced his readers to the authors and arguments in his books. For the first year or so his paradoxers were historical figures. When he turned his attention to his nineteenth century compatriots, his columns carried the allure of exposés.

At home De Morgan cast his paradoxers' net more widely. For years he and Sophia had eagerly soaked up stories about the world of William Frend, and De Morgan began recording what they had heard of the Spitalfields Mathematical Society; of Maseres, who led Frend into mathematics; and of Higgins, whose ideas had so intrigued Sophia and infuriated Augustus. All were included in the *Budget of Paradoxes* that Sophia published as a freestanding volume in 1872. The book's twentieth-century editor valued it as a "delicious satire,"[16] but for De Morgan it was something more. In a wonderful mishmash of metaphors, he once explained that "in reading an old mathematician you will not read his riddle unless you plough with his heifer; you must see with his light if you want to know how much he saw."[17] Ploughing with the heifers of paradoxers brought to life the extraordinary variety of people among whom De Morgan lived.

As the parents were drinking in stories about the family's past, their son George was following his father's mathematical footsteps. William had confused and disappointed his father by leaving UCL to study at the Academy of Arts, but George thrived in his father's classes. Cambridge was the next step forward for many similarly successful students, but George was a De Morgan who would never consider studying in an Anglican institution. Instead, he moved through various teaching positions in London until by 1866 he had secured the position of mathematics teacher at the University College School that he and both of his brothers had attended.

As George was working his way into this position, he was also looking for a community in which to discuss mathematical ideas. In 1864, he and his friend Arthur Cowper Raynard decided to form a society, similar to the Astronomical but focused on Mathematics. At first they were thinking of a school group to be called either the "London University Mathematics Society" or the "University College Mathematical Society." By the time of their first regular meeting, however, the group had expanded its vision to become the London Mathematical Society, or LMS.[18]

Augustus De Morgan was very pleased when the group's founding young men asked him to become the first chair of the LMS. Their choice for vice-chair was T. Archer Hirst, who was about to take up a position as the first professor of applied mathematics at UCL. Hirst had earned his PhD from the University of Marbourg in Germany, where mathematics was carried on within the rigorous tradition initiated by Cauchy. Hirst, who once confided to his diary that he found De Morgan to be a "dry pedantic pedant,"[19] had little patience with the rich definition of the mathematics of reason that the older man pursued, but De Morgan was willing to recognize the influx of continental ideas that Hirst represented. In the opening speech he delivered to the group of twenty-seven mathematical afficionados on January 16, 1865, he called for a society in which those who like Hirst were versed in the highest branches of pure mathematics would also consider "the inclinations and pursuits of those engaged on the more elementary branches."[20] Mathematics was a multifarious subject, wide enough to include both the meaningful subject he pursued and the abstract formal one Hirst had learned.

De Morgan grounded his call for a broad view of mathematics in a sense of his historical moment. The variety of approaches he saw around him constituted evidence that mathematics was in a stage similar to that of seventeenth-century Europe, in which a wide variety of people were developing pieces of the whole that later became calculus. Before Newton and Leibniz synthesized their results, no one could see how their individual investigations related to each other. De Morgan saw the same to be true of mathematics in 1865. "All these subjects," from esoteric continental studies of canonical algebra and elliptic transcendentals to the most reasoned analyses of geometry, "are thriving in their own way, and are laying the foundation of something great to be discovered by and by," he asserted.[21] As long as the LMS provided a place where people engaged in all of these mathematical interests communicated with each other, it would serve as the locus of mathematical progress.

De Morgan's hope that the LMS would recognize that "our subject is really rather a wide one"[22] reflected his sense of the field he had spent a lifetime pursuing. Standing comfortably before a sympathetic group that his son had brought together, he opened up the vision of a reasoned mathematics that had supported all of his researches. In the second part of his speech, he asked that "the Analysis of the necessary *Matter* of Thought," that is, mathematics, always be accompanied by "the Analysis of the necessary *Laws* of Thought," that is, logic. He pointed to the importance of history for those who wanted to move the subject forward. Everyone casts the past to fit their version of the present, he observed, but if anyone wants "to have his own researches guided in the way which will best lead him to success," he must recognize "the curious ways" that mathematics has evolved.[23] Studying language was equally important as a way to free thought from the constraints of too narrowly defined words. Finally, he advocated acknowledging the deeply human nature of mathematical ideas, the power of simple "commonsense" to point people toward the most interesting more advanced problems.[24] Including considerations of logic, history, language, and simple human reason in the society's discussions was the best way to ensure that the subject continued to grow into all of its rich possibilities.

De Morgan was highly respected and his reputation was a large part of the reason the LMS trebled in size within the year. But by 1865 his view of mathematics as the ultimate expression of human reason was at best "idiosyncratic," and it did not last long.[25] The direction of change may be seen in the fate of the "four-color axiom" he had long seen as representing the cutting edge of human conceptual progress. In an 1860 review of one of Whewell's books, De Morgan had published Guthrie's claim as "a fundamental and axiomatic position—meaning one which cannot be made to depend upon anything more simple and fundamental."[26] He acknowledged it might seem strange to claim axiomatic status for an arcane observation that he himself had not suspected until he was well into his forties. But the man who had spent a lifetime teaching geometry to beginners was well aware that despite claims of self-evidence, actually coming to *see* the absolute truth of geometrical axioms may require considerable effort. "That a geometrical axiom may be doubted, and even denied, by those who have not a clear conception of its terms, we know to be true," he explained. Year after year, he pointed out, students rebelled when they were first told that two lines could not enclose a space, and he himself remembered drawing "two straight lines upon the ground, in thought," and following them further and further until "he made them meet again at the antipodes, and inclose a gore of the sphere."[27] But, he went on, with education, all of these rebellions ended in insight and everyone who truly understood the meaning of parallel lines came to recognize the necessary truth of the parallel postulate.

De Morgan was convinced that the same would happen to anyone who truly engaged with the four-color axiom. This fact had not been recognized until the middle of the nineteenth century, but now that it had been uncovered, those who considered it could come to know its truth with the same certainty that they knew the parallel postulate. For De Morgan that such a straightforward and practical axiom had just been discovered was marvelous evidence of mathematical progress. He exulted in being able to stand at the cutting edge of human understanding.

Despite his excitement, few saw De Morgan's four-color axiom the way he did. In 1878 Arthur Cayley reopened the subject in a meeting of the LMS. The striking aspect of Cayley's presentation was that the claim that four colors suffice to color a map ceased to be an "axiom" and became instead a "statement" that required a "solution."[28] This choice of words marked a significant change in the fortunes of Guthrie's observation. That four colors suffice to color a map was never again a "fact" or an "axiom"; rather it was the "four-color problem" the "four-color conjecture," or the "four-color theorem." This change in terms was fundamental. The new words signaled the overthrow of the human element that stood in the center of De Morgan's mathematics of reason. It was no longer enough for a person to spend a couple of hours experimenting with colored pencils to establish a mathematical proposition as a fact.

Well into the twentieth century mathematicians tried to establish in their terms what De Morgan had seen so clearly in his. It is highly doubtful that he would have been impressed by the way their efforts were effectively showing that his "axiom" depended "on anything I see more clearly," but the new group set little store by his form of insightful understanding. Within a decade of his death, his vision had evaporated so completely that his most successful mathematics student, J. J. Sylvester, could simply dismiss him with the comment, "He did not write Mathematics, he wrote about Mathematics."[29] For Sylvester, mathematical truth could not be validated by the convictions of people who experienced it as insight. It had to be established through a process of mathematical reason that could be codified. From this point of view, all of De Morgan's concerns with meaning were obfuscations that distracted from the essence of the subject. Real mathematics stood pure and absolute above all of the human contingencies of language, reason, and personal insight that bound De Morgan's work to the ultimate truths of the universe.

There was a strong generational element in the move away from De Morgan's mathematics of reason. Even as the aging man stood before the fledgling society his son had called together, his friends were being overtaken by mortality: Boole died in the year before his LMS address, Hamilton was to die later in that year, Whewell in the year after that. De Morgan himself recognized that much of his audience did not agree with the view of mathematics he was painting, but he declared, "I have entire faith in the future."[30] De Morgan's children were entering their twenties. It was time to pass the torch of reason to a new generation.

In a counterfactual world, the next generation of the family of reason would have been headed by William Kingdom Clifford. He entered Trinity

College, Cambridge as a deeply committed Anglican in 1863, and then followed the familiar trajectory of falling away from Anglicanism because its doctrines did not accord with his understanding of reason. He graduated from Cambridge in 1866 as second wrangler and second Smith's Prizeman, was elected to membership in the LMS on May 21, 1866, and four years later succeeded De Morgan as professor of mathematics at UCL. Like De Morgan and Frend before him, Clifford was a consummate teacher who devoted himself to the implications of reason in his classes and writings. Despite all of this, he did not become friends with De Morgan and did not marry one of the De Morgan daughters. Thus one of the De Morgan children would have to carry the torch forward.

Aside from his name, William Frend De Morgan would seem a poor choice for this role. Neither a theologian nor a mathematician, he was a rebellious son who deeply disappointed his father by leaving UCL after only one year.[31] Augustus reluctantly accepted his eldest's decision to pursue an alternative education at the Academy of Arts, but William did not thrive there either. Instead, he became friends with William Morris and Edward Burne-Jones, decided he was not really a painter, and once again dropped out of school to pursue the debonair life of a bohemian artist, while designing tiles, stained glass, and furniture for Morris & Co.

In 1866, changing circumstances in his natal family began putting pressure on William's carefree lifestyle. The troubles began in November of that year, when his father became involved in a controversy about religious neutrality at UCL. The issue centered on who should be named to the chair of mental philosophy and logic. Several months of searching identified two candidates for this position: the Unitarian minister, James Martineau, who held the position of professor of mental and moral philosophy and political economy at the Unitarian Manchester New College, and the young, religiously neutral Croom Robertson. The senate of the university recommended that the more experienced Martineau be offered the position, but the Council, which had the power to make the appointment, decided to hire Robertson instead. The politics behind this decision were complex. At one extreme stood those who found any public religious commitment to be a violation of the secular character of UCL and therefore supported Robertson; at the other stood those who saw the Council to have allowed Martineau's religious convictions to enter into a hiring decision and therefore to have violated the principle of religious neutrality. De Morgan was a member of the latter group, and as the London papers exploded into a discussion of the rights and wrongs of the situation, he wrote a detailed letter explaining his position and announcing his determination to resign at the end of the year if Robertson were hired.

De Morgan undoubtedly hoped that the weight of his decades of service to the college would carry the day, but he was wrong; rather than changing their considered decision, the Council simply accepted his resignation. For the next six months, he continued to teach his classes while the Council's decision to hire Robertson moved through the bureaucracy until it was finally ratified in a special meeting of the Proprieters. Too proud and angry to accept support from his former students and colleagues, De Morgan was alone when, at the end of the term, he packed up all of his teaching materials and left the college to which he had devoted so much of his life. Sophia tried to comfort him by pointing out all of the "strong and honest minds" he had cultivated through the years, but her support was not enough.[32] De Morgan had built his entire life around what he understood to be the secular principles of UCL, and "every one who saw him observed the change which had passed over him" when he saw his college to be violating those principles.[33]

De Morgan's resignation from UCL was hardly the end of the family's troubles, however. Just four months later, in October 1867, the De Morgans' mathematical son George died of tuberculosis. Augustus claimed to "bear it well," bolstered by "a strong and practical conviction of a better and higher existence" that "reduces the whole thing to emigration to a country from which there is no way back," but he may have been protesting too much.[34] Within a year, a stroke left him "so prostrated that it was evident he never again would be equal to sustained effort."[35] From then on, William found himself called upon to act as a major support for his mother.

The first order of business was to move the family around the corner to a more affordable house at 6 Merton Road. While Sophia busied herself with the challenge of fitting furniture into the smaller space, Augustus set to work designing the layout of his new library. He had to sell a considerable number of books to finance the move, but more than three thousand volumes remained to be sorted and put away. "They *shall* all go in, and I will put them all in myself,"[36] he insisted, as he found an appropriate place for each one. It took several months to set his world in order, but when it was done, the now feeble old man again nestled into his library. There he returned to the ways of his forefathers and focused "a good deal" of his reasoning attention on "reading the Greek Testament, and comparing the different versions and translations."[37] His creative energies fading, De Morgan joined Lindsey's and Frend's search for truth in the Bible.

Even so, Augustus's new library oasis could no more protect him from tragedy than any of his other ones had. Soon after George's death, Chrissy began coughing in troubling ways. In response, Edward took her on an ocean trip, ostensibly as an adventure but actually in hopes that the warmth of southern climes would restore her to health. The journey was a great success for Edward, who was so taken by South Africa that he decided to stay, but Chrissy returned to London sicker than ever. In a desperate attempt to find healing air, Sophia and William took her to the seaside in Bournemouth, but Augustus was far too frail and miserable to accompany them out of town. "Say something to my mother!" William pleaded from his dying sister's bedside, but his father could find nothing to say.[38]

After burying Chrissy, Sophia and William returned to London, where they found Augustus "so weak that he had that day fallen on the floor, and was unable to rise without help."[39] Seven months later, on March 18, 1871, Augustus De Morgan died and was buried with Alice and George in the family plot in Kensal Green. Sophia then sold his library to the University of London, which scattered his books like ashes through their collections.[40]

The following year, Sophia gathered De Morgan's notes and articles into *A Budget of Paradoxes,* a popular book which was most recently reissued in 2015. Nonetheless, money was tight. Annie's marriage to Dr. R. E. Thompson in 1872 removed one dependent from the household, which meant Sophia and Mary could move to another, yet smaller house on Cheyne Row, Chelsea. Once settled in her new address, Sophia turned her attention to establishing the righteousness of her husband's causes in a *Memoir of Augustus De Morgan.* Finding the best way to present his life's work entailed considerable negotiation with family and friends, but by 1882 she was finally satisfied that she had succeeded in explaining her husband's life of reason. Sophia then turned to herself. Working with Mary to draw her memories together occupied her through the last decade of her life.

Outside of Cheyne Row, the family's troubles continued. In 1877 Edward died falling from a horse in South Africa, and in 1884 Annie died of the same disease that had taken Alice, George, and Chrissie. Sophia's communions with the dead sustained her through all of these losses. Until the very end of her life, she remained as warmly interested in everyone around her as had been the father she so adored. She became actively involved in the movement against vivisection, and from the time of its founding in 1882 she continued her explorations of the afterlife as a member of the Society of Psychical Research in

London. In 1895, at the age of eighty-seven, Sophia De Morgan died in her sleep. She was buried with Augustus, Alice, and George in Kensal Green.

Despite all of the tragedies that surrounded her immediate family, Sophia remained a believer in progress who always pronounced the world to be greatly improved since her childhood. But as she was writing her *Memoir of Augustus De Morgan,* William Clifford — the mathematical thinker who did not marry a De Morgan daughter — was doing everything he could to drive a stake through the heart of the reason that had supported her and the family for generations. Clifford spent his undergraduate years at Trinity College, where Whewell was master, and emerged determined to destroy the power of reason that was the foundation of Tory Anglicanism. His chosen weapon was non-Euclidean geometry. This kind of geometry, represented, for example, by the longitude lines on a globe, is one in which the parallel postulate does not hold. Navigators had for centuries been plotting their courses in ways that acknowledged the intersection of all longitudinal straight lines at the poles, but they saw no reason to question whether the space we live in might be similarly curved. Clifford, however, did. He recognized that all claims for the transcendent power of human reason rested on the assumption that in our space parallel lines do *not* intersect at some far distant poles. The peculiarity of this position lay in the fact that it was a claim to know what happened at an infinite distance, which could never be directly experienced. And so, from at least the time of Locke, coming to see that the parallel postulate was absolutely true was the experience that established human's power to truly know the universe. De Morgan's account of the student who moved from the erroneous belief that two lines could enclose a space to the wonderful recognition of the absolute truth of the parallel postulate was a universally recognized story of personal insight that established the transcendent power of reason.

For Clifford, however, this educational narrative was a tale of indoctrination rather than insight. Recognizing that there were alternatives to the parallel postulate returned power to the student, whose original conviction that two lines could enclose a space might actually be true. Even more satisfying was the way the simple *possibility* that the space we live in could be non-Euclidean destroyed the claims of absolute truth that had upheld Anglican — and Unitarian — theology since the time of Newton. Clifford was thrilled to be thus freed from the religion that he found so oppressive. He gleefully adopted the word "agnostic" to express his conviction that non-Euclidean geometry had cast doubt on the very existence of God.[41]

The non-Euclidean possibility also destroyed all claims that the practice of reason could lead to true understanding. "The idea of the Universe, the Macrocosm, the All, as subject of human knowledge, and therefore of human interest, has fallen to pieces," Clifford proclaimed triumphantly.[42] And so, while Sophia was commemorating her husband's life, Clifford was leading a charge against the understanding of reason that had powered and shaped the lives of her family for over a century. It would be decades before his challenge to the absolute truth of Euclidean geometry would become mainstream, but his work pointed to a future in which human experiences of reasoned insight could no longer be used to support claims of truth.

There is little reason to think that William De Morgan engaged with non-Euclidean geometry, but his educational trajectory suggests that he did not need Clifford to persuade him that the experience of accepting the fifth postulate did not capture the essence of his being. In the difficult years that surrounded his father's death, he was finding his way in ceramics. In 1872, he moved to a house just three doors down from his mother on Cheyne Row and set up a pottery studio. It took him several years and many mistakes to master the complicated processes involved in making pots and tiles. But when he succeeded in taming his medium, the ebullient imagination that was relegated to the margins of his father's reasoned world exploded in images of adventuring ships, fire-breathing dragons, contemplative mermaids, and imps peering out from among flowers. The colors he finally mastered flashed in exuberant peacock's tails, deep blue oceans, and bees that positively buzz with redness. He easily matched the playfulness his father expressed in his doodles; in fact, Charles Dodgson, better known as Lewis Carroll, is said to have written *The Hunting of the Snark* in response to the De Morgan tiles he had installed in his college rooms.[43] But William also recognized and respected the deeper regularities underlying his quirky sense of humor. The symmetries that supported his father's logical thinking hold his imps in place, control the rolls of his dolphins, shape the flights of his dragons, and structure whole walls covered with carnations, roses, daisies and swans. William did not claim that he was practicing reason when he created his designs, but the intense experiences of insight that inspired his father's logical researches leap directly from his pieces.

In 1887 a forty-eight-year-old William emulated his grandfather and startled his friends with a decision to get married. Evelyn Pickering was fifteen years his junior and had already established herself as a painter at the time of their marriage. The talented new member of the family was in many ways a suitable heiress to Sophia's world of strong women. On her canvases, resilient sea maidens stand united among ocean waves, and powerful women direct thunderstorms. But the younger woman also struggled against the limitations of the world that Sophia's generation defined for her. In others of her paintings, women are crushed in captivity and weep by the waters of Babylon. Evelyn and Sophia may have disagreed about the proper role of women in society, but they saw eye to eye in other ways. Sophia's ways of understanding pain are reflected in Evelyn's later works, where spirits pull away from exhausted bodies, and Christ rises from a graveyard supported by angels. In her work, Evelyn expanded upon the reflections on matter and spirit that her mother-in-law had begun.

William's and Evelyn's art may thus be seen as expressions of their inheritance, but neither of them felt the need to reason their way to the validity of the worlds they depicted. Their late nineteenth-century world was very different from the eighteenth-century one in which the Lindseys and the Frends had grounded their vision of a truly Christian Church in the power of reason. It was almost equally different from the early nineteenth-century world in which Augustus and Sophia had used their reason to explore powerful new understandings of mathematics, logic, and life after death. In William's and Evelyn's world the nature, power, and limits of reason were being actively redefined. They responded in and through their work. William's tiles and Evelyn's canvases stand as portals into the transcendent understandings that their forebears had entered through reason.

# *Epilogue*

In January 1839, when Alice De Morgan was just six months old, a new couple moved onto Gower Street. The newcomers considered themselves Unitarians, though they were more socially conservative than the De Morgans and had married in an Anglican church. Like De Morgan, the head of household was Cambridge-educated and had as a student been introduced to and challenged by the ideas of men like William Whewell and John Herschel. After completing his studies, he too had decided not to become a clergyman, and by the time he and his wife moved to Gower Street he was as involved in the scientific community as De Morgan was, serving as the secretary of the Geological Society. To complete the connection, at the end of that year, both families had sons they named William.

It is pleasant to think of these two young men on the rise raising their hats to each other on Gower Street and sharing the ups and downs of life with their infant Williams. However, there is no reason to believe that they ever did so. In fact, in all the mounds of personal and professional correspondence that each of them left behind, neither ever mentions the other. Despite the thickets of connection that would seem to bind Augustus De Morgan and Charles Darwin together, including three years as close neighbors, it is as if the two men lived in two completely different worlds.

That difference may be seen in the two families' responses to the urban street they shared for three years. What was for the De Morgans a wonderfully

convenient neighborhood in the midst of an intellectually vibrant environment was for the Darwins a filthy road in a Dickensian city that they escaped
as soon as possible. The urban/rural preference that separated De Morgan
from Darwin ran very deep. It was in fact so visceral for both men that in the
final decades of their lives, each was essentially unable to travel to the world
of the other.

Fundamentally different views of the world lay behind the two men's differently focused agorophobias; their refusals to leave their familiar haunts was
an insistence on remaining in touch with what was for them the most real.
Darwin's move to the country was a return to a natural world that for him
comprised existence; De Morgan's world was to be found in the words of the
reasoning forebears who were gathered around him in the books of his library.
Darwin may well have spent as much time in his Down library as De Morgan
did in his London one, but the truth the naturalist was seeking was always
the truth of an external world that was but weakly described by the words in
books. There was nothing weak in the written words De Morgan worked with,
however; all of his life was spent shining the light of reason into a world that
was wholly embedded in them.

The difference that divided De Morgan from Darwin had distinguished the
family of reason from Darwin's maternal Wedgwood and paternal Darwin ancestors for generations. Lindsey's friendship with Priestley meant that he was
separated by at most one degree from Darwin's forebears, but otherwise their
worlds did not overlap. Darwin's ancestors were provincial natural philosophers who stood at the forefront of the Industrial Revolution. Priestley's work
on oxygen fit in well with their concerns, but they were politely uninterested
in his pursuit of reason. The contrast between these interests points to a fundamental split that divided the classical sciences of mathematics and astronomy from the more modern, empirical sciences like physics or chemistry. From
the seventeenth through the nineteenth centuries, these two traditions both
claimed the mantle of science, but they were not the same.[1] Mathematics and
astronomy might be described as meditative studies, in which knowledge con-
stitutes understanding. In the empirical sciences knowledge could be equated
to power. Members of the family of reason all found ways to ground their lives
in the contemplative side of this split. The doctors and potters of Charles Darwin's family thrived on the other, pragmatic side. And the difference between
the two groups continued in the thought and lives of their Victorian offspring.

Throughout their lives, De Morgan and Darwin lived in two complementary worlds that were all but invisible to each other. But over the course of

their shared nineteenth century, the developing strength of scientific ways of knowing was shifting the balance of power. Darwin's *Origin of Species* may serve as a convenient stand-in for the myriad forces that were challenging the intellectual world of the family of reason. In Darwin's evolutionary theory, Locke's faculty of reason was reduced to a chance variation in a process that was focused on survival rather than understanding. In a Darwinian world, reason might still differentiate humans from animals, but there is no reason to believe that practicing it would lead to deep insights into the transcendent order of the universe.

The *Origin* may be useful as an indicator, but the watershed that it marks had been building since before De Morgan and Darwin were apparently unaware of each other on Gower Street. Over the course of their lives, the De Morgans contended with a host of forces that threatened to overturn their essential conviction that humans were able to reason their way to a true understanding of the world. De Morgan pushed back against Cauchy's efforts to replace reason with rigor in mathematics, against Laplace's efforts to transform reason into probabilistic calculation, against Hamilton's and Mansel's efforts to separate logic from human thinking, and against a scientific community that was determined to control his wife's investigations into a spirit world. All of these efforts may be seen as attempts to defend the value of human agency against powerful alienating forces. And from a twenty-first century perspective, it may seem as if all of them failed.

However, the truths of our convictions lie less in the words we use to describe them than in the lives we lead to fulfill them, and by this measure the family's efforts to lead lives of reason have lasted. The Lindseys' insistence on speaking the truth that reason revealed to them is still manifest in the headquarters of the English Unitarian Church that stands on the site of their Essex Street Chapel. Students at Cambridge still identify with William Frend, whose bust watches over readers in the library of Jesus College, and the Society of Antiquaries of London still awards the Frend Medal in honor of a namesake who devoted his life to the study of the archaeological and material remains of the early Christian Church. Students in the mathematics department at UCL still gather in meetings of the Augustus De Morgan Society, and his bust gazes on all who enter the library of the University of London. Women don't tend to be memorialized in these ways, but Sophia's spirit certainly rejoices with every poverty-stricken family who finds some form of support, every person who finds a way out of mental illness, and every woman pursuing her education at the Royal Holloway College, which absorbed the Bedford School for Girls.

In these ways and more, the family's activities in support of a world in which all are able to follow their reason continue to reverberate.

At the same time, examining their lives reveals the kinds of internal tensions that lurk behind the commitments to reason that these monuments commemorate. Frend may have seen himself as engaging London's Jewish community in reasoned discussion, but the enduring friendships that resulted were grounded in open-hearted human contact, not reason. He reveled in endless conversations, but they do not seem to have changed anyone's mind. Lindsey, for his part, devoted himself to creating a religion that was universal because reasoned. Nonetheless, even as he offered his insistence on the humanity of Jesus as a way to bring Jews and Moslems into the fold, he refused to recognize that it also cut the vast majority of his countrymen out of the discussion entirely. De Morgan spent his life opening the insights of his exclusive Cambridge education to the young men of London, but he was a very rigid thinker. Under pressure from Sophia and his experiences with Alice, he agreed that girls deserved an education, but he never thought they should be included in his world. As these examples suggest, even as the family talked of opening their communities to the diversity of the world's people, empowering the disenfranchised, and lifting up the dispossessed, their insistence that reason is the only way to come to truth had its downsides. Not only was reasoned discussion not always the way to bring people together, insisting that it was could also serve to discredit those who approached the world differently.

Much has changed since the Lindseys founded their chapel, Frend fought his battles, and the De Morgans pursued their researches. Nonetheless, finding ways to embrace diversity is a challenge as pressing today as it was when Lindsey was reaching out to the Jews of London. Standing up against the threat of authoritarian political forces is as urgent in the early twenty-first century as it was when Frend was tried and banished in the late eighteenth. Creating a reasoning citizenship is as essential to us as it was when De Morgan was teaching the marginalized young men of London and Sophia was raising her children to tell the truth. Our ways of dividing the world may have effectively dismembered the conceptual matrix in which their understandings were embedded, but the power of the reason that is common to all of humanity forms the bedrock of the democratic systems of government they bequeathed to us. Following their lives, which were shaped by an unswerving commitment to cultivating, sharing, and defending the power of reason, may serve at once as a comfort and a goad, an inspiration and a warning as we negotiate our own very challenging times.

# Notes

*BIBLIOGRAPHIC ABBREVIATIONS*

## Archives

CUL—Cambridge University Library, Department of Manuscripts and University Archives

DWL—Dr. Williams's Library, London, Letters from Archdeacon Francis Blackburne to the Reverend Theophilus Lindsey, MS 12.52

NJ—Sophia Frend De Morgan, Nursery Journal, 1839–1842 (DMF_MS_0024), De Morgan Collection, London

OBL—Oxford, Bodleian Libraries, Archive of the Noel, Byron, and Lovelace Families

RS:HS—Royal Society, London, Herschel Papers

RHUL—Royal Holloway University of London, Bedford College Archives

SHL—Senate House Library, University of London

TCL—Archives of Trinity College, Cambridge University

UCL—University College London Archives

## People

FB—Francis Blackburne

LNB—Lady Noel Byron

MF—Mary Frend

SF—Sophia Frend

SEDM—Sophia Elizabeth De Morgan

TL—Theophilus Lindsey

WF—William Frend

WRH—Sir William Rowan Hamilton (Irish)

## INTRODUCTION

1. OBL, box 71, folios 102–3.
2. ADM's description of the conversation in UCL, MS Add. 7, p. 39.
3. Sophia Elizabeth De Morgan, *Memoir of Augustus De Morgan* (London: Longmans, Green 1882), 88.
4. Psalm 15:2, 4.
5. For an analysis of the various meanings of reason in the eighteenth century, see Arthur Lovejoy, "The Parallel of Deism and Classicism," in *Essays in the History of Ideas* (Baltimore: Johns Hopkins University Press, 1948), 78–98.

## 1. FAITH IN REASON

1. Catharine Cappe and Mary Cappe, *Memoirs of the Life of the Late Mrs. Catharine Cappe* (Boston: Wells and Lily, 1824), 5.
2. Nikolaus Peysner, *Yorkshire: The North Riding* (Harmondsworth: Penguin Books, 1966), 290.
3. Sheldon Rothblatt, "The Student Sub-culture and the Examination System in Early 19th Century Oxbridge," *The University in Society,* vol. 1 (Princeton: Princeton University Press, 1974), 248.
4. D. A. Winstanley, *Unreformed Cambridge* (Cambridge: University Press, 1935), 200–201.
5. Winstanley, *Unreformed Cambridge,* 202.
6. The classic text for the "Holy Alliance" between science and liberal theology is Margaret C. Jacob, *The Newtonians and the English Revolution: 1689–1720* (New York: Gordon and Breach, 1976).
7. Francis Blackburne, *The Works, Theological and Miscellaneous* (Cambridge: Benjamin Flower, 1804), I: iv n.
8. Blackburne, *Works,* I: vi.
9. Blackburne, *Works,* I: vi.
10. Blackburne, *Works,* I: vi.
11. Blackburne, *Works,* I: iv–v.
12. Cappe and Cappe, *Memoirs,* 80.
13. FB to TL, Jul 1756. DWL 12.52 (14).
14. FB to TL, May 1757. DWL 12.52 (46).
15. FB to TL, Nov 15, 1756, in Thomas Belsham, *Memoirs of the Late Reverend Theophilus Lindsey, M.A. Including a Brief Analysis of His Works* (London: J. Johnson, 1812), 494.
16. FB to TL, Mar 8, 1757. DWL 12.52 (37).
17. Catharine Cappe, "Memoir of the Late Rev. Theophilus Lindsey, A.M," *Monthly Repository of Theology and General Literature* 3 (1808): 638.
18. Cappe and Cappe, *Memoirs,* 97.
19. Cappe, "Lindsey," 638.
20. Belsham, *Memoirs of Lindsey,* 7.
21. Belsham, *Memoirs of Lindsey,* 7.

22. This meant that when the incumbent died, Lindsey could take the position himself or bestow it on whomever else he might like. Cappe, "Lindsey," 638.
23. John Locke, *The Reasonableness of Christianity: As Delivered in the Scriptures,* in Higgins-Biddel, Clarendon Edition of the Works of John Locke (Oxford: Clarendon Press, 1999), 169–70.
24. Locke, *Reasonableness,* 169–70.
25. Locke, *Reasonableness,* 168.
26. Locke, *Reasonableness,* 23.
27. The Thirty-Nine Articles remain the basic statement of faith of the Anglican Church.
28. [Francis Blackburne], *The Confessional or A Full and Free Inquiry into the Right, Utility, Edification, and Success, of Establishing Systematical Confessions of Faith and Doctrine in Protestant Churches,* 2nd ed. (London: S. Bladon, 1767), 225. The reference here is to John 8:32, "Then you shall know the truth and the truth shall set you free."
29. [Blackburne], *Confessional,* 224–25.
30. Clarke quoted in [Blackburne], *Confessional,* 205. For a fuller treatment of Clarke's position, see John Gascoigne, *Cambridge in the Age of the Enlightenment* (Cambridge: Cambridge University Press, 1989), 115–23.
31. [Blackburne], *Confessional,* 213.
32. [Blackburne], *Confessional,* 223.
33. [Blackburne], *Confessional,* 234.
34. [Blackburne], *Confessional,* 102.
35. Belsham, *Memoirs of Lindsey,* 10.
36. Lindsey to the Countess of Huntington, 1755, quoted in G. M. Ditchfield, *Theophilus Lindsey: From Anglican to Unitarian* (London: Dr. Williams's Trust, 1998), 10.
37. FB to TL, Jan 2, 1756. DWL 12.52 (1). Much of this letter is reprinted in John Towill Rutt, ed., *Memoirs and Correspondence of Joseph Priestley* (London: Society for the Diffusion of Useful Knowledge, 1817–32), I: 83n.
38. FB to TL, Jan 2, 1756. DWL 12.52 (1).
39. FB to TL, Jan 6, 1756. DWL 12.52 (1*).
40. FB to TL, Mar 26, 1756. DWL 12.52 (7).
41. FB to TL, Mar 19, 1756. DWL 12.52 (6).
42. FB to TL, Feb 10, 1756. DWL 12.52 (4).
43. FB to TL, Jan 2, 1756. DWL 12.52 (1).
44. FB to TL, Mar 26, 1756. DWL 12.52 (7).
45. FB to TL, Mar 19, 1756. DWL 12.52 (6). The reference here is to John Jones, who published a pamphlet, "Free and Candid Disquisitions Relating to the Church of England and the Means of Advancing Religion Therein." Ditchfield, *Lindsey,* 9.
46. FB to TL, May 2, 1756. DWL 12.52 (9). Blackburne's understanding of his situation is rooted in Matthew 12: 43–45.
47. FB to TL, Jul 20, 1756. DWL 12.52 (15).
48. FB to TL, Jan 2, 1756. DWL 12.52 (1).
49. FB to TL, Mar 19, 1756. DWL 12.52 (6).
50. FB to TL, Mar 11, 1757. DWL 12.52 (38).

51. FB to TL, Jan 2, 1756. DWL 12.52 (1). Blackburne wrote this in the first person: "That I must fulfill with all my heart."
52. FB to TL, Jan 2, 1756. DWL 12.52 (1).
53. Locke, *Reasonableness,* 169–70.
54. FB to TL, Jan 2, 1756. DWL 12.52 (1).
55. FB to TL, May 2, 1756. DWL 12.52 (9).
56. FB to TL, Jan 25, 1757. DWL 12.52 (28).
57. FB to TL, Mar 1757. DWL 12.52 (42).
58. FB to TL, Feb 10, 1756. DWL 12.52 (4).
59. FB to TL, Mar 26, 1756. DWL 12.52 (7).
60. FB to TL, Mar 1757. DWL 12.52 (47).
61. FB to TL, Feb 23, 1759. DWL 12.52 (60).
62. FB to TL, Jul 3, 1759. DWL 12.52 (66).
63. Fb to TL, Dec 18, 1759. DWL 12.52 (72).

## 2. BREAKING AWAY

1. FB to TL, Feb 1, 1763. DWL 12.52 (73).
2. Thomas Belsham, *Memoirs of the Late Reverend Theophilus Lindsey, M.A. Including a Brief Analysis of His Works* (London: J. Johnson, 1812), 11.
3. FB to TL, Mar 29, 1763. DWL 12.52 (80)
4. For a somewhat antiquated but straightforward overview of the development of anti-Trinitarian ideas in Europe and America, see Earl Morse Wilbur, *Our Unitarian Heritage* (Boston: Beacon Press, 1925). The entire text of this work can also be found on the web at http://online.sksm.edu/ouh.
5. Lindsey, Apology. Quoted in Belsham, *Memoirs of Lindsey,* 22.
6. Lindsey, Apology. Quoted in Belsham, *Memoirs of Lindsey,* 13.
7. In G. M. Ditchfield, *Theophilus Lindsey: From Anglican to Unitarian* (London: Dr. Williams's Trust, 1998), 3n11, Ditchfield lays out the career of the man who took the position Lindsey rejected as a "neat illustration" of the probable shape of the road Lindsey did not travel. "Dodgson was an exact contemporary of Lindsey at St John's college, Cambridge; in 1755 he succeeded Lindsey as rector of Kirby Wiske, through the patronage of Northumberland." Upon accepting the offer Lindsey declined, Dodgson went on to become bishop of Ossory in 1765, and then of Elphin in 1775.
8. Catharine Cappe and Mary Cappe, *Memoirs of the Life of the Late Mrs. Catharine Cappe* (Boston: Wells and Lily, 1824), 106.
9. FB to TL, Feb 1, 1763. DWL 12.52 (73); FB to TL, Apr 30, 1763. DWL 12.52 (84); FB to TL, May 9, 1763. DWL 12.52 (85).
10. Cappe and Cappe, *Memoirs,* 95.
11. Cappe and Cappe, *Memoirs,* 95.
12. Cappe and Cappe, *Memoirs,* 95.
13. Cappe and Cappe, *Memoirs,* 95.
14. Cappe and Cappe, *Memoirs,* 97.
15. Cappe and Cappe, *Memoirs,* 92.

16. Cappe and Cappe, *Memoirs,* 93.

17. Cappe and Cappe, *Memoirs,* 104.

18. Catharine Cappe, "Memoir of Lindsey," quoted in Belsham, *Memoirs of Lindsey,* 29–30.

19. Catherine Cappe, *Observations on Charity Schools, Female Friendly Societies, and Other Subjects Connected with the Views of the Ladies Committee* (York: W. Blanchard, 1805), ii.

20. Cappe and Cappe, *Memoirs,* 99.

21. Cappe and Cappe, *Memoirs,* 93.

22. The pamphlets are listed in *Gentleman's Magazine* 41 (1771): 405; 42 (1772): 263. See also a "Short View of the Controversy" by John Disney, 1773.

23. Lindsey, Apology. Quoted in Belsham, *Memoirs of Lindsey,* 32.

24. Belsham, *Memoirs of Lindsey,* 33.

25. For Priestley's life and work, see Robert Schofield, *The Enlightenment of Joseph Priestley: A Study of His Life and Work from 1733 to 1773* (University Park: Pennsylvania State University Press, 1997); Robert Schofield, *The Enlightened Joseph Priestley: A Study of His Life and Work from 1773 to 1804* (University Park: Pennsylvania State Press, 2004); Isabel Rivers and David L Wykes, eds., *Joseph Priestley, Scientist, Philosopher, and Theologian* (Oxford: Oxford University Press, 2009).

26. John Towill Rutt, ed., *Memoirs and Correspondence of Joseph Priestley* (London: Society for the Diffusion of Useful Knowledge, 1817–32), I: 82.

27. John Locke, *The Reasonableness of Christianity: As Delivered in the Scriptures,* in Higgins-Biddel, Clarendon Edition of the Works of John Locke (Oxford: Clarendon Press, 1999), 169–70.

28. Belsham, *Memoirs of Lindsey,* 35.

29. Cappe and Cappe, *Memoirs,* 123.

30. For the Feather's Tavern Petition, see John Gascoigne, *Cambridge in the Age of the Enlightenment* (Cambridge: Cambridge University Press, 1989), and John Gascoigne, "Anglican Latitudinarianism and Political Radicalism in the Late Eighteenth Century," *History* 71 (1986): 22–38. Contemporary treatments are to be found in Francis Blackburne, *The Works, Theological and Miscellaneous* (Cambridge: Benjamin Flower, 1804). For Jebb, see John Disney, *The Works, Theological, Medical, Political, and Miscellaneous, of John Jebb, MD, F.R.S. with Memoirs of the Life of the Author* (London: T. Cadell, 1787). For Lindsey's role in the movement as well as a one-sided presentation of the debate in Parliament, see Belsham, *Memoirs of Lindsey,* 46–61.

31. [Francis Blackburne], *The Confessional or A Full and Free Inquiry into the Right, Utility, Edification, and Success, of Establishing Systematical Confessions of Faith and Doctrine in Protestant Churches,* 2nd ed. (London: S. Bladon, 1767), x.

32. [Blackburne], *Confessional,* xi.

33. TL to John Jebb, Mar 3, [1772], in Rutt, *Priestley Memoir,* 160n.

34. Belsham, *Memoirs of Lindsey,* 62.

35. Rutt, *Priestley Memoir,* 160.

36. JP to TL, Mar 9, 1772, in Rutt, *Priestley Memoir,* 161. The Lindsey side of this correspondence has disappeared. I am inferring his response from Priestley's reaction.

37. FB to TL, May 13, [1773]. DWL 12.52 (100).

38. Blackburne, *Works,* I: xlviii.

39. FB to TL, Oct 13, 1775. DWL 12.52 (102).

40. Belsham, *Memoirs of Lindsey,* 77–78.

41. Cappe and Cappe, *Memoirs,* 94.

42. FB to TL, Jan 2, 1756 (1). DWL 12.52 (1).

43. Theophilus Lindsey, *A Farewell Address to the Parishioners of Catterick* (London, 1774), 10. The last quoted phrase is in italics in the original.

44. Cappe and Cappe, *Memoirs,* 144.

45. Hannah Lindsey to Catharine Harrison, Dec 5, 1773, in Cappe and Cappe, *Memoirs,* 143.

46. Belsham, *Memoirs of Lindsey,* 79.

47. Cappe and Cappe, *Memoirs,* 146–47.

48. For Priestley's natural philosophical thinking, see John G. McEvoy, "Joseph Priestley, 'Aerial Philosopher': Metaphysics and Methodology in Priestley's Thought, 1772–1781," *Ambix* 25 (1978): 1–55, 93–116, 153–75; John G. McEvoy, "Joseph Priestley, 'Aerial Philosopher': Metaphysics and Methodology in Priestley's Thought, 1772–1781," *Ambix* 26 (1979): 16–38.

49. Samuel Clarke, William Burgh, and Theophilus Lindsey, *The Book of Common Prayer Reformed According to the Plan of the Late Dr. Samuel Clarke: Together with the Psalter or Psalms of David* (London: J. Johnson, 1774).

50. Rutt, *Priestley Memoir,* 225–26.

51. Belsham, *Memoirs of Lindsey,* 111.

52. Belsham, *Memoirs of Lindsey,* 92.

53. I have chosen this cross section of women for their relevance to this family story. For a somewhat longer list, see G. M. Ditchfield, "Hannah Lindsey and Her Circle: The Female Element in Early English Unitarianism," *Enlightenment and Dissent* 26 (2010): 54–79.

54. Ditchfield, "Hannah Lindsey," 67.

55. Ditchfield, "Hannah Lindsey," 62.

56. Belsham, *Memoirs of Lindsey,* 144.

57. Belsham, *Memoirs of Lindsey,* 144.

58. Belsham, *Memoirs of Lindsey,* 114.

59. Lindsey quoted in Belsham, *Memoirs of Lindsey,* 114.

60. Belsham, *Memoirs of Lindsey,* 115.

61. Belsham, *Memoirs of Lindsey,* 115.

62. Francis Blackburne, "Answer to the Question: Why Are You Not a Socinian?," in Blackburne, *Works,* I: cxxi.

63. FB to Wiche, Sep 9, 1783. DWL 12.45.

### 3. EDUCATION OF AN ANGLICAN

1. Memoir of William Frend [1826]. CUL, MS Add. 2886, misc. folder 285.

2. Memoir of William Frend [1826]. CUL, MS Add. 2886, misc. folder 285. George's drinking partner was Dr Richard Farmer, who became master of Emmanuel College,

Cambridge, from 1775 to 1797, and canon of Canterbury Cathedral from 1782 to 1788. Farmer carried his hard-drinking habits with him as he rose in stature.

3. Stella Corpe, "Canterbury Freemen 1700–1750" (final diss. in local history, University of Kent, 1982).

4. Memoir of William Frend [1826]. CUL, MS Add. 2886, misc. folder 285.

5. Memoir of William Frend [1826]. CUL, MS Add. 2886, misc. folder 285.

6. William Frend, *A Sequel to the Account of the Proceedings in the University of Cambridge against the Author of a Pamphlet Entitled Peace and Union* (London: The Author, 1795), 102–3 Frida Knight credits Six the elder with the invention of the maximum and minimum thermometer. Frida Knight, *University Rebel* (London: Victor Gollancz, 1971), 65n.

7. [Augustus De Morgan], "William Frend, Esq," *Christian Reformer or Unitarian Magazine and Review* 8, no. 90 (Jun 1841): 373; [De Morgan], "William Frend, Esq," 373.

8. WF to Harriet Frend, Mar 23, 1831. CUL, MS Add. 7886, folder 3, 76.

9. The personal nature of this admission process is typical and reflected in the historical sources; what we know of the Cambridge education is primarily to be found in personal memories. In Frend's case the most direct of these are in Frend, *Sequel;* the description of Paley in A Christian [William Frend], "Anecdotes of Dr. Paley," *Universal Magazine* 4 (1805): 415; and various memories that show up in his personal correspondence. Very useful, because written by a man who went through Christ's just seven years after Frend, is Henry Gunning, *Reminiscences of the University, Town and County of Cambridge from the Year 1780,* 2 vols. (London: George Bell, 1855). An early attempt to bring together the bewildering array of contemporary descriptions of the eighteenth-century Cambridge experience is Christopher Wordsworth, comp., *Social Life at the English Universities in the Eighteenth Century* (Cambridge: Deighton, Bell, 1874), but the result is so undigested as to be all but incomprehensible. The radically revised and abridged version, Christopher Wordsworth and R. Brimley Johnson, *The Undergraduate* (London: Stanley Paul, 1928) is considerably more accessible, and D. A. Winstanley, *Unreformed Cambridge* (Cambridge: University Press, 1935) is even more so. John Gascoigne, *Cambridge in the Age of the Enlightenment* (Cambridge: Cambridge University Press, 1989) opens the world of modern treatments.

10. Gunning, *Reminiscences,* I: 322.

11. Gunning, *Reminiscences,* I: 42–43.

12. Gunning, *Reminiscences,* I: 14–15.

13. George Pryme, quoted in Winstanley, *Unreformed Cambridge,* 205.

14. A Christian [William Frend], "Paley," 510. This piece has been widely attributed to Frend. The attribution is a bit delicate because Frend and Paley overlapped at Cambridge for at most two terms, but other unimpeachable sources, e.g., the letter to his daughter Harriet quoted in note 8, above, show Frend remembering Paley with great clarity and affection.

15. Wordsworth and Johnson, *Undergraduate,* 31.

16. Wordsworth and Johnson, *Undergraduate,* 32.

17. Gunning, *Reminiscences,* I: 7–8.

18. William Paley, *The Principles of Moral and Political Philosophy* (Boston: W. H. Whitaker, 1841), 141–42.

19. M. L. Clarke, *Paley: Evidences for the Man* (Toronto: University of Toronto Press, 1974), 20, n35.

20. After Jebb's reforms, students had simply to affirm their membership in "the Church of England as by law established." William Frend, *Considerations on the Oaths Required by the University of Cambridge at the Time of Taking Degrees* (London: J. Deighton, 1787), 44.

21. Winstanley, *Unreformed Cambridge*, 329. For a fuller discussion of Jebb's educational efforts, see John Disney, *The Works, Theological, Medical, Political, and Miscellaneous, of John Jebb, MD, F.R.S. with Memoirs of the Life of the Author* (London: T. Cadell, 1787) and Gascoigne, *Enlightenment Cambridge*, 202–5.

22. Frend, *Sequel,* 103. I have not seen the portrait. My description is from Knight, *University Rebel,* 31.

23. A Christian [William Frend], "Paley," 415.

24. John Locke, *An Essay Concerning Human Understanding,* ed. Alexander Campbell Fraser (New York: Dover Publications, 1959), I: 116.

25. Locke, *Essay,* II: 350. In the original, "increase" does not have a final *d.*

26. Frend, *Sequel,* 103 His slightly younger classmate was given the same message. Gunning, *Reminiscences,* I: 15.

27. Frend, *Sequel,* 103.

28. My description of the mathematical curriculum is taken from Gunning who went through Christ's College with Parkinson seven years after Frend. Gunning, *Reminiscences,* I: 14.

29. Isaac Newton to Richard Bentley, Dec 10, 1692, in Richard Thayer, ed., *Newton's Philosophy of Nature: Selections from His Writings* (Mineola, N.Y.: Dover Publications, 2005), 46.

30. Locke, *Essay,* II: 407.

31. Locke, *Essay,* II: 407.

32. Locke, *Essay,* II: 342.

33. Gillbert Wakefield, *Memoirs of the Life of Gilbert Wakefield, B. A., Late Fellow of Jesus College, Cambridge,* comp. and ed. John Towill Rutt and Arnold Wainewright (London: J. Deighton, 1792), 100, 101.

34. Locke, *Essay,* II: 306.

35. Locke, *Essay,* II: 306.

36. Locke, *Essay,* II: 309.

37. Locke, *Essay,* II: 309.

38. Locke, *Essay,* II: 310.

39. Winstanley, *Unreformed Cambridge,* 47.

40. Gunning, who kept his Acts in 1787, debated in the spring on readings from "the second and third sections of Newton, and Paley on utility" and the following fall on "the ninth and eleventh sections of Newton, and the Credibility of Miracles." Gunning, *Reminiscences,* I: 81, 86.

41. Locke, *Essay,* II: 126–27.

42. I am taking class size numbers from Wakefield, *Memoirs,* 100 For the eighteenth-century examination experience, see Winstanley, *Unreformed Cambridge,* 48–57; An-

drew Warwick, *Masters of Theory: Cambridge and the Rise of Mathematical Physics* (Chicago: University of Chicago Press, 2003), 52–58. For a student's perception of the examinations in the mid-1780s, see Gunning, *Reminiscences.*

43. For the Smith's Prize exam, see Warwick, *Masters of Theory,* 5–8.

44. *Dictionary of National Biography from the Earliest Times to 1900.* London: Oxford University Press [1959–1960], s.v. "William Frend." I have not seen any independent confirmation of this offer, but this source claims that the position would have earned Frend an extraordinary £2,000 per year, with a retirement pension of £800 per year for life.

45. George Frend to WF, Feb 23, 1779. CUL, MS Add. 7886, folder 2,42.

46. WF to MF, Jan 7, 1784. CUL, MS Add., misc. folder T45.

47. Neil McKendrick, John Brewer, and J. H. Plumb, *The Birth of a Consumer Society: The Commercialization of Eighteenth-Century England* (Bloomington: Indiana University Press, 1982).

48. WF to MF, [May] 1784. CUL, MS Add., misc. folder T64.

49. WF to MF, Feb 10, 1784. CUL, MS Add., misc. folder T46.

50. WF to MF, Jan 7, 1784. CUL, MS Add., misc. folder T45.

51. WF to MF, Feb 10, 1784. CUL, MS Add., misc. folder T46.

52. WF to MF, [May] 1784. CUL, MS Add., misc. folder T64.

53. MF to WF, May 2, 1784. SHL, MS 913B/3/1.

54. MF to WF, May 2, 1784. SHL, MS 913B/3/1. WF to MF, Feb 10, 1784. CUL, MS Add., misc. folder T46.

55. Gunning, *Reminiscences,* I: 26.

56. WF to MF, Jul 7, 1784. CUL, MS Add., misc. folder T48.

57. WF to MF, [Jun(?), 1784]. CUL, MS Add., misc. folder T63.

58. WF to MF, [soon after Nov 22, 1784]. CUL, MS Add., misc. folder T49.

## 4. THE ROAD TO UNITARIANISM

1. WF to MF, Nov 30, 1785. CUL, MS Add., misc. folder T50.

2. MF to WF, [Dec 18, 1785]. SHL, MS 913B/3/1.

3. MF to WF, Apr 15, 1786. CUL, MS Add., misc. folder T51.

4. Laqueur emphasizes the lower-class aspects of the movement in Thomas Walter Laqueur, *Religion and Respectability: Sunday Schools and Working-Class Culture, 1780–1850* (New Haven: Yale University Press, 1976). Cliff sees more clearly the kind of middle-class impetus that seems to have supported the Maddingly project in Philip B Cliff, *The Rise and Development of the Sunday School Movement in England: 1780–1980* (Surrey, Eng.: National Christian Education Council, 1986). Though both works recognize the eighteenth-century precursors of the movement, neither mentions the Cambridge project with which Frend was involved.

5. The official nature of Frend's position is unclear. There is a long De Morgan family tradition, reflected in much of the secondary literature, that Frend was the vicar in Maddingly. However, although records of Frend's ordination to and resignation from the ministry of Long Stanton are in the Cambridge archive, there is nothing there about

Maddingly. What is more, in George Dyer, *Memoirs of the Life and Writing of Robert Robinson* (London: G. G. and J. Robinson, 1796), Frend's friend and contemporary George Dyer mentions Frend only at Long Stanton. In addition, the impressively complete list of the vicars of Maddingly that begins in the fourteenth century did not include Frend until it was amended at the insistence of Frida Knight, who believed that Frend was the vicar there.

6. WF to LNB, 1818. Reprinted in Sophia Elizabeth De Morgan, *Threescore Years and Ten: Reminiscences of the Late Sophia Elizabeth De Morgan to Which Are Added Letters to and from Her Husband the Late Augustus De Morgan, and Others,* ed. Mary De Morgan (London: Richards Bentley and Son, 1895), 105–6n.

7. WF to LNB, 1818. Reprinted in De Morgan, *Reminiscences,* 106n.

8. Henry Gunning, *Reminiscences of the University, Town and County of Cambridge from the Year 1780,* 2 vols. (London: George Bell, 1855), I: 12.

9. WF to LNB, 1818. Reprinted in De Morgan, *Reminiscences,* 106n.

10. MF to WF, Dec 18, 1785. SHL, MS 913B/3/1.

11. WF to MF, Nov 30, 1785. CUL, MS Add., misc. folder T50.

12. MF to WF, Dec 18, 1785. SHL, MS 913B/3/1.

13. Gunning, *Reminiscences,* I: 108n.

14. WF to MF, Jul 15, 1786. CUL, MS Add., misc. folder T53.

15. WF to MF, Jul 15, 1786. CUL, MS Add., misc. folder T53.

16. WF to MF, Jul 15, 1786. CUL, MS Add., misc. folder T53.

17. WF to MF, Jul 25, 1786. CUL, MS Add., misc. folder T54.

18. WF to MF, Jul 25, 1786. CUL, MS Add., misc. folder T54.

19. WF to MF, Aug 18, 1786. CUL, MS Add., misc. folder T56.

20. WF to MF, Jul 15, 1786. CUL, MS Add., misc. folder T53.

21. WF to MF, Jul 15, 1786. CUL, MS Add., misc. folder T53.

22. WF to MF, [Aug] 21, [1786]. CUL, MS Add., misc. folder T56.

23. WF to MF, Aug 10, 1786. CUL, MS Add., misc. folder T55.

24. WF to MF, Jun 29, 1786. CUL, MS Add., misc. folder T52.

25. WF to MF, Aug 18, 1786. CUL, MS Add., misc. folder T56.

26. WF to MF, Aug 18, 1786. CUL, MS Add., misc. folder T56.

27. WF to MF, Sep 8, 1786. CUL, MS Add., misc. folder T59.

28. MF to WF, Dec 18, 1785. SHL, MS 913B/3/1.

29. WF to MF, [Oct 9], 1786. CUL, MS Add., misc. folder T57.

30. WF to MF, [Oct 9], 1786. CUL, MS Add., misc. folder T57.

31. William Frend, *A Sequel to the Account of the Proceedings in the University of Cambridge against the Author of a Pamphlet Entitled Peace and Union* (London: The Author, 1795), 104.

32. Thomas Rees, ed. and trans., *The Racovian Catechism, with Notes and Illustrations, Translated from the Latin* (London: Longman, Hurst, Rees, Orme, and Brown, 1818), votes of Parliament in frontmatter.

33. Frend, *Sequel,* 104.

34. William Frend, *Thoughts on Subscription to Religious Tests* (St Ives: T. Bloom, 1788), 25.

35. John Gascoigne, *Cambridge in the Age of the Enlightenment* (Cambridge: Cambridge University Press, 1989), 227.

36. WF to LNB, Nov 15, 1835. OBL, box 71, folios 206–7.

37. William Frend, *Animadversions on the Elements of Christian Theology by the Reverend George Pretyman, D.D., F.R.S., Lord Bishop of London* (London: The Author, 1800), 45–46.

38. *The Book of Common Prayer, and Administration of the Sacraments and Other Rites and Ceremonies of the Church, According to the Use of the Church of England, Together with the Psalter or Psalms of David* (Oxford: Clarendon Press, 1799), 36–37.

39. Quoted in Frida Knight, *University Rebel* (London: Victor Gollancz, 1971), 58.

40. A Member of the Senate [William Frend], *Considerations on the Oaths Required by the University of Cambridge at the Time of Taking Degrees, and on Other Subjects Which Relate to the Discipline of the Seminary* (London: J. Deighton, 1787), 8.

41. A Member of the Senate [William Frend], *Considerations*, 9.

42. A Member of the Senate [William Frend], *Considerations*, 22.

43. Frend, *Thoughts*, 23.

44. Frend, *Thoughts*, i.

45. Frend, *Thoughts*, i.

46. [William Frend], *An Address to the Members of the Church of England and to Protestant Trinitarians in General, Exhorting Them to Turn from the False Worship of Three Persons, to the Worship of the One True God* (London: J. Johnson, 1788), 7.

47. Mark 4:22.

48. [Frend], *Address*, 7.

49. [Frend], *Address*, 9.

50. [Frend], *Address*, 7.

51. [Frend], *Address*, 9–10.

52. [Frend], *Address*, 6.

53. [William Frend], *Appendix to Thoughts on Subscription* (St Ives: T. Bloom, 1789), 27–28.

54. [William Frend], *A Second Address to the Members of the Church of England and to Protestant Trinitarians in General, Exhorting Them to Turn from the False Worship of Three Persons, to the Worship of One True God* (London: J. Johnson, 1789), 12.

55. [Frend], *Appendix*, 27–28.

56. Frend, *Sequel*, 107. I have not been able to fix the amount of Frend's income after the loss of his tutorship. In his *Appendix* he notes that the tutorship he lost was worth £150—a fine living wage at the time—and it is tempting to conclude from his calculation in terms of thirds, that his ministry and his fellowship were each worth the same amount.

57. Frend, *Thoughts*, i.

## 5. EXERCISING REASON

1. TL to Tayleur, Dec 31, 1787. Quoted in H. McLachlin, *Letters of Theophilus Lindsey* (Manchester: At the University Press, 1920), 127.

2. TL to WF, Jun 20, 1788. CUL, MS Add., 7886, folder 6, 144.

3. TL to WF, Jun 20, 1788. CUL, MS Add., 7886, folder 6, 144.

4. John Disney, *The Works, Theological, Medical, Political, and Miscellaneous, of John Jebb, MD, F.R.S. with Memoirs of the Life of the Author* (London: T. Cadell, 1787), 155.

5. William Pitt, "Debate on Mr. William Pitt's Motion for a Reform in Parliament," *The Parliamentary History of England from the Earliest Period to the Year 1803,* vol. 22 (Mar 27, 1781–May 7, 1782), 1417.

6. Pitt, "Debate," 1417.

7. Disney, *Jebb,* 215–16.

8. Frend's correspondence with Reynolds and Hammond is to be found in Frida Knight, ed., *Letters to William Frend from the Reynolds Family of Little Paxton and John Hammond of Fenstanton: 1793–1814* (Cambridge: Cambridge Antiquarian Record Society, 1974).

9. George Dyer, *Memoirs of the Life and Writing of Robert Robinson* (London: G. G. and J. Robinson, 1796), 315–16.

10. Now the St Andrews Street Baptist Church.

11. I have gathered the basics of Robinson's biography from the *Oxford Dictionary of National Biography* (Oxford University Press, 2004), s.v. "Robinson, Robert," by John Stevens.

12. Dyer, *Robinson,* 136.

13. Dyer, *Robinson,* 121.

14. Robert Robinson, *A Plea for the Divinity of Our Lord Jesus Christ: In a Pastoral Letter Addressed to a Congregation of Protestant Dissenters, at Cambridge* (Cambridge: Fletcher and Hodson, 1776) Theophilus Lindsey, "Advertisement," in *An Examination of Mr Robinson of Cambridge's Plea for the Divinity of Our Lord Jesus Christ* (London: J. Johnson, 1789).

15. William Frend, *A Sequel to the Account of the Proceedings in the University of Cambridge against the Author of a Pamphlet Entitled Peace and Union* (London: The Author, 1795), 107.

16. Frida Knight, *University Rebel* (London: Victor Gollancz, 1971) locates Frend's summer months with the Jews before his first European tour. I see it as having occurred after his return from that trip, when he was determined to "search at the fountainhead" for clarity about his religion. In either case, we are guessing about the date of an occasion that he mentioned decades later as having occurred in "a summer's vacation." In 1784 he was traveling in Wales; 1785 is the summer Knight prefers, but it seems early for him to have realized he needed Hebrew; in 1786 he was in Europe. That leaves 1787 and/or 1788 as my best guess for when he lived with the Jews. In 1789 he was again in Europe; by 1790 his reputation as an exceptional Hebraist was already strong enough that he was assigned to translate the Pentateuch for the Unitarian retranslation project.

17. For Lord George Gordon, see David S. Katz, *The Jews in the History of England* (Oxford: Clarendon Press, 1994), 303–11.

18. For more on the history of the Bevis Mark Synagogue, see Richard D. Barnett and Abraham Levy, *The Bevis Marks Synagogue* (London: Society of Heshaim, 1970). For its architecture, see Sharman Kadish, *Bevis Marks Synagogue: A Short History of the Building and an Appreciation of Its Architecture,* illus. Barbara Bowman, Derek Kendall,

photographer, Survey of the Jewish Built Heritage in the United Kingdom & Ireland (London: English Heritage, 2001).

19. J. van den Berg, "Priestley, the Jews and the Millennium," in *Sceptics, Millenarians and Jews,* ed. David S. Katz and Jonathan I. Israel (1990), 256–74; David B. Ruderman, *Jewish Enlightenment in an English Key* (Princeton: Princeton University Press, 2000), 170–79.

20. WF, [On Jews] "Undated autograph manuscript by 'Follower of Jesus.'" CUL, MS Add. 7886, misc. folder 300.

21. WF, [On Jews] "Undated autograph manuscript by 'Follower of Jesus'" CUL, MS Add. 7886, misc. folder 300.

22. Frend, *Sequel,* 102.

23. Frend, *Sequel,* 107 Frend made this statement in the first person.

24. WF to MF, [?] 1789. CUL, MS Add., misc. folder T60.

25. WF to MF, [?] 1789. CUL, MS Add., misc. folder T60.

26. WF to MF, Jun 17, 1789. CUL, MS Add., misc. folder T61.

27. WF to MF, Sep 14, 1789. CUL, MS Add., misc. folder T62.

28. TL to WF, Nov 14, 1789. CUL, MS Add., misc. folder T149.

29. Memoir of William Frend [1826]. CUL, MS Add. 2886, misc. folder 285. This does not clearly date Mary's death, but the lack of further letters between them suggests that it happened at about the time of his return from his second European tour.

30. John 1:1–2, 14.

31. TL to WF, May 31, 1790. CUL, MS Add. 7886, folder 6, 155.

32. Joseph Priestley to WF, Sep 12, 1790, in John Towill Rutt, ed., *Memoirs and Correspondence of Joseph Priestley* (London: Society for the Diffusion of Useful Knowledge, 1817–32), II: 82.

33. Joseph Priestley to WF, Sep 12, 1790, in Rutt, *Priestley Memoir,* 82.

34. "Committees for Repeal of the Test and Corporation Acts: Minutes, 1788–9 (nos. 61–99)," *Committees for Repeal of the Test and Corporation Acts: Minutes 1786–90 and 1827–28* (1978), p. 36, http://www.british-history.ac.uk/report.asp?compid=38779.

35. Declaration of the Rights of Man, Aug 1789, http://www.yale.edu/lawweb/avalon/rights of.htm.

36. TL to WF, Nov 14, 1789. CUL, MS Add., misc. folder T149.

37. The full text of Price's address is to be found at http://oll.libertyfund.org/Texts/LFBooks /Sandoz0385/HTMLs/LoveOfCountry.html.

38. TL to WF, Mar 10, 1790. CUL, MS Add., misc. folder T150.

39. Joseph Priestley, *Familiar Letters Addressed to the Inhabitants of Birmingham, in Refutation of Several Charges Advanced Against the Dissenters* (Birmingham: J. Thompson, 1790), II: 13.

40. For Burke, see Richard Bourke, *Empire and Revolution: The Political Life of Edmund Burke* (Princeton: Princeton University Press, 2018).

41. This wonderful simile is attributed to A. J. Grieve in *Political Sermons of the American Founding Era: 1730–1805,* 2nd ed., 2 vols., foreword by Ellis Sandoz (Indianapolis: Liberty Fund, 1998). http://oll.libertyfund.org/Texts/LFBooks/Sandoz0385/HTMLs /0018_Pt05_Part4.html#hd_lf018.1.head.105.

42. Edmund Burke, *Reflections on the Revolution in France,* ed. Frank M. Turner (New Haven: Yale University Press, 2003), 29.

43. As entrées into the literature on the role of language in Burke's work, see Steven Blakemore, *Burke and the Fall of Language* (Hanover: University Press of New England, 1988); John Turner, "Burke, Paine and the Nature of Language," *Yearbook of English Studies* 19 (1989): 36–53. For the role of language in political discussion more generally, see Olivia Smith, *The Politics of Language* (Oxford: Clarendon Press, 1984); James A. Epstein, *Radical Expression: Political Language, Ritual, and Symbol in England, 1790–1850* (New York: Oxford University Press, 1994).

44. Burke quoted in Turner, "Burke and Paine," 45–46. It was Mary Wollstonecraft who was so irritated by a work with "no first principles to refute."

45. "Richard Price's Reply to Burke," in Edmund Burke, *Reflections on the Revolution in France,* ed. J. C. D. Clark (Stanford: Stanford University Press, 2001), 424–25.

46. Joseph Priestley, *Letters to the Right Honourable Edmund Burke, Occasioned by His Reflections on the Revolution in France* (Dublin, 1791), iii, Eighteenth Century Collections Online.

47. Thomas Paine, *Political Writings,* ed. Bruce Kuklick (Cambridge: Cambridge University Press, 2000), 89–90.

48. Paine, *Political Writings,* 91.

49. Joseph Priestley to TL, Jul 15, 1791, in Gillbert Wakefield, *Memoirs of the Life of Gilbert Wakefield, B.A. Late Fellow of Jesus College, Cambridge,* comp. and ed. John Towill Rutt and Arnold Wainewright (London: J. Deighton, 1792), 124–25.

50. E. P. Thompson, *The Making of the English Working Class* (New York: Pantheon Books, 1964), 17.

51. Thompson, *English Working Class,* 18.

52. Thompson, *English Working Class,* 113–14.

## 6. TRIALS IN CAMBRIDGE

1. I am getting these specific Cambridge events from Henry Gunning, *Reminiscences of the University, Town and County of Cambridge from the Year 1780,* 2 vols. (London: George Bell, 1855), 251–53.

2. William Frend, *Peace and Union Recommended to the Associated Bodies of Republicans and Anti-Republicans* (St Ives, 1793), 44, Eighteenth Century Collections Online.

3. Frend, *Peace and Union,* 5.

4. Frend, *Peace and Union,* 23.

5. Frend, *Peace and Union,* 17.

6. Frend, *Peace and Union,* 26.

7. Frend, *Peace and Union,* 30.

8. Frend, *Peace and Union,* 34.

9. Frend, *Peace and Union,* 45.

10. John Johnstone, *The Works of Samuel Parr, LL.D. Prebendary of St. Paul's, Curate of Hatton &. With Memoirs of His Life and Writings and Selection from His Correspondence* (London: Longman, Rees, Orme, Brown & Green, 1828), I: 447–48.

11. Frend, *Peace and Union,* 47–49 Gunning reproduced this appendix in full in Gunning, *Reminiscences,* I: 278–80, but he there replaced the word "sconced" with the word "scotched." In Augustus De Morgan, *A Budget of Paradoxes,* ed. David Eugene Smith (Chicago: Open Court, 1915), I: 198, De Morgan pointed to this as a mistake in Gunning. Since Frida Knight used "scotched" as well (Knight, *University Rebel,* 122), I suspect it may be found in some versions of the pamphlet. My copy uses the word "sconced," so I have done so as well.

12. The Scheldt is an estuary that has considerable strategic importance in the Netherlands. In 1792 the revolutionaries in France opened the Scheldt to trade, which upset a previous status quo that had favored the English.

13. Frend, *Peace and Union,* 48.

14. Frend, *Peace and Union,* 48.

15. Frend, *Peace and Union,* 49.

16. Frend, *Peace and Union,* 49.

17. Frend, *Peace and Union,* 19.

18. Johnstone, *Parr,* I: 447–48; Gunning, *Reminiscences,* I: 278–80.

19. I am taking this discussion of seditious libel from John Barrell and Jon Mee, eds., "Introduction," in *Trials for Treason and Sedition, 1792* (London: Pickering and Chatto, 2006–7), xiv.

20. William Frend, *An Account of the Proceedings in the University of Cambridge against William Frend, M.A. Fellow of Jesus College, Cambridge, for Publishing a Pamphlet, Intitled Peace and Union, &c. Containing the Proceedings in Jesus College, the Trial in the Vice-Chancellor's Court, and in the Court of Delegates* (Cambridge: Benjamin Flower, 1793), x.

21. Frend, *Account,* xx.

22. Frend, *Account,* xxii.

23. Frend, *Account,* xxvii.

24. Frend, *Account,* xxvi.

25. Frend, *Account,* xxvi.

26. Mary Milner, *The Life of Isaac Milner, D.D., F.R.S.* (London: John W. Parker, 1842), 61.

27. I owe this insight to John Gascoigne, *Cambridge in the Age of the Enlightenment* (Cambridge: Cambridge University Press, 1989), 251.

28. UCL, MS. Add. 7. The proposed grace is in Latin. An English translation would read: "Since, from the generous donations of numerous members of the Academy, donations given for the purpose of raising funds to cover the expenses of those men who have formed a Society for the abolition of the slave trade, it seems appropriate that this goal, which has been sought by individual persons in private, would be approved by all men publicly: let it therefore please the Senate that, in this spirit, forty pounds be obtained from the community chest, and placed in the hands of our most esteemed prochancellor, as well as those of Master Farish, senior procurator, and of Master Dixon, to be transported by these men to an appointed meeting of the aforementioned Society, to be held at London."

29. Gunning, *Reminiscences,* I: 162.

30. The *Statute De Concionibus* reads: "Prohibemus ne quisquam in concione aliqua, in

loco communi tractando, in lectionibus publicis, seu aliter publice infra universita-
tem nostram quicquam doceat, tractet, vel defendat contra religionem seu ejusdem
aliquam partem in regno nostro publica authoritate receptam et stabilitam, aut contra
aliquem statum authoritatem dignitatem seu gradum vel ecclesiasticum vel civilem
hujus nostri regni vel Angliae vel Hiberniae." (We prohibit anyone from teaching anything or conducting oneself or defending oneself in any public meeting, in a public
place, in public lectures, or otherwise publicly within our university, against religion or
to anything pertaining to it what has been received and established by public authority
in our kingdom; or against any rank, authority, office or position, be it ecclesiastical or
civil, of our kingdom, be it England or Ireland.)

31. Frend, *Peace and Union,* l–li.

32. For more on student antics, see Gunning, *Reminiscences,* I: 272–75.

33. The story is told in De Morgan, *Budget,* I: 198–99. De Morgan reveled in identifying the
unruly arsonists as the to-be lord chancellor, John Singleton Copley, Baron Lyndhurst;
the to-be bishop, Herbert Marsh; and the to-be chief justice, Sir William Rough.

34. Frend, *Account,* 1.

35. Frend, *Account,* 88, 89.

36. Frend, *Account,* 89–90.

37. Frend, *Account,* 161.

38. Frend, *Account,* 161.

39. Milner, *Life of Milner,* 98.

40. Frend, *Account,* 169.

41. Frend, *Account,* 176.

42. Frend, *Account,* 173.

43. Frend, *Account,* xlvii.

44. Johnstone, *Parr,* I: 447.

45. One of the many humorous twists of Frend's trials and appeals is his objection to the
decision to expel him from college because "the sentence of removal from college . . .
is clearly inconsistent with that which requires the constant residence of the master
and fellows." Frend, *Account,* xxiv.

46. Frend, *Account,* xlvii.

47. Frend, *Account,* xlii.

48. Frend, *Account,* 91.

49. Johnstone, *Parr,* I: 448.

50. Frend, *Account,* xlii.


*7. TRIALS IN LONDON*

1. Christopher Hibbert, Ben Weinreb, et al., *The London Encyclopedia* (Bethesda, Md.:
Adler and Adler, 1986), s.v. "Population," 613.

2. Joseph Gerrald died of consumption in New South Wales in 1796. William Skirving
died of dysentery in New South Wales in 1796. Thomas Muir escaped from Botany
Bay on an American ship in 1796, and after a series of remarkable adventures made his
way to France where he died in 1799. Palmer sailed out of New South Wales in 1800,

but after a series of misadventures he died of dysentery on the Mariana Island of Guguan in 1802.

3. For more comprehensive considerations of the role of language in this period, see Olivia Smith, *The Politics of Language* (Oxford: Clarendon Press, 1984); James A. Epstein, *Radical Expression: Political Language, Ritual, and Symbol in England, 1790–1850* (New York: Oxford University Press, 1994).

4. "The Charge Delivered by the Right Honourable Sir James Eyre … to the Grand Jury," reprinted in Jack W. Marken and Burton Pollin, eds., *Uncollected Writings by William Godwin* (Gainesville, Fl.: Scholars' Facsimiles and Reprints, 1968), 4.

5. "Charge Delivered," reprinted in Marken and Pollin, *Uncollected Godwin,* 134.

6. William Frend, *An Account of the Proceedings in the University of Cambridge against William Frend, M.A. Fellow of Jesus College, Cambridge, for Publishing a Pamphlet, Intitled Peace and Union, &c. Containing the Proceedings in Jesus College, the Trial in the Vice-Chancellor's Court, and in the Court of Delegates* (Cambridge: Benjamin Flower, 1793), vii–viii.

7. Marken and Pollin, *Uncollected Godwin,* 152.

8. Richard Reynolds to WF, Jul 31, 1795, in Frida Knight, ed., *Letters to William Frend from the Reynolds Family of Little Paxton and John Hammond of Fenstanton, 1793–1814* (Cambridge: Cambridge Antiquarian Record Society, 1974), 14.

9. William Frend, *Scarcity of Bread. A Plan for Reducing the High Price of This Article, in a Letter Addressed to W. Devaynes* (London, 1795), 2–3.

10. Richard Reynolds to WF, Jul 31, 1795, in Knight, *Frend Letters,* 14.

11. E. P. Thompson, *The Making of the English Working Class* (New York: Pantheon Books, 1964), 145.

12. The Diary of Joseph Farington, quoted in Nicholas Roe, *Wordsworth and Coleridge: The Radical Years* (Oxford: Clarendon Press, 1988), 153.

13. Diary of Joseph Farington, quoted in Roe, *Wordsworth and Coleridge,* 153.

14. Francis Place, *The Autobiography of Francis Place,* ed. Mary Thale (Cambridge: Cambridge University Press, 1972), 187.

15. Eusebia [Mary Hays], *Cursory Remarks on an Enquiry into the Expedience and Propriety of Public or Social Worship* (London: T. Knott, 1792).

16. WF to Mary Hays, Apr 16, 1792, in A. F. Wedd, ed., *The Love-Letters of Mary Hays, 1779–1780* (London: Methuen, 1925), 220.

17. WF to Mary Hays, Apr 16, 1792, in Wedd, *Hays Letters,* 220.

18. For Hays, see Gina Luria Walker, *Mary Hays, 1759–1843: The Growth of a Woman's Mind* (Burlington, Vt.: Ashgate, 2006); Marilyn L Brooks, ed., *The Correspondence (1779–1843) of Mary Hays, British Novelist* (Lewiston, N.Y.: Edwin Mellen Press, 2004).

19. WF to MF, [soon after Nov 22, 1784]. CUL, MS Add., misc. folder T49.

20. At the time Frend seems to have investigated whether the rules against married fellows might be lifted. T. Jones to WF, Dec 18, 1895. CUL, MS Add. 7886, T118.

21. Mary Hays, *Memoirs of Emma Courtney,* introd. by Eleanor Ty, Oxford World Classics (Oxford: Oxford University Press, 1996), 139.

22. A copy of the Godwin-Hays correspondence is to be found in Brooks, *Hays's Correspondence,* 363–470.

23. Hays, *Memoirs,* 121–22.

24. Hays, *Memoirs,* 78.

25. Hays, *Memoirs,* 71. The quotation is from Alexander Pope, "Eloisa to Abelard," 1:66.

26. Hays, *Memoirs,* 99.

27. Hays, *Memoirs,* 71. The source of this line is Alexander Pope, "The first satire of the second book of Horace, Imitated," 1:128.

28. Hays, *Memoirs,* 97.

29. Hays, *Memoirs,* 90n.

30. Hays, *Memoirs,* 140.

31. Hays, *Memoirs,* 143.

32. Hays, *Memoirs,* 106.

33. Helena Bergmann, *A Revised Reading of Mary Hays' Philosophical Novel* Memoirs of Emma Courtney *(1796)* (Lewiston, N.Y.: Edwin Mellen Press, 2011).

34. The identity of Harley was still a mystery in 1925, when Hays's great-niece edited her love letters: Wedd, *Hays Letters.*

## 8. REASONING IN UNEASY TIMES

1. William Frend, *A Sequel to the Account of the Proceedings in the University of Cambridge against the Author of a Pamphlet Entitled Peace and Union* (London: The Author, 1795), v–vi.

2. Frend, *Sequel,* 112.

3. Frend, *Sequel,* 119.

4. Frend, *Sequel,* 119. I do not know the "absurd regulation" to which Frend is referring.

5. Frend, *Sequel,* 119.

6. Frend, *Sequel,* 119.

7. WF to LNB, Nov 15, 1835. OBL, box 71, folios 206–7.

8. William Frend, *The Principles of Algebra* (London: J. Davis, 1796), iv.

9. The literature on Jesuit science is vast. A helpful entrée is John W. O'Malley, ed., *The Jesuits: Cultures, Sciences and the Arts, 1540–1773* (Toronto: University of Toronto Press, 1999). For mathematics see also Peter Dear, *Discipline and Experience: The Mathematical Way in the Scientific Revolution* (Chicago: University of Chicago Press, 1995).

10. The challenge is easily seen in the convolutions that characterize all efforts to reduce these centuries to a mathematically coherent narrative line. Cf. Carl Boyer, *A History of Mathematics* (New York: John Wiley and Sons, 1968); Morris Kline, *Mathematical Thought from Ancient to Modern Times* (New York: Oxford University Press, 1972). Carl Boyer engages the problem directly in Carl Boyer, "Analysis: Notes on the Evolution of a Subject and a Name," *The Mathematics Teacher* 47 (1954): 450–62.

11. Frend, *Principles,* x.

12. Frend, *Principles,* vi.

13. Frend, *Principles,* viii.

14. Frend, *Principles,* xiii.

15. Frend, *Principles,* xii.

16. Frend, *Principles,* x.

17. Frend, *Principles,* x.

18. Frend, *Principles,* xi–xii.

19. James Wood, *The Elements of Algebra: Designed for the Use of Students in the University,* 3rd ed. (Cambridge, 1801), 37, 26.

20. For a further discussions of the issues that surrounded this choice, see Helena M. Pycior, *Symbols, Impossible Numbers, and Geometric Entanglements* (Cambridge: Cambridge University Press, 1997), 307–16, and Kevin C. Knox, "The Negative Side of Nothing: Edward Waring, Isaac Milner and Newtonian Values," in *From Newton to Hawking: A History of Cambridge University's Lucasian Professors of Mathematics,* ed. Kevin C. Knox and Richard Noakes (Cambridge: Cambridge University Press, 2003), 205–40.

21. Thomas Belsham, *Memoirs of the Late Reverend Theophilus Lindsey, M.A. Including a Brief Analysis of His Works* (London: J. Johnson, 1812), 433n.

22. Francis Maseres, appendix to William Frend, *The Principles of Algebra: Or the True Theory of Equations Established on Mathematical Demonstration. Part the Second* (London: J. Davis, 1799), 418.

23. Frend, *True Theory,* x.

24. William Frend, *A Letter to the Vice-Chancellor of the University of Cambridge* (Cambridge: Benjamin Flower, 1798), 2.

25. For an overview of these attempts to establish the fundamental theorem of algebra, see Kline, *Mathematical Thought,* 597 ff.

26. Frend, *Letter to Vice-Chancellor,* 2–3.

27. Frend, *Letter to Vice-Chancellor,* 2–3.

28. Frend, *Letter to Vice-Chancellor,* 2.

29. Frend, *True Theory,* 2. The reference is to *Paradise Lost,* i:302: "Thick as the autumnal leaves that strew the brooks in Vallombrosa."

30. Frend, *True Theory,* 119.

31. [Robert Woohouse], "Review of Frend, *Principles of Algebra,*" *Monthly Review; or Literary Journal Enlarged* 22 (1797): 441.

32. [Robert Woodhouse], "Review of Frend, Principles of Algebra: Or the True Theory of Equations Established by Mathematical Demonstration,'" *Monthly Review; or Literary Journal Enlarged* 33 (1800): 183.

33. Letter from Woodhouse to Maseres, quoted in the obituary of William Frend embedded in [Augustus De Morgan], "Report of the Council of the Society to the Twenty-Second Annual General Meeting, Feb 11, 1842," *Memoirs of the Royal Astronomical Society* 13 (1843): 462. [Augustus De Morgan], "Report of the Council of the Society to the Twenty-First Annual General Meeting, February 12, 1841," *Memoirs of the Royal Astronomical Society* 12 (1842): 429–73.

34. William Frend, *The Effect of Paper Money on the Price of Provisions* (London: J. Ridgway, 1801), 20.

35. Frend, *Paper Money,* 26.

36. Frend, *Paper Money,* 12.

37. Frend, *Paper Money,* 26–27.

38. Quoted in Nicholas Roe, *Wordsworth and Coleridge: The Radical Years* (Oxford: Clarendon Press, 1988), 153.

39. WF to LBN, Feb 27, 1829. OBL box 71, folio 147.

40. TL to WF, [1800?]. CUL, Add. 7886[170].

41. Quoted in Frida Knight, ed., *Letters to William Frend from the Reynolds Family of Little Paxton and John Hammond of Fenstanton: 1793–1814* (Cambridge: Cambridge Antiquarian Record Society, 1974), 193.

42. William Frend, *Animadversions on the Elements of Christian Theology by the Reverend George Pretyman, D.D., F.R.S., Lord Bishop of London* (London: The Author, 1800), 45–47.

43. Frend saw his first Chinese abacus after this book was published. William Frend, *Evening Amusements: Or, The Beauty of the Heavens Displayed* (London: J Mawman, 1804–1822), (1807), 232. Hereafter *EA.*

44. William Frend, *Tangible Arithmetic: Or, The Art of Numbering Made Easy by Means of an Arithmetical Toy* (London: J. Mawman, 1806), advertisement to the 2nd edition.

45. Isaac Newton, *Opticks or A Treatise of the Reflections, Refractions, Inflections & Colours of Light* (New York: Dover Publications, 1979), Query 28, 370.

46. Isaac Newton, *The Principia,* trans. Andrew Motte, Great Minds (New York: Prometheus Books, 1995), General Scholium, 440–41.

47. Frend, *EA* (1804): 183.

48. Frend, *EA* (1804): 154–56, 137–38.

49. Frend, *EA* (1805): 87.

50. Frend, *EA* (1804): 169–70.

51. Frend, *EA* (1804): 183.

52. Frend, *EA* (1805): 199–200.

53. Frend, *EA* (1806): 167.

54. Frend, *EA* (1808): 73.

55. Frend, *EA* (1808): 74.

56. Frend, *EA* (1807): 4–6.

57. Frend, *EA* (1806): 2.

58. Hannah Lindsey to Sarah Blackburne, Aug 13, 1803. Letter reprinted in Sophia Elizabeth De Morgan, *Threescore Years and Ten: Reminiscences of the Late Sophia Elizabeth De Morgan to Which Are Added Letters to and from Her Husband the Late Augustus De Morgan, and Others,* ed. Mary De Morgan (London: Richards Bentley and Son, 1895), 14n.

59. William Frend, *Patriotism; or The Love of Our Country: An Essay Illustrated by Examples from Antient and Modern History; Dedicated to the Volunteers of the United Kingdom* (London: J. Mawman, 1804).

60. Quoted in Peter Spence, *The Birth of Romantic Radicalism: War, Popular Politics and English Radical Reformism, 1800–1815* (Oxford: Clarendon Press, 1996), 59.

61. William Frend, "A Monthly Retrospect of Public Affairs; or a Christian's Survey of the Political World," *Monthly Repository of Theology and General Literature* 3 (Apr 1808): 215.

62. Frend, "Christian's Survey," in *Monthly Repository* 3 (Mar 1808): 159.

63. Frend, "Christian's Survey," in *Monthly Repository* 3 (Apr 1808): 215.

64. Frend, "Christian's Survey," in *Monthly Repository* 3 (Apr 1808): 215.

65. Frend, "Christian's Survey," in *Monthly Repository* 3 (Jan 1808): 47–48.

66. Frend, "Christian's Survey," in *Monthly Repository* 3 (Jan 1808): 47–48.

67. Frend, "Christian's Survey," in *Monthly Repository* 3 (Apr 1808): 215.

68. Frend, "Christian's Survey," in *Monthly Repository* 3 (Jan 1808): 48.

69. Frend, "Christian's Survey," in *Monthly Repository* 4 (Jul 1809): 404.

70. Frend, *EA* (1809): 228–30.

71. Frend, *EA* (1807): 4–6.

72. Frend, *EA* (1810): iii–iv.


*9 . HEIRESS*

1. Hannah Lindsey to FB, Sep 1796. CUL, MS Add. 7886, folder 5, 131.

2. G. M. Ditchfield, "Hannah Lindsey and Her Circle: The Female Element in Early English Unitarianism," *Enlightenment and Dissent* 26 (2010): 62.

3. Hannah Lindsey to FB, Nov 13, 1798. CUL, MS Add. 7886, folder 5, 135.

4. Hannah Lindsey to FB, Jan 5, 1805. CUL, MS Add., misc. folder T142.

5. FB to WF, Nov 14, 1807. CUL, MS Add. 7886, folder 1, 8.

6. Hammond to WF, Apr 3, 1808, in Frida Knight, ed., *Letters to William Frend from the Reynolds Family of Little Paxton and John Hammond of Fenstanton, 1793–1814* (Cambridge: Cambridge Antiquarian Record Society, 1974), 79.

7. There is some question about the date of Sophia's birth. Mary De Morgan says 1809, but I am following Frida Knight, who says 1808, because it agrees with the entry in the little "family book" De Morgan wrote (UCL, MS Add. 7) and is more compatible with the Frends' marriage date.

8. Sophia Elizabeth De Morgan, *Threescore Years and Ten: Reminiscences of the Late Sophia Elizabeth De Morgan to Which Are Added Letters to and from Her Husband the Late Augustus De Morgan, and Others,* ed. Mary De Morgan (London: Richards Bentley and Son, 1895), 71.

9. De Morgan, *Reminiscences,* 67.

10. Frend borrowed this phrase from an elderly woman who used it when she first saw his infant daughter. It so completely captured Frend's view of the essence of a child that he told this story twice to Lady Bryon: WF to LNB, Aug 17, 1815. OBL, box 71, folios 120–21; WF to LNB, Aug 16, 1816. OBL, box 71, folios 133–34.

11. This is a brief summary of a complicated process. For more detail, see J. Anne Hone, *For the Cause of Truth: Radicalism in London, 1796–1821* (Oxford: Clarendon Press, 1982), 148–61.

12. William Frend, "A Monthly Retrospect of Public Affairs; or a Christian's Survey of the Political World," *Monthly Repository of Theology and General Literature* 5 (Apr 1810): 210.

13. Dudley Miles, *Francis Place: The Life of a Remarkable Radical, 1771–1854* (New York: St. Martin's, 1988), 29.

14. De Morgan, *Reminiscences,* 8.

15. Mary De Morgan, "Introductory Memoir of S. E. De Morgan," in De Morgan, *Reminiscences,* xxvi.

16. WF to Lady Byron, Oct 10, 1818. OBL, box 71, folios 139–41. Reprinted in De Morgan, *Reminiscences,* 104n.

17. De Morgan, *Reminiscences,* 6–7.

18. De Morgan, *Reminiscences,* 31–38.

19. De Morgan, *Reminiscences,* 20.

20. WF to LNB, Aug 16, 1816. OBL, box 71, folios 133–34.

21. WF to Dawson Turner, Sep 2, 1820. M. E. Grenander Department of Special Collections and Archives, University at Albany, State University of New York, MS 016.8.

22. De Morgan, *Reminiscences,* 89.

23. De Morgan, *Reminiscences,* xxvii.

24. WF to SF, 1821. CUL, MS Add. 7887, 34.

25. Psalm 119:169.

26. Frances Caroline Frend, Aug 31, 1810; Harriet Ann Frend, Jan 15, 1813; William Wandesforde Frend (always called Richard) Jul 20, 1815; Alicia Frend, Jan 27, 1817; Henry Tyrwitt Frend (dob unknown); Alfred Blackburne Frend (dob unknown).

27. Sarah Frend to SF, Jun 2, 1824. CUL, MS Add. 7887, 13.

28. Sarah Frend to SF, May 13, [1824]. CUL, MS Add. 7887, 11.

29. WF to SF, [1824]. CUL, MS Add. 7887, 36.

30. WF to SF, [1824]. CUL, MS Add. 7887, 36.

31. De Morgan, *Reminiscences,* 104, 105.

32. De Morgan, *Reminiscences,* 104, 105.

33. De Morgan, *Reminiscences,* 105.

34. Quoted on the AIM25 website: http://www.aim25.ac.uk/cgi-in/vcdf/detail?coll_id=51 59&inst_id=51&nv1=search&nv2=.

35. De Morgan, *Reminiscences,* 106.

36. De Morgan, *Reminiscences,* 97.

37. De Morgan, *Reminiscences,* 207.

38. De Morgan, *Reminiscences,* 24. An elderly Sophia located this plantation in either Jamaica or Georgia, but I am assuming it is the educational experiment on a West Indian plantation William Frend described in a letter to Lady Byron, Oct 10, 1818. OBL, box 71, folios 139–41.

39. De Morgan, *Reminiscences,* 97–98.

40. De Morgan, *Reminiscences,* 97–98.

41. De Morgan, *Reminiscences,* 93–94.

42. De Morgan, *Reminiscences,* 96.

43. De Morgan, *Reminiscences,* 96.

44. Joyce Godwin, *The Theosophical Enlightenment* (Albany: State University of New York Press, 1994), 76.

45. Quoted in *Oxford Dictionary of National Biography* (Oxford: Oxford University Press, 2004), s.v. "Taylor, Thomas," by Andrew Louth.

46. De Morgan, *Reminiscences,* 62.

47. De Morgan, *Reminiscences,* 54–56; Godwin, *Theosophical Enlightenment,* 25.

48. De Morgan, *Reminiscences,* 60.

49. Godwin, *Theosophical Enlightenment,* 82.

50. Godwin, *Theosophical Enlightenment,* 80.

51. De Morgan, *Reminiscences,* 54–56; Godwin, *Theosophical Enlightenment,* 25.

52. De Morgan, *Reminiscences,* 43–44.

53. De Morgan's position at this point seems to have been in flux because he must have subscribed to the Thirty-Nine Articles in order to take his degree. The sticking point for him seems to have been the public performance of subscription that would have been required to accept a fellowship.

54. Sophia Elizabeth De Morgan, *Memoir of Augustus De Morgan* (London: Longmans, Green, 1882), 20.

### 10. SON OF INDIA

1. ADM family notebook. UCL, MS Add. 7.

2. ADM to WRH, Jul 17, 1864, in Robert Perceval Graves, *Life of Sir William Rowan Hamilton* (1889), III: 612.

3. ADM family notebook. UCL, MS Add. 7.

4. The last name was also recorded as Tirville and Tivill.

5. Christianna was the daughter of a Danish missionary in Cuddalore.

6. ADM family notebook. UCL, MS Add. 7.

7. ADM family notebook. UCL, MS Add. 7.

8. ADM family notebook. UCL, MS Add. 7.

9. ADM family notebook. UCL, MS Add. 7.

10. ADM family notebook. UCL, MS Add. 7.

11. Among these was Augustus's brother George, who married Josephine Coghill, the third daughter of Sophia née Dodson and Josiah Coghill.

12. ADM to WRH, Jul 17, 1864, in Graves, *Hamilton,* III: 612–13.

13. ADM family notebook. UCL, MS Add. 7.

14. ADM to WRH, Aug 21, 1857, in Graves, *Hamilton,* III: 525.

15. Sophia Elizabeth De Morgan, *Memoir of Augustus De Morgan* (London: Longmans, Green, 1882), 22.

16. ADM to WRH, Jul 17, 1864, in Graves, *Hamilton,* III: 612–13.

17. De Morgan, *Memoir,* 108.

18. ADM to G. B. Airy, Aug 31, 1854. CUL, RGO 6234.

19. ADM to WRH, Aug 21, 1857, in Graves, *Hamilton,* III, 524–25.

20. ADM to WRH, Jul 17, 1864, in Graves, *Hamilton,* III, 612–13.

21. ADM to WRH, May 4, 1857, in Graves, *Hamilton,* III, 515–16.

22. Augustus revisited and debunked these stories in his notebook on the De Morgan family history. UCL, MS Add. 7.

23. De Morgan, *Memoir,* 4.

24. De Morgan, *Memoir,* 9.

25. De Morgan, *Memoir,* 10.

26. De Morgan, *Memoir,* 3.

27. Sophia interprets this image as "a gigantic father having the contents of his pocket picked by a crowd of dwarfish children, one meaning of which I understand to be, to represent the properties of magnitude analysed by the aid of number." De Morgan, *Memoir,* 66.

28. De Morgan, *Memoir,* 5–6.
29. De Morgan, *Memoir,* 5, 6.
30. De Morgan, *Memoir,* 8.

## *11. READING MAN*

1. Peter Searby, *A History of the University of Cambridge: 1750–1870* (Cambridge: Cambridge University Press, 1997), 61, 63.
2. Sophia claimed that his rooms were "over the gateway" (Sophia Elizabeth De Morgan, *Memoir of Augustus De Morgan* [London: Longmans, Green, 1882], 13), but they were on the first floor of Q entry.
3. Mr. Parsons's master's was from Oriel College, Oxford.
4. I have here changed De Morgan's verb form, from "forage" to "foraging." ADM to Rev. W. Heald, Aug 21, 1869, in De Morgan, *Memoir,* 393.
5. ADM to Rev. W. Heald, Aug 21, 1869, in De Morgan, *Memoir,* 393.
6. De Morgan, *Memoir,* 16.
7. ADM to WRH, Apr 1, 1858, in Robert Perceval Graves, *Life of Sir William Rowan Hamilton* (1889), III: 353–54.
8. De Morgan, *Memoir,* 14.
9. Quoted in De Morgan, *Memoir,* 13.
10. William Wordsworth, *The Prelude, Or, Growth of a Poet's Mind: An Autobiographical Poem* (England: Edward Moxon, 1850), 58.
11. A. Rupert Hall, *Philosophers at War: The Quarrel between Newton and Leibniz* (Cambridge: Cambridge University Press, 2002).
12. [Robert Woodhouse], "Review of J. L. LaGrange, Théorie des Fonctions Analytiques; &c. i.e. The Theory of Analytical Functions; Containing the Principles of the Differential Calculus, Divested of All Reference to Infinitely Small or Evanescent Quantities, Limits, or Fluxions, and Reduced to the Algebraical Analysis of Finite Quantities," *Monthly Review; or Literary Journal, Enlarged* 28 (1799): 483.
13. Michael S. Mahoney, "Changing Canons of Mathematical and Physical Intelligibility in the Later 17th Century," *Historia Mathematica* 11 (1984): 417–23; Michael Mahoney, "Infinitesimals and Transcendent Relations: The Mathematics of Motion in the Late Seventeenth Century," in *Reappraisal of the Scientific Revolution,* ed. David C Lindberg and Robert S Westman (Cambridge: Cambridge University Press, 1990), 461–92.
14. The literature on Descartes's *Geometry* is far too large to enter here. A remarkably clear and helpful entrée into his thinking is to be found in H. J. M. Bos, "The Structure of Descartes' Géometrie," in *Lectures in the History of Mathematics* (Washington, D.C.: American Mathematical Society, 1991), 37–57.
15. Isaac Newton, "Prime and Ultimate Ratios from *Principia,* Book 1," in *A Source Book in Mathematics, 1200–1800,* ed. D. J. Struik (Cambridge: MIT Press, 1969), 299–300.
16. Leibniz's infinitesimals have not been generally used since the early eighteenth century, but in 1966 Abraham Robinson showed that such a system could be developed following rigorous modern standards. Abraham Robinson, *Non-Standard Analysis* (Princeton: Princeton University Press, 1966).

17. For a cogent presentation of the powers of Leibnizian symbology in historical context, see H. J. M. Bos, "Differentials, Higher-Order Differentials and the Derivative in the Leibnizian Calculus," *Archive for the History of the Exact Sciences* 14 (1974): 1–90.

18. This quotation, which is generally attributed to d'Alembert, has become so ubiquitous as to transcend footnotes. For one interpretation of eighteenth-century French thinking about the nature of mathematics, see Joan L. Richards, "Historical Mathematics in the French Eighteenth Century," *Isis* 97 (2006): 700–713.

19. Judith V. Grabiner, *The Origins of Cauchy's Rigorous Calculus* (Cambridge: MIT Press, 1981).

20. Robert Woodhouse, *The Principles of Analytic Calculation* (Cambridge, 1803), xxv.

21. These words are from [Robert Woodhouse], "Review of Lacroix, S. F., Traité du calcul différentie, &c.; i.e. A Treatise on the Differential and Integral Calculus," *Monthly Review; or Literary Journal* 31 (1800): 494–95. He repeated the point, though less succinctly, in the preface to Woodhouse, *Principles*.

22. Woodhouse, *Principles,* ii.

23. Again these words are from [Woodhouse], "Lacroix, Pt 1," 494–95, but the point is repeated less succinctly in the preface to Woodhouse, *Principles*.

24. Charles Babbage to John Herschel, Aug 10, 1814, quoted in Anthony Hyman, *Charles Babbage: Pioneer of the Computer* (Princeton: Princeton University Press, 1982), 32.

25. These men and their immediate followers, including notably George Biddell Airy and William Whewell, were central to the development of English science and mathematics throughout the first six decades of the nineteenth century, and the literature surrounding them is immense. A readable introduction to the group and their influence that emphasizes the Baconian nature of their thinking is Laura J. Snyder, *The Philosophical Breakfast Club* (New York: Broadway Books, 2011). Balancing that position with an emphasis on industrial thinking is William J. Ashworth, "Memory, Efficiency, and Symbolic Analysis: Charles Babbage, John Herschel, and the Industrial Mind," *Isis* 87 (1996): 629–53. An emphasis on class thinking is to be found in Harvey Becher, "Radicals, Whigs and Conservatives: The Middle and Lower Classes in the Analytical Revolution at Cambridge in the Age of Aristocracy," *British Society for the History of Science* 28 (1995): 405–26.

26. William Frend, "A Monthly Retrospect of Public Affairs; or a Christian's Survey of the Political World," *Monthly Repository of Theology and General Literature* 6 (1811): 572.

27. For the Milner version of the story, see Mary Milner, *The Life of Isaac Milner, D.D., F.R.S.* (London: John W. Parker, 1842), 463–82.

28. Charles Babbage, *Passages from the Life of a Philosopher,* ed. Martin Campbell-Kelly (New Brunswick, N.J.: Rutgers University Press, 1994), 20, 21.

29. Titles ranged from "Remarks on the Theory of Analytical Developments" to "On Trigonometrical Functions of Different Orders."

30. Phillip C. Enros, "The Analytical Society, 1812–1813: Precursor of the Revival of Cambridge Mathematics," *Historia Mathematica* 10 (1983): 41.

31. Quoted in Joan L. Richards, "Rigor and Clarity: Foundations of Mathematics in France and England: 1800–1840," *Science in Context* 4 (1991): 312. Note: In this article, I erroneously attributed the notes to Babbage rather than to Peacock.

32. For a fuller discussion of the issues that divided Lacroix from his English translators see: Richards, "Rigor and Clarity," 297–319. For different analyses of the dynamics within the Analytical Society, see Becher, "Radicals, Whigs and Conservatives" and Ashworth, "Industrial Mind."

33. De Morgan, *Memoir,* 12.

34. ADM to WRH, Apr 26, 1863, in Graves, *Hamilton,* III: 590.

35. Andrew Warwick, *Masters of Theory: Cambridge and the Rise of Mathematical Physics* (Chicago: University of Chicago Press, 2003), 77. Warwick's sweeping book is the best modern treatment of the development of the nineteenth-century Tripos. The best nineteenth-century treatment is W. W. Rouse Ball, *A History of the Study of Mathematics at Cambridge* (Cambridge: Cambridge University Press, 1889). The nineteenth-century Tripos changed enough over time that it can be difficult to zero in on the experience at a particular moment. University records are so sparse that historians are left trying to construct the experience through piles of incidental memories, which range from the florid extravagances of someone trying to sell books in A Trinity Man [J. M. F. Wright], *Alma Mater; or Seven Years at the University of Cambridge* (London: Black, Young, and Young, 1827), to the admittedly vague recollections of an eighty-six-year-old George Biddell Airy in George Biddell Airy, "On the Earlier Tripos of the University of Cambridge," *Nature,* 24 (Feb 1887): 397–99.

36. Technically it was not until November 1827 (De Morgan took the Tripos in January of that year) that the university ruled that all questions as well as answers were to be written, but the members of the first and second classes (which included De Morgan) had been working from written examinations since at least 1802. J. W. L. Glaisher, "The Mathematical Tripos," *Nature,* 2 (Dec 1886): 104.

37. A Trinity Man [J. M. F. Wright], *Alma Mater,* 93.

38. A Trinity Man [J. M. F. Wright], *Alma Mater,* 96.

39. A Trinity Man [J. M. F. Wright], *Alma Mater,* 101.

40. A Trinity Man [J. M. F. Wright], *Alma Mater,* 104.

41. Thomas Thorp to ADM, Jun 3, 1824. SHL, MS 913A/2/9.

42. De Morgan, *Memoir,* 15.

43. De Morgan, *Memoir,* 11.

44. De Morgan, *Memoir,* 15.

## 12. LAYING THE GROUNDWORK

1. Sophia Elizabeth De Morgan, *Threescore Years and Ten: Reminiscences of the Late Sophia Elizabeth De Morgan to Which Are Added Letters to and from Her Husband the Late Augustus De Morgan, and Others,* ed. Mary De Morgan (London: Richards Bentley and Son, 1895), 17.

2. J. L. E. Dryer and H. H. Turner, *History of the Royal Astronomical Society* (London: Royal Astronomical Society, 1923), 14, 7. David Miller attributes a resurgent interest in the physical sciences in early nineteenth-century England to this group in David Philip Miller, "The Revival of the Physical Sciences in Britain, 1815–1840," *Osiris* 2nd ser., 2 (1986): 113. He focuses on their relationship to the Royal Society in David Philip

Miller, "Between Hostile Camps: Sir Humphry Davy's Presidency of the Royal Society of London," *British Journal for the History of Science* 16 (1983): 1–47. William Ashworth emphasizes their ties to business and the emerging world of industrial capitalism in William Ashworth, "The Calculating Eye: Baily, Herschel, Babbage and the Business of Astronomy,'" *British Journal for the History of Science* 27 (1994): 409–41.

3. For a rich description of the world of Joseph Banks, see Richard Holmes, *The Age of Wonder* (New York: Vintage Books, 2008).

4. Banks quoted in J. L. Heilbron, "A Mathematicians' Mutiny, with Morals," in *World Changes: Thomas Kuhn and the Nature of Science,* ed. Paul Horwich (Cambridge: MIT Press, 1993), 87, 88. The relationship between mathematicians and natural philosophers was an issue from the very beginning of the Royal Society. Cf. Steven Shapin, "Robert Boyle and Mathematics: Reality, Representation, and Experimental Practice," *Science in Context* 2 (1988): 23–58.

5. John Herschel, "Baily Obituary," *Notes and Records of the Royal Astronomical Society* 6, no. 10 (Nov 8, 1844): 101.

6. For a lively presentation of Harrison's story, see Dava Sobel, *Longitude: The True Story of a Lone Genius Who Solved the Greatest Scientific Problem of His Day* (New York: Walker, 1995).

7. Mary Croarken, "Human Computers in Eighteenth- and Nineteenth-Century Britain," in *The Oxford Handbook of the History of Mathematics,* ed. Eleanor Robson and Jacqueline Stedall (Oxford: Oxford University Press, 2009), 375–403.

8. William Frend, *Evening Amusements: Or, The Beauty of the Heavens Displayed* (London: J. Mawman, 1804–22), (1820): 5–6.

9. Dryer and Turner, *History RAS,* 56.

10. Dryer and Turner, *History RAS,* 9.

11. Herschel, "Baily," 97.

12. Dryer and Turner, *History RAS,* 5.

13. Miller, "Revival," 130–31 On the enlightenment perception of calculation as reason, see Lorraine Daston, "Enlightenment Calculations," *Critical Inquiry* 21 (1994): 182–202.

14. Dryer and Turner, *History RAS,* 14.

15. Quoted in Marcus Adler, "Memoir of the Late Benjamin Gompertz," *Assurance Magazine and Journal of the Institute of Actuaries* 13 (1867): 9.

16. WF to LNB, Feb 11, 1829. OBL, box 71, folio 149.

17. For a fuller analysis of the kinds of implications that attended Babbage's machines in the early nineteenth century, see Simon Schaffer, "Babbage's Intelligence; Calculating Machines and the Factory System," *Critical Inquiry* 21 (1994): 203–27; Ashworth, "Calculating Eye."

18. ADM to WRH, Jul 17, 1864, in Robert Perceval Graves, *Life of Sir William Rowan Hamilton* (1889), III: 613.

19. Dryer and Turner, *History RAS,* 14.

20. Quoted in Adler, "Gompertz," 9.

21. Herschel, "Baily," 93.

22. I gleaned this date from Herschel, "Baily," but modern sources date the eclipse at May 25, 585 BC. For the kinds of issues that may lie behind the discrepancy, see Matthew

Stanley, "Predicting the Past: Ancient Eclipses, and Airy, Newcomb and Huxley on the Authority of Science," *Isis* 103 (2012): 254–77.

23. For a discussion of the astronomical work being done in India, see Simon Schaffer, "The Asiatic Enlightenment of British Astronomy," in *The Brokered World: Go-Betweens and Global Intelligence,* ed. Simon Schaffer (Sagamore Beach: Watson Publishing, 2009), 49–104.

24. Dryer and Turner, *History RAS,* 7.

25. Quoted in Dryer and Turner, *History RAS,* 38–39.

26. For the life of Briggs, see Evans Bell, *Memoir of General John Briggs of the Madras Army* (London: Chatto and Windus, 1885).

27. Briggs laid out the position that it was essential for Englishmen respect their Indian soldiers, which included learning to address them in their native languages, in John Briggs, *Letters Addressed to a Young Person in India; Calculated to Afford Instruction for His Conduct in General, and More Especially in His Intercourse with the Natives* (London: John Murray, 1828).

28. Bell, *Briggs,* 103.

29. Kapil Raj, "Relocating," in *Relocating Modern Science: Circulation and the Construction of Knowledge in South Asia and Europe, 1650–1900* (New York: Palgrave Macmillan, 2007), 98.

30. Bell, *Briggs,* 103.

31. Sophia Elizabeth De Morgan, *Memoir of Augustus De Morgan* (London: Longmans, Green, 1882), 21.

32. For the founding of the London University, see Negley Harte and John North, *The World of UCL: 1828–1990* (London: UCL, 1991). Sophia gave her version in De Morgan, *Memoir,* 22–28. Sophia claimed that the university's origins lay in a series of articles her father had published under the pen name Civis, "somewhere about 1819" "in a monthly periodical edited by Mr. John Thirlwall," which she was, however, unable to find in the British Museum. As far as I know, no one else has found this periodical either. Hugh Hale Bellot, *University College London, 1826–1926* (London: University of London Press, 1929), 26, has instead attributed to Frend a series of articles on "The London University" published by "Civis" in *Panoramic Miscellany* in 1826. Numerous internal clues, like the inclusion of a loving remembrance of Paley's teaching, support Bellot's attribution.

33. "The London College," *The Times,* no. 12672 (Jun 6, 1825): 4.

34. Harte and North, *World of UCL,* 26.

35. "London College."

36. F. A. Cox, "The London College," *The Times,* no. 12677 (Jun 11, 1825): 3.

37. Bellot, *University College London,* 56.

38. Civis, "The London University," *Panoramic Miscellany* (1826): 506.

39. Civis, "London University," 507.

40. WF to LNB, Dec 29, 1828. OBL, box 71, folio 144.

41. William Frend, *Animadversions on the Elements of Christian Theology by the Reverend George Pretyman, D.D., F.R.S., Lord Bishop of London* (London: The Author, 1800), 45–47.

42. WF to LNB, Dec 29, 1828. OBL, box 71, folio 144.

43. WF, [On Jews] "Undated autograph manuscript by 'Follower of Jesus.'" CUL, Add. 7886³⁰⁰.

44. WF, [On Jews] "Undated autograph manuscript by 'Follower of Jesus.'" CUL, Add. 7886³⁰⁰.

45. Thomas Campbell. Quoted in Bellot, *University College London,* 23.

46. Quoted in Adrian Rice, "Inspiration or Desperation? Augustus De Morgan's Appointment to the Chair of Mathematics at London University in 1828," *British Journal for the History of Science* 30 (1997): 268.

47. Harte and North, *World of UCL,* 31.

48. Quoted in Rice, "Inspiration," 266.

49. WF to ADM. CUL, MS Add. 7887 T14. For a more detailed discussion of the process that led to De Morgan's being offered this position, see Rice, "Inspiration."

50. De Morgan, *Memoir,* 28.

51. Rice, "Inspiration," 267.

52. Ronald Anderson, "Augustus De Morgan's Inaugural Lecture of 1828," *Mathematical Intelligencer* 25, no. 3 (2006): 20.

53. Anderson, "De Morgan's Inaugural Lecture," 20.

54. Anderson, "De Morgan's Inaugural Lecture," 20.

55. Anderson, "De Morgan's Inaugural Lecture," 19.

56. Anderson, "De Morgan's Inaugural Lecture," 22–23.

57. Anderson, "De Morgan's Inaugural Lecture," 23.

58. Anderson, "De Morgan's Inaugural Lecture," 21.

59. Louis Pierre Marie Bourdon, *Elements of Algebra,* trans. Augustus De Morgan (London, 1828).

60. Anderson, "De Morgan's Inaugural Lecture," 20.

61. [Augustus De Morgan], "Study of Natural Philosophy," *Quarterly Journal of Education* 3 (Jan–Apr 1832): 60.

62. [De Morgan], "Study of Natural Philosophy," 60–61.

63. [De Morgan], "Study of Natural Philosophy," 61.

64. Anderson, "De Morgan's Inaugural Lecture," 25.

65. Anderson, "De Morgan's Inaugural Lecture," 26.

## 13. UNITARIAN WOMEN

1. WF to LNB, Apr 14, 1829. OBL, box 71, folio 150.

2. WF to Miss Milbank, [1806–7]. OBL, box 71, folios 1–6.

3. WF to LNB, Apr 17, 1816. OBL, box 71, folio 15.

4. WF to LNB. Feb 18, 1816. OBL, box 71, folio 132.

5. WF to LNB, May 25, 1816. OBL, box 71, folio 126–28.

6. LNB to WF, May 19, 1816. OBL, box 71, folios 21–22.

7. My discussion of the social situation in the English countryside relies heavily on E. J. Hobsbawm and George Rude, *Captain Swing* (London: Lawrence and Wishart, 1969), 1–93.

8. WF to LNB, Mar 13, 1830. OBL, box 71, folios 55–56.

9. WF to LNB, Mar 13, 1830. OBL, box 71, folios 55–56.

10. For the Swing Riots, see Hobsbawm and Rude, *Captain Swing.*

11. WF to LNB, Dec 25, 1830. OBL, box 71, folios 162–63.

12. Sidney Pollard, "Dr William King of Ipswich: A Cooperative Pioneer," *Cooperative College Papers* 6 (1959): 17.

13. LNB to WF, Jun 29, 1829. OBL, box 71, folio 51.

14. Pollard, "King," 21.

15. Henry Brougham, *Practical Observations on the Education of the People Addressed to the Working Classes and Their Employers* (London: Richard Taylor, 1825), 2.

16. Brougham, *Observations,* 1.

17. Brougham, *Observations,* 5.

18. Brougham, *Observations,* 32. The quotation is Ephesians 4:14.

19. Steven Shapin and Barry Barnes, "Science, Nature and Control: Interpreting Mechanics' Institutes," *Social Studies of Science* 7 (1977): 33n4.

20. Brougham, *Observations,* 6.

21. Brougham, *Observations,* 17.

22. For one such interpretation that may also serve as an entrée into this literature, see James Secord, *Victorian Sensation: The Extraordinary Publication, Reception and Secret Authorship of Vestiges of the Natural History of Creation* (Chicago: University of Chicago Press, 2000).

23. LNB to WF, Jun 29, 1829. OBL, box 71, folio 51.

24. Sophia Elizabeth De Morgan, *Threescore Years and Ten: Reminiscences of the Late Sophia Elizabeth De Morgan to Which Are Added Letters to and from Her Husband the Late Augustus De Morgan, and Others,* ed. Mary De Morgan (London: Richards Bentley and Son, 1895), 146–47.

25. WF to LNB, Oct 21, 1831. OBL, box 71, folios 172–73. LNB to SF, Sep 28, 1834. OBL, box 67, folios 56–57. In the longer run the goal was to set up a Mechanics Institute.

26. LNB to SF, Sep 28, 1834. OBL, box 67, folios 56–57.

27. WF to LNB, Apr 29, 1831. OBL, box 71, folio 167.

28. In Sophia Elizabeth De Morgan, *Memoir of Augustus De Morgan* (London: Longmans, Green, 1882), 41, this move took place in 1830, but I think that's incorrect. I'm taking my chronology from WF to LNB, Aug 24, 1831. OBL, box 71, folio 69, where he declares they are about to move to London.

29. De Morgan, *Reminiscences,* 110.

30. De Morgan, *Reminiscences,* 108–9.

31. De Morgan, *Reminiscences,* 152.

32. De Morgan, *Reminiscences,* 152.

33. De Morgan, *Reminiscences,* 153.

34. De Morgan, *Reminiscences,* 154–55.

35. De Morgan, *Reminiscences,* 155.

36. Roger Cooter, *The Cultural Meaning of Popular Science: Phrenology and the Organization of Consent in Nineteenth-Century Britain* (Cambridge: Cambridge University Press, 1984), 296. For more information on the London Institution where Sophia and

her family heard Spurzheim's lecture, see J. N. Hays, "Science in the City: The London Institution, 1819–40," *British Journal for the History of Science* 7, no. 2 (Jul 1974):
146–62.

37. For an analysis of the elements of Spurzheim's success in England, see John van Wyhe,
"The Diffusion of Phrenology through Public Lecturing," in *Science in the Marketplace:
Nineteenth-Century Sites and Experiences,* ed. Aileen Fyfe and Bernard Lightman (Chi-
cago: University of Chicago Press, 2007).

38. De Morgan, *Reminiscences,* 175.

39. De Morgan, *Reminiscences,* 169–71.

40. De Morgan, *Reminiscences,* 165–66.

41. De Morgan, *Reminiscences,* 165–66.

42. De Morgan, *Reminiscences,* 165–66.

43. De Morgan, *Reminiscences,* 166.

44. De Morgan, *Reminiscences,* 155.

45. De Morgan, *Reminiscences,* 160.

46. De Morgan, *Reminiscences,* 163.

47. De Morgan, *Reminiscences,* 133.

48. LNB to SF, Mar 18, De Morgan, *Reminiscences,* 201n.

49. De Morgan, *Reminiscences,* 197.

50. De Morgan, *Reminiscences,* 195.

51. LNB to SF, Jun 8, 1835. OBL, box 67, folios 62–63.

52. The Children's Friend Society had 2,588 subscribers, almost twice the number of any
other charity on the list of "Societies with one Subscription List" appended to F. K.
Prochaska, *Women and Philanthropy in Nineteenth-Century England* (Oxford: Clarendon Press, 1980).

53. De Morgan, *Reminiscences,* 166.

54. LNB to SF, Nov 2, [1834]. OBL, box 67, folios 58–59.

55. Maria Edgeworth, *Helen* (London: R. Bentley, 1834), 63.

56. Edgeworth, *Helen,* 396.

57. Edgeworth, *Helen,* 45–46.

58. LNB to SF, Jul 20, 1835. OBL, box 67, folio 68.

59. LNB to SF, Nov 2 [1834]. OBL, box 67, folios 58–59.

60. LNB to SF, Jun 8, 1835. OBL, box 67, folios 62–63.

61. *Monthly Repository,* 2nd ser., no 64 (Apr 1832): 259. (The author of this poem was Sarah
Flower Adams. Frances E Mineka, *The Dissidence of Dissent: The Monthly Repository,
1806–1838* [Chapel Hill: University of North Carolina Press, 1944], 401.)

62. There is a growing literature about Rammohun Roy. I have used Dermot Killingley,
*Rammohun Roy in Hindu and Christian Tradition* (Newcastle upon Tyne: Grevatt &
Grevatt, 1993) as an introductory overview. For a treatment of his political views in
the Indian context, see C. A. Bayly, "Rammohan Roy and the Advent of Constitu-
tional Liberalism in India, 1800–30," *Modern Intellectual History* 4, no. 1 (2007): 25–41.
For his interactions with the Unitarians in England, see Lynn Zastoupil, "Defining
Christians, Making Britons: Rammohun Roy and the Unitarians," *Victorian Studies*
44, no. 2 (Winter 2002): 215–43; Lynn Zastoupil, *Rammohun Roy and the Making of*

*Victorian Britain* (New York: Palgrave Macmillan, 2010); and Clare Midgley, "Transoceanic Commemoration and Connections between Bengali Brahmos and British and American Unitarians" (2011).

63. I have taken this précis of Rammohun Roy's life from Killingley, *Rammohun Roy*, 6–7.

64. Zastoupil, *Rammohun Roy*, 1.

65. Zastoupil, *Rammohun Roy*, 1–3.

66. De Morgan, *Reminiscences*, 91–92.

67. Zastoupil, *Rammohun Roy*, 30.

68. Killingley, *Rammohun Roy*, 46.

69. De Morgan, *Memoir*, 91–92.

70. De Morgan, *Reminiscences*, 91–92.

71. LNB to SF, Aug 24, [1838]. OBL, box 67, folios 69–70.

72. LNB to SF, Sep 16, 1835. OBL, box 67, folios 71–72.

73. LNB to SF, Sep 16, 1835. OBL, box 67, folios 71–72.

74. LNB to SF, Sep 16, 1835. OBL, box 67, folios 71–72.

75. LNB to SF. Oct 18, 1835. OBL, box 67, folios 74–75.

76. LNB to WF, Oct 23, 1835. OBL box 71, folios 82–83.

77. LNB to WF, Oct 23, 1835. OBL, box 71, folios 82–83.

78. LNB to WF, Oct 23, 1835. OBL, box 71, folios 82–83.

79. WF to LNB, Nov 15, 1835. OBL, box 71, folios 206–7.

80. De Morgan, *Reminiscences*, 92–93.

## 14. GENTLEMAN OF REASON

1. Evans Bell, *Memoir of General John Briggs of the Madras Army* (London: Chatto and Windus, 1885), 106.

2. Bell, *Briggs*, 106.

3. Bell, *Briggs*, 106.

4. Sophia Elizabeth De Morgan, *Memoir of Augustus De Morgan* (London: Longmans, Green, 1882), 32.

5. ADM to Thomas Coates, Aug 15, 1830. UCL, SDUK papers, Augustus De Morgan.

6. Sophia Elizabeth De Morgan, *Memoir*, 42. Charles Babbage, *Reflections on the Decline of Science in England, and on Some of Its Causes* (London: B. Fellowes, 1830); Jack Morrell and Arnold Thackray, *Gentlemen of Science* (Oxford: Clarendon Press, 1981), 48. See also "The Great Battle," in Laura J. Snyder, *The Philosophical Breakfast Club* (New York: Broadway Books, 2011), 128–57.

7. Morrell and Thackray, *Gentlemen*, 19–20, 25–28. For further reflections on the implications of this conception of scientific practitioners as gentlemen, see "The Paradoxes of Gentility," in James Secord, *Victorian Sensation: The Extraordinary Publication, Reception and Secret Authorship of Vestiges of the Natural History of Creation* (Chicago: University of Chicago Press, 2000), 403–36. For a more theoretical consideration of the challenges natural philosophers have faced in understanding their roles, see Lorraine Daston and H. Otto Sibum, "Introduction: Scientific Persona and Their Histories," *Science in Context* 16 (2003): 1–8.

8. For the complex relationship between mathematics and science in the English seventeenth century, see the chapter "Certainty and Civility," in Steven Shapin, *A Social History of Truth: Civility and Science in Seventeenth-Century England* (Chicago: University of Chicago Press, 1994), 310–54. For the more contemporary episode in the ongoing saga, the expulsion of mathematicians from the Royal Society in the 1780s, see J. L. Heilbron, "A Mathematicians' Mutiny, with Morals," in *World Changes: Thomas Kuhn and the Nature of Science,* ed. Paul Horwich (Cambridge: MIT Press, 1993), 81–130.

9. UCL, MS Add. 7.

10. ULC, MS Add. 7.

11. Negley Harte and John North, *The World of UCL: 1828–1990* (London: UCL, 1991), 30.

12. Sophia Elizabeth De Morgan, *Memoir,* 29.

13. The exact figures for 1828–29 were 641; for 1829–30 this number fell to 630. Harte and North, *World of UCL,* 45.

14. I am taking my figures for the Trinity Fellows stipend from: A Trinity Man [J. M. F. Wright], *Alma Mater; or Seven Years at the University of Cambridge* (London: Black, Young, and Young, 1827), II: 167.

15. Sophia Elizabeth De Morgan, *Memoir,* 30.

16. For a fuller discussion of the issues surrounding Pattison's dismissal, see "Importing the New Morphology," in Adrian Desmond, *The Politics of Evolution: Morphology, Medicine, and Reform in Radical London* (Chicago: University of Chicago Press, 1989), 25–100.

17. Sophia Elizabeth De Morgan, *Memoir,* 35.

18. Technically, it seems that De Morgan's objection was less to the Council itself making this decision, and more that they had appointed a committee to do so in a way that De Morgan did not think constituted a "competent tribunal." Cf. the additional letter he wrote to the Warden of the London University reprinted in Sophia Elizabeth De Morgan, *Memoir,* 38.

19. Sophia Elizabeth De Morgan, *Memoir,* 39.

20. Sophia Elizabeth De Morgan, *Memoir,* 39.

21. WF to ADM, Aug 5, 1831. CUL, Add. 7887[19]. This letter was sent from Stoke Newington before the Frends left for Hastings.

22. Sophia Elizabeth De Morgan, *Memoir,* 28.

23. Sophia Elizabeth De Morgan, *Memoir,* 60.

24. Henry Brougham, *Practical Observations on the Education of the People Addressed to the Working Classes and Their Employers* (London: Richard Taylor, 1825), 10.

25. For a comprehensive analysis of the development of science publishing in this period, see "Steam Reading" in Secord, *Victorian Sensation,* 41–76.

26. ADM to Thomas Coates, Mar 30, 1827. UCL, SDUK papers. For the paper on statics, see Adrian Rice, "Inspiration or Desperation? Augustus De Morgan's Appointment to the Chair of Mathematics at London University in 1828," *British Journal for the History of Science* 30 (1997): 270.

27. Augustus De Morgan, *On the Study and Difficulties of Mathematics* (Chicago: Open Court, 1910).

28. A fourth edition was published by Open Court in 1943.

29. Charles Knight, *Passages of a Working Life during a Half Century: With a Prelude of Early Reminiscences* (London: Bradbury & Evans, 1864), 163.

30. Knight, *Passages,* II: 184.

31. Knight, *Passages.*

32. Knight, *Passages,* II: 182.

33. Knight, *Passages,* II: 201 My comparison of the *Penny Cyclopædia* to Diderot's *Encyclopédie* runs counter to the dismissive treatments it has received among recent historians of science. In Richard Yeo, "Reading Encyclopedias: Science and the Organization of Knowledge in British Dictionaries of Arts and Sciences, 1730–1850," *Isis* 82 (1991): 24–49, Yeo mentions the *Penny Cyclopædia* only as an example of a work directed at the lower classes. James Secord classifies it among England's cheap books that entailed "cramming an amazing quantity of print into a small space," to produce "treatises" that "were often forbiddingly technical." Secord, *Victorian Sensation,* 48–49.

34. Knight, *Passages,* II: 209.

35. Sophia Elizabeth De Morgan, *Memoir,* 50.

36. A list of most of his articles is to be found in the "List of Writings" appended to Sophia Elizabeth De Morgan, *Memoir.*

37. Knight, *Passages,* II: 209.

38. Knight, *Passages,* II: 218.

39. Kevin Lambert, "A Natural History of Mathematics: George Peacock and the Making of English Algebra," *Isis* 104 (2013): 278–302. The historical literature focused on the development of English ideas of algebra in the early decades of the nineteenth century is enormous. As an entrée in the literature I recommend following the footnotes in this excellent and relatively recent article.

40. George Peacock, "Report on the Recent Progress and Present State of Certain Branches of Analysis," *Report of the Third Meeting of the British Association for the Advancement of Science* (1833): 99.

41. [Augustus De Morgan], "Review of George Peacock, *A Treatise on Algebra,*" *Quarterly Journal of Education* 9 (1835): 311.

42. [De Morgan], "Peacock's *Algebra,*" 293.

43. [De Morgan], "Peacock's *Algebra,*" 92.

44. [De Morgan], "Peacock's *Algebra,*" 310.

45. John Warren, *Treatise on the Geometrical Representation of the Square Roots of Negative Quantities* (Cambridge: J. Smith, 1828).

46. [De Morgan], "Peacock's *Algebra,*" 309.

47. WF to ADM, Jun 22, 1836. CUL, Add. 7787[29]. For further development of the meanings of "art" and "science" in 1830s discussions of mathematics, see Joan L. Richards, "The Art and the Science of British Algebra: A Study in the Perception of Mathematical Truth," *Historia Mathematica* 7 (1980): 229–367.

48. A number of different scholars have considered the relations of history, natural history, and mathematics in this period, including Joan L. Richards, "Augustus De Morgan, the History of Mathematics, and the Foundations of Algebra," *Isis* 78 (1987): 6–30; Adrian Rice, "Augustus De Morgan: Historian of Science," *History of Science* 34

(1996): 201–40; Lambert, "Peacock's Mathematics"; and Marie-Jose Durand-Richard, "L'École Algebrique Anglaise: Les Conditions Conceptuelles et Institutionnelles d'un Calcul Symbolique comme Fondement de la Connaissance," in *L'Europe Mathematique: Histoires, Mythes, Identites,* ed. Catherine Goldstein, Jeremy Gray, and Jim Ritter (Paris: Éditions de la Maison des sciences de l'homme, 1996), 445–77.

## 15. REASONING AMONG THE STARS

1. Sophia Elizabeth De Morgan, *Memoir of Augustus De Morgan* (London: Longmans, Green, 1882), 22. That Frances was the piano player is suggested by a tongue-in-cheek question Sophia posed to Augustus in an addendum to WF to ADM, Jul 24, 1836, CUL, Add. 7887[32]. "When did you hear Fanny … say she wished you would not play with her?"

2. ADM to WF, Sep 1, 1834. Sophia Elizabeth De Morgan, *Memoir,* 78–79.

3. ADM to WF, Sep 1, 1834. Sophia Elizabeth De Morgan, *Memoir,* 80.

4. WF to ADM, Aug 17, 1835. CUL, Add. 7887[24].

5. WF to ADM, Oct 6, 1834. CUL, Add. 7887[23].

6. WF to ADM, Aug 17, 1835. CUL, Add. 7887[24].

7. Baily's protégé and De Morgan's friend, William Stratford, became the superintendent of the *Nautical Almanac* in 1831 and held the position until his death in 1854.

8. Sophia Elizabeth De Morgan, *Memoir,* 48.

9. John Herschel, "Baily Obituary," *Notes and Records of the Royal Astronomical Society* 6, no. 10 (Nov 8, 1844): 96–97.

10. Herschel, "Baily," 119–20.

11. Sophia Elizabeth De Morgan, *Memoir,* 45.

12. For a discussion of the different way Robert Boyle used his home for scientific work, see Steven Shapin, "The House of Experiment in Seventeenth-Century England," *Isis* 79 (1988): 373–404. For Baily's house on Tavistock Place, see L. H. Horton-Smith, *The Baily Family of Thatcham and Later of Speen and Newbury, All in the County of Berkshire* (Leicester: W. Thornley & Son, 1951). The house is now gone, but Francis Street and Bailey [sic] Street remain in the neighborhood of Tavistock Place.

13. Sophia Elizabeth De Morgan, *Threescore Years and Ten: Reminiscences of the Late Sophia Elizabeth De Morgan to Which Are Added Letters to and from Her Husband the Late Augustus De Morgan, and Others,* ed. Mary De Morgan (London: Richards Bentley and Son, 1895), 47.

14. Sophia Elizabeth De Morgan, *Memoir,* 45.

15. John Herschel, *A Preliminary Discourse on the Study of Natural Philosophy,* fascimile of 1830 ed. (Chicago: University of Chicago Press, 1987), 4–5.

16. Herschel, *Preliminary Discourse,* 7.

17. Herschel, *Preliminary Discourse,* 16.

18. Rebekah Higgitt, "Astronomers against Newton," *Endeavour* 28 (2004): 20–24.

19. Higgitt, "Astronomers against Newton," 21.

20. Higgitt, "Astronomers against Newton," 22.

21. Herschel, "Baily," 118–19.

22. Herschel, "Baily," 100.

23. Ken Alder, *The Measure of All Things: The Seven-Year Odyssey and Hidden Error That Transformed the World* (New York: Free Press, 2002).

24. Herschel, "Baily," 104.

25. Sophia Elizabeth De Morgan, *Memoir,* 106.

26. Sophia Elizabeth De Morgan, *Memoir,* 119.

27. Charles Knight, *Passages of a Working Life during a Half Century: With a Prelude of Early Reminiscences* (London: Bradbury & Evans, 1864), II: 210.

28. Sophia Elizabeth De Morgan, *Memoir,* 50.

29. Sophia Elizabeth De Morgan, *Memoir,* 107.

30. Sophia Elizabeth De Morgan, *Memoir,* 66–67.

31. Knight, *Passages,* II: 64.

32. [Augustus De Morgan], "Halley's Comet," in *British Almanac and Companion,* Society for the Diffusion of Useful Knowledge (London: SDUK, 1835), 5–15.

33. WF to ADM, Oct 8, 1835. CUL, Add. 7887[27].

34. WF to ADM, Sep 24, 1835. CUL, Add. 7887[26].

35. WF to ADM, Oct 8, 1835. CUL, Add. 7887[27].

36. ADM to WRH, Aug 25, 1853, in Robert Perceval Graves, *Life of Sir William Rowan Hamilton* (1889), III: 300.

37. WF to LNB, Oct 11, 1835. OBL, box 71, folios 203–5.

38. WF to ADM, Sep 24, 1835. CUL, Add. 7887[26].

39. *Monthly Notices of the Royal Astronomical Society* 3, no. 18 (Dec 11, 1835): 138, 139.

40. *Monthly Notices of the Royal Astronomical Society* 3, no. 18 (Dec 11, 1835): 138, 139.

41. WF to LNB, Oct 11, 1835. OBL, box 71, folios 203–5.

42. WF to LNB, Oct 11, 1835. OBL, box 71, folios 203–5.

43. "The Comet," UCL, MS Add. 163. In *The Comet,* Rowbotham infuriates Augustus with his astronomical ignorance. The reference is to John Rowbotham, who is most often remembered as John Ruskin's tutor, but was also a fellow of the RAS. De Morgan wrote the obituary that appeared in *Memoirs of the Astronomical Society* 16 (1847): 511.

44. "The Comet," UCL, MS Add. 163.

45. Quotes in the next few paragraphs from "The Comet," UCL, MS Add. 163.

46. Sophia Elizabeth De Morgan, *Memoir,* 48.

47. Sophia Elizabeth De Morgan, *Memoir,* 48.

48. David Philip Miller, "The Revival of the Physical Sciences in Britain, 1815–1840," *Osiris,* 2nd ser., vol. 2 (1986): 107–34; William Ashworth, "The Calculating Eye: Baily, Herschel, Babbage and the Business of Astronomy,'" *British Journal for the History of Science* 27 (1994): 409–41; William J. Ashworth, "Memory, Efficiency, and Symbolic Analysis: Charles Babbage, John Herschel, and the Industrial Mind," *Isis* 87 (1996): 629–53.

49. Herschel, "Baily," 101.

50. A. De Morgan, "Preface" to Baily, "Journal of a Tour in Unsettled Parts of North America," iii.

51. James South, *Charges against the President and Councils of the Royal Society* (London: R. Clay, 1830).

52. Michael Hoskin, "Astronomers at War: South v. Sheepshanks," *Journal for the History of Astronomy* 20, no. 3 (Oct 1989): 178.

53. Hoskin, "Astronomers at War," 182. After South's death, this enormous lens was presented to Dunsink Observatory in Trinity College, Dublin.

54. Hoskin, "Astronomers at War," 180.

55. Letter from the Reverend R. Sheepshanks to the Editor of *The Times,* Dec 4, 1838. Reprinted in Richard Sheepshanks, *A Letter to the Board of Visitors of the Greenwich Royal Observatory* (London: G. Barclay, Castle St. Leicester Sq, 1860), 20.

56. Sophia Elizabeth De Morgan, *Memoir,* 63–64.

57. Hoskin, "Astronomers at War."

58. Quoted in Hoskin, "Astronomers at War," 185.

59. The Knights of the Trinitie, UCL, MS Add. 163.

60. For this group's experience, see George Tate, *The History of the Borough, Castle and Barony of Alnwick* (Alnwick: Henry Hunter Blair, 1866), 372.

61. "Knights," UCL, MS Add. 163. South's wealth came from marriage.

62. "Knights," UCL, MS Add. 163.

63. Notes in the next two paragraphs from "The Comet," UCL, MS Add. 163.

64. Sophia Elizabeth De Morgan, *Reminiscences,* 111–32.

65. ADM to SF, [Aug] 8, [1835]. SHL, MS 913A/1/1.

## 16. EXPANDING CONSCIOUSNESS

1. WF to ADM, Aug 17, 1835. CUL, Add. 7887[24].

2. ADM to SF, [1835] Fri Evening. SHL, MS 913A/1/1.

3. ADM to WF, [1835]. Reprinted in Sophia Elizabeth De Morgan, *Threescore Years and Ten: Reminiscences of the Late Sophia Elizabeth De Morgan to Which Are Added Letters to and from Her Husband the Late Augustus De Morgan, and Others,* ed. Mary De Morgan (London: Richards Bentley and Son, 1895), 128n.

4. LNB to WF, Mar 26, 1836. OBL, box 71, folio 86.

5. WF to LNB, Mar 28, 1836. OBL, box 71, folio 210.

6. WF to Harriet Frend. Feb 8, 1831. CUL, Add. 7886[75].

7. I copied these lyrics from http://www.worshipmap.com/lyrics/messiahtext.html.

8. 1 Corinthians 15:36.

9. 1 Corinthians 15:42–44.

10. WF to Harriet Frend, Feb 8, 1831. CUL, Add. 7886[75].

11. WF to Harriet Frend, Feb 8, 1831. CUL, Add. 7886[75].

12. WF to LNB, Mar 28, 1836. OBL, box 71, folio 210.

13. WF to LNB, Mar 28, 1836. OBL, box 71, folio 210.

14. For an introduction to the cemetery culture of this period, see James Stevenson Curl, *Victorian Celebration of Death* (Gloucestershire: Sutton Publishing, 2000). For more on Victorian views of death, see Pat Jalland, *Death in the Victorian Family* (New York:

Oxford University Press, 1996); Michael Wheeler, *Death and the Future Life in Victorian Literature and Theology* (New York: Cambridge University Press, 1990). The Frend family was very restrained in marking Harriet's grave with just a simple flat stone.

15. WF to ADM, Jun 9 1836. CUL, Add. 7887[28].

16. WF to ADM, Jun 9 1836. CUL, Add. 7887[28].

17. ADM to SF, Jun 16, [1836]. SHL, MS 913A/1/1.

18. WF to ADM, Jul 12 1836. CUL, Add. 7887[31].

19. WF to ADM, Jul 12, 1836. CUL, Add. 7887[31]. WF to ADM, Jul 24, 1836. CUL, Add. 7887[32].

20. WF to ADM, Jul 12, 1836. CUL, Add. 7887[31].

21. Sophia Elizabeth De Morgan, *Reminiscences,* 98.

22. Sophia Elizabeth De Morgan, *Reminiscences,* 178.

23. LNB to SF, Jun 15, 1836. OBL, box 67, folios 77–78.

24. Francis Blackburne, *No Proof in the Scriptures of an Intermediate State of Happiness or Misery Between Death and the Resurrection* (London: S Blaidon, 1756).

25. LNB to SF, Jul 13, [1836]. OBL, box 67, folios 79–80.

26. LNB to SF, [1836]. OBL, box 67, folio 87.

27. LNB to SF, Jul 13, [1836]. OBL, box 67, folios 79–80.

28. LNB to SF, Jul 13, [1836]. OBL, box 67, folios 79–80.

29. George Berkeley, *The Analyst: Or, a Discourse Addressed to an Infidel Mathematician. Wherein It Is Examined Whether the Object, Principles, and Inferences of the Modern Analysis Are More Distinctly Conceived, or More Evidently Deduced, Than Religious Mysteries and Points of Faith,* 2nd ed. (London: J. and R. Tonson and S. Draper in the Strand, 1754), 59.

30. Robert Woodhouse, *The Principles of Analytic Calculation* (Cambridge, 1803), xviii.

31. "Berkeley," s.v. *Penny Cyclopædia* (1835).

32. Augustus De Morgan, *Formal Logic, or, The Calculus of Inference, Necessary and Probable,* ed. A. E. Taylor (London: Open Court Company, 1926), 31.

33. Sophia Elizabeth De Morgan, *Memoir of Augustus De Morgan* (London: Longmans, Green, 1882), 165.

34. "Berkeley," s.v. *Penny Cyclopædia* (1835).

35. [Augustus De Morgan] "Report of the Council of the [Royal Astronomical] Society to the Twenty-Second Annual General Meeting, February 11, 1842" *Royal Astronomical Society* 12 (1842): 463.

36. William Whewell, "Thoughts on the Study of Mathematics as a Part of a Liberal Education," in *On the Principles of English University Education* (London: John W. Parker, 1838), 138.

37. Whewell, "Thoughts," 140.

38. Whewell, "Thoughts," 163.

39. Augustus De Morgan, *The Connexion of Number and Magnitude: An Attempt to Explain the Fifth Book of Euclid* (London: Taylor and Walton, 1836).

40. [ADM] "Limit," s.v. *Penny Cyclopædia* 13 (1839).

41. Whewell, "Thoughts," 149.

42. Augustin-Louis Cauchy, *Cauchy's* Cours d'analyse*: An Annotated Translation,* ed. and trans. Robert E Bradley and Edward C Sandifer (New York: Springer, 2009), 7.

43. Cauchy, *Cours d'analyse,* 2.

44. Cauchy, *Cours d'analyse,* 3 I have diverged from the Bradley and Sandifer translation for the second half of this sentence. They rendered Cauchy's "ni donner pour sanction à la morale des théorèmes d'algebre ou de calcul integral" as "nor to make moral judgements the theorems of algebra or of integral calculus."

45. Bruno Belhoste, *Cauchy: Un mathématicien légitimiste au XIXe siècle* (Paris: Belin, 1985), 78–84.

46. Whewell, "Thoughts," 149.

47. Augustus De Morgan, *The Differential and Integral Calculus,* Library of Useful Knowledge (London: Baldwin and Cradock, 1842).

48. [ADM] "Interpretation" s.v. *Penny Cyclopædia* 13 (1839).

49. This quotation is from De Morgan's article "Limit" in the *Penny Cyclopædia,* 13 (1839). He made the same point, though less neatly in his *Calculus:* "Our object is to show that there is no great refinement or abstruseness in the nature of the fundamental ideas of the science; but that they do, in fact, suggest themselves in cases which occur in common life." Augustus De Morgan, *Calculus,* 27.

50. Augustus De Morgan, *Calculus,* 27.

51. Cauchy, *Cours d'analyse,* 2.

52. Augustus De Morgan, "On Divergent Series and Various Points of Analysis Connected with Them" (read March 4, 1844)," *Transactions of the Cambridge Philosophical Society* VIII (1849): 183. For De Morgan's discussion of this particular series, see Augustus De Morgan, *Calculus,* 224. For a fuller development of De Morgan's views of divergent series, see Joan L. Richards, "Augustus De Morgan, the History of Mathematics, and the Foundations of Algebra," *Isis* 78 (1987): 6–30.

53. WF to LNB, Mar 28, 1836. OBL, box 71, folio 210.

54. SF to ADM, postscript to WF to ADM, Sep 2, 1836. CUL, Add. 7887[30]. This letter, without Sophia's postscript, is reprinted in Sophia Elizabeth De Morgan, *Reminiscences,* n136–40.

55. ADM to SF. [Sep 1836]. SHL, MS 913/2/1.

## *17. HOME ON GOWER STREET*

1. ADM to SF. UCL, MS Add. 7.

2. SF to ADM, Sep 6, 1836. UCL, MS Add. 7.

3. SF to ADM, Sep 6, 1836. UCL, MS Add. 7.

4. SF to ADM, Sep 6, 1836. UCL, MS Add. 7.

5. Sophia Elizabeth De Morgan, *Memoir of Augustus De Morgan* (London: Longmans, Green, 1882), 72–73.

6. LNB to SF, Feb 6 1837. OBL, box 67, folios 83–84.

7. Sophia Elizabeth De Morgan, *Memoir,* 86.

8. ADM to Charles Babbage. Apr 17, 1837. British Library, Add. MS 37190, folio 107.

9. Augustus De Morgan, *The Connexion of Number and Magnitude: An Attempt to Explain*

*the Fifth Book of Euclid* (London: Taylor and Walton, 1836) in Senate House Library Special Collections [Rare] AWF Dem.

10. Sophia Elizabeth De Morgan, *Memoir,* 88.

11. Sophia Elizabeth De Morgan, *Memoir,* 88.

12. Afterwards, the house was renumbered to become 35 Gower Street.

13. WF to SEDM, Sep 26 1837. CUL, Add. 7887³³.

14. Hugh Hale Bellot, *The University of London: A History* (Bristol: Western Printing Services, 1969), 2–4.

15. Augustus De Morgan, *Thoughts Suggested by the Establishment of the University of London* (London: Taylor and Walton, 1837).

16. Augustus De Morgan, *Thoughts,* 3.

17. Augustus De Morgan, *Thoughts,* 6.

18. Augustus De Morgan, *Thoughts,* 8–9; cf. Matthew 6:6.

19. Augustus De Morgan, *Thoughts,* 8–9.

20. Augustus De Morgan, *Thoughts,* 10.

21. Augustus De Morgan, *Thoughts,* 15.

22. Augustus De Morgan, *Thoughts,* 15.

23. Augustus De Morgan, *Elements of Arithmetic* (London: John Taylor, 1830).

24. For a thorough treatment of the curriculum Augustus developed in these notebooks, see Adrian Rice, "What Makes a Great Mathematics Teacher?: The Case of Augustus De Morgan," *American Mathematical Monthly* 106, no. 6 (Jun–Jul 1999): 534–52.

25. Augustus De Morgan, *Thoughts,* 7.

26. Sophia Elizabeth De Morgan, *Memoir,* 101.

27. Hugh Hale Bellot, *University College London, 1826–1926* (London: University of London Press, 1929), 83.

28. [Sir John William] Lubbock and [John Elliot Drinkwater] Bethune, *On Probability,* published under the Superintendence of the Society for the Diffusion of Useful Knowledge (London: Baldwin and Cradock, 1830).

29. Augustus De Morgan, *An Essay on Probabilities, and on Their Application to Life Contingencies and Insurance Offices,* in *The Cabinet Cyclopædia,* ed. Dionysius Lardner (London: Longman, Orme, Brown, Green, & Longmans, and John Taylor, 1838), 13.

30. Pierre-Simon Laplace, *Philosophical Essay on Probabilities,* 5th ed., trans. Andrew I. Dale (New York: Springer-Verlag, 1995), 1.

31. Laplace, *Probabilities,* 12.

32. [Augustus De Morgan], "Theory of Probabilities," in *Encyclopædia Metropolitana,* ed. Hugh James Rose, Edward Smedley, and Henry John Rose (London: B. Fellowes, 1838), 394.

33. Augustus De Morgan, *Essay on Probabilities,* 7.

34. [De Morgan], "Probabilities," 396–97. For the phrase "moral probability," see Lorraine Daston, *Classical Probability in the Enlightenment* (Princeton: Princeton University Press, 1988). As an entrée into a vast literature on the history of probability, see Gerd Gigerenzer, *The Empire of Chance: How Probability Changed Science and Everyday Life* (New York: Cambridge University Press, 1989).

35. ADM to his mother, 1836, in Sophia Elizabeth De Morgan, *Memoir,* 144.

36. Augustus De Morgan, *Essay on Probabilities,* 7.

37. For an introduction to the correspondence between De Morgan and Lovelace, see Christopher Hollins, Ursula Martin, and Adrian Rice, *Ada Lovelace: The Making of a Computer Scientist* (Oxford: Bodleian Library, 2018)

38. Sophia Elizabeth De Morgan, *Memoir,* 90.

39. Sophia Elizabeth De Morgan, *Threescore Years and Ten: Reminiscences of the Late Sophia Elizabeth De Morgan to Which Are Added Letters to and from Her Husband the Late Augustus De Morgan, and Others,* ed. Mary De Morgan (London: Richards Bentley and Son, 1895), 184. Sophia remembered writing this article for "a periodical (of which only two numbers appeared) called, I think, *The Monthly Chronicle,*" but I have been unable to find it in *The Monthly Chronicle: A National Journal of Politics, Literature, Science, and Art,* seven volumes of which were published by Longman, Orme, Brown, Green and Longmans between 1838 and 1841.

40. Sophia Elizabeth De Morgan, *Memoir,* 93.

41. Sophia Elizabeth De Morgan, *Memoir,* 91.

42. Sophia Elizabeth De Morgan, *Reminiscences,* 188–89.

43. Sophia Elizabeth De Morgan, *Reminiscences,* 190.

44. Sophia Elizabeth De Morgan, *Reminiscences,* 188–89.

45. Sophia Frend De Morgan, Nursery Journal, 1839–1842. DMF_MS_0024 (hereafter NJ).

46. Maria Edgeworth and Richard Lovell Edgeworth, *Practical Education* (New York: Harper & Brothers, 1835), x.

47. Edgeworth and Edgeworth, *Practical Education,* vii.

48. NJ. The single exception is "Conscientiousness," which she judged to be "Moderate."

49. NJ. [Thursday] Apr 30, [1840].

50. Isaac Watts, "Discourse on the Education of Children and Youth," in *The Improvement of the Mind. To Which Is Added a Discourse on the Education of Children and Youth,* in *The Improvement of the Mind. To Which is Added a Discourse on the Education of Children and Youth* (Boston: Thomas, 1812), 313.

51. Edgeworth and Edgeworth, *Practical Education,* 133.

52. NJ. Monday, [Jan 20, 1840].

53. NJ. Tuesday, Wednesday, & Thursday [Jan 21–24, 1840].

54. NJ. [Monday, July] 26, [1841].

55. LNB to SEDM, Mar [1840]. OBL, box 67, folios 104–5.

56. NJ. Sunday, Jan 5, [1840].

57. NJ. Apr 9, [1841].

58. NJ. Yesterday, Thursday [Apr 15, 1840].

59. NJ. [Wednesday] Dec 2, 1840.

60. NJ. [Feb] 10, [1841].

61. NJ. [Monday] Aug 9, [1841]. Conolly was the resident physician at the Middlesex County Pauper Lunatic Asylum at Hanwell. John Conolly, *An Inquiry Concerning the Indications of Insanity,* reprinted from 1830 (London: Dawsons of Pall Mall, 1964).

62. NJ. [Sunday, Jul] 25, [1841].

63. NJ. Sunday, Aug 15, 1841.

64. NJ. Saturday, [Jan 1840].

65. NJ. Jul 4, 1841.

66. NJ. Dec 30, 1840.

67. NJ. Sunday, Feb 15, [1840].

68. NJ. Dec 10, 1840.

69. NJ. [Wednesday], Dec 2, 1840.

70. NJ. [Dec] 18, [1840].

71. NJ. Dec 7, 1841.

72. For more on Victorian views of lying in children, see "Lies and Imagination," in Sally Shuttleworth, *The Mind of the Child: Child Development in Literature, Science and Medicine, 1840–1900* (Oxford: Oxford University Press, 2010), 60–74.

73. NJ. Aug 29, [1841].

74. NJ. [Tuesday, Mar] 30, [1841].

75. WF to LNB, Jan 7, 1838. OBL, box 71, folios 114–15.

76. WF to LNB, Apr 22, 1839. OBL, box 71, folios 116–7.

77. Sophia Elizabeth De Morgan, *Memoir,* 109.

78. NJ. [Feb] 14, [1841].

79. Sophia Elizabeth De Morgan, *Reminiscences,* 84 Sophia seems here to have confused the Psalm 19, which opens with "The heavens declare the glory of God," with Psalm 103, which contains the line "As for man, his days are as grass."

80. [Augustus De Morgan], "Report of the Council of the Society to the Twenty-First Annual General Meeting, February 12, 1841," *Memoirs of the Royal Astronomical Society* 12 (1842): 465.

81. [De Morgan], "Report, 1841," 465.

82. NJ. Mar 26, [1841].

83. NJ. Mar 26, [1841].

84. NJ. Mar 26, [1841].

85. Sophia Elizabeth De Morgan, *Memoir,* 94.


*18. REARING YOUNG SEEDLINGS*

1. A. M. W. Stirling, *William De Morgan and His Wife* (New York: H. Holt, 1922), 39.

2. Stirling, *William De Morgan,* 42.

3. Stirling, *William De Morgan,* 41–42.

4. Stirling, *William De Morgan,* 42.

5. NJ. [Feb] 10, [1841].

6. Draft letter from ADM to his mother, [no date]. SHL, MS 913A/2/12.

7. Stirling, *William De Morgan,* 47.

8. Stirling, *William De Morgan,* 45.

9. Sophia Elizabeth De Morgan, *Memoir of Augustus De Morgan* (London: Longmans, Green, 1882), 116.

10. For the demise of the Children's Friend Society, see Geoff Blackburn, *The Children's Friend Society* (Western Australia: Access Press, 1993) For Sophia's responses, see Sophia Elizabeth De Morgan, *Threescore Years and Ten: Reminiscences of the Late Sophia*

*Elizabeth De Morgan to Which Are Added Letters to and from Her Husband the Late Augustus De Morgan, and Others,* ed. Mary De Morgan (London: Richards Bentley and Son, 1895), 202–3.

11. For an expansive treatment of these developments, see Alison Winter, *Mesmerized: Powers of Mind in Victorian Britain* (Chicago: University of Chicago Press, 1998).

12. NJ. [Apr] 4, [1841].

13. NJ. May 23, [1841].

14. For the development of nineteenth-century therapeutic techniques, see Charles Rosenberg, "The Therapeutic Revolution," in *The Therapeutic Revolution: Essays in the Social History of American Medicine,* ed. Morris Vogel and Charles Rosenberg (Philadelphia: University of Pennsylvania Press, 1979), 3–26. For a more comprehensive treatment, see John Harley Warner, *The Therapeutic Perspective: Medical Practice, Knowledge, and Identity in America, 1820–1885* (Cambridge: Harvard University Press, 1986). Both of these works focus primarily on America, but developments in England were similar.

15. NJ. May 23, [1841].

16. NJ. May 23, [1841].

17. NJ. May 23, [1841].

18. Stirling, *William De Morgan,* 48.

19. Stirling, *William De Morgan,* 60.

20. C. D. [Sophia De Morgan], *From Matter to Spirit. The Result of Ten Years' Experience in Spirit Manifestations. Intended as a Guide to Enquirers,* with an introduction by A. B. [Augustus De Morgan] (London: Longman, Green, Longman, Roberts & Green, 1863), 43–46.

21. ADM to William Heald, [1849], in Sophia Elizabeth De Morgan, *Memoir,* 206.

22. For a description of the interactions between these publications and the development of English science in this period, see "Steam Reading" in James Secord, *Victorian Sensation: The Extraordinary Publication, Reception and Secret Authorship of Vestiges of the Natural History of Creation* (Chicago: University of Chicago Press, 2000), 41–76.

23. Sloan Evans Despeaux and Adrian C Rice, "Augustus De Morgan's Anonymous Reviews for *The Athenaeum*: A Mirror of a Victorian Mathematician," *Historia Mathematica* 43 (2016): 148–71. For a list of De Morgan's almost one thousand reviews, see http://smcse.city.ac.uk/doc/cisr/web/athenaeum/reviews/contributors/contributor files/DEMORGAN,Augustus.html.

24. Sophia Elizabeth De Morgan, *Memoir,* 105.

25. ADM to WRH, Sep 27, 1852, in Robert Perceval Graves, *Life of Sir William Rowan Hamilton* (1889), III: 415.

26. Sophia Elizabeth De Morgan, *Memoir,* 105.

27. ADM to John Herschel, Apr 8. RS:HS 6.203.

28. ADM to John Herschel, Sep 15, 1865, in Sophia Elizabeth De Morgan, *Memoir,* 333.

29. Ronald Anderson, "Augustus De Morgan's Inaugural Lecture of 1828," *Mathematical Intelligencer* 25, no. 3 (2006): 19.

30. Augustus De Morgan, *Arithmetical Books from the Invention of Printing to the Present Time Being Brief Notices of a Large Number of Works Drawn up from Actual Inspection* (London: Taylor and Walton, 1847), vi.

31. Augustus De Morgan, *Arithmetical Books,* vi.

32. Augustus De Morgan, *Arithmetical Books,* ii.

33. Augustus De Morgan, *Arithmetical Books,* 13.

34. Augustus De Morgan, *Arithmetical Books,* 51.

35. Augustus De Morgan, *Arithmetical Books,* 55.

36. ADM to Sir John Herschel, May 11, 1844. RS:HS 6.205. De Morgan's example of an obscure equation was $(a + b > -1)^{c + d\sqrt{-1}}$.

37. Augustus De Morgan, *Arithmetical Books,* xxvii.

38. ADM to WRH, Oct 11, 1849, in Graves, *Hamilton,* III: 283.

39. John Locke, *An Essay Concerning Human Understanding,* 29th ed. (London: Thomas Tegg, 1841), 349.

40. Richard Whately, *Elements of Logic: Comprising the Substance of the Article in the Encyclopædia Metropolitana* (London: John W. Parker, 1848), xi.

41. Whately, *Logic,* xxxiii–xxxiv.

42. Whately, *Logic,* xiii.

43. Whately, *Logic,* 11.

44. Whately, *Logic,* 20n.

45. James Van Evra, "Richard Whately and Logical Theory," in *Handbook of the History of Logic,* ed. Dov M Gabbay and John Woods (Amsterdam: Elsevier, 2008), 75–91.

46. Augustus De Morgan, *On the Study and Difficulties of Mathematics* (Chicago: Open Court, 1910), 212n. For a detailed analysis of the development of De Morgan's ideas of logic in the 1830s, see Maria Panteki, "French 'Logique' and British 'Logic': On the Origins of Augustus De Morgan's Early Logical Inquiries, 1805–1835," *Historia Mathematica* 30 (2003): 278–340.

47. Augustus De Morgan, *On the Study and Difficulties of Mathematics* (Chicago: Open Court, 1910), 220–21. For the historical relations of logic and mathematics, see Ian Mueller, "Greek Mathematics and Greek Logic," in *Ancient Logic and Its Modern Interpretations,* ed. John Corcoran (Boston: Reidel, 1974).

48. Sir William Hamilton, "Logic: The Recent English Treatises on That Science," in *Discussions on Philosophy and Literature, Education and University Reform,* introd. by Robert Turnbull (New York: Harper & Brothers, 1853), 130.

49. Hamilton, "Logic," 137.

50. Sir William Hamilton, "On the Study of Mathematics as an Exercise of Mind," in *Discussions on Philosophy and Literature, Education and University Reform,* introd. by Robert Turnbull (New York: Harper & Brothers, 1853), 302.

51. Whately, *Logic,* xiii.

52. Augustus De Morgan, *First Notions of Logic (Preparatory to the Study of Geometry)* (London: Taylor and Walton, 1840).

53. [ADM], "Syllogism," s.v. *Penny Cyclopædia.*

54. Augustus De Morgan, *First Notions,* 4.

55. Augustus De Morgan, "On the Foundation of Algebra, No. I" (read Dec 9, 1839), *Transactions of the Cambridge Philosophical Society* 7 (1842): 173–88; Augustus De Morgan, "On the Foundation of Algebra, No II" (read Nov 29, 1841), *Transactions of the Cambridge Philosophical Society* 7 (1842): 287–300; Augustus De Morgan, "On the

Foundation of Algebra, No. III" (read Nov 27, 1843), *Transactions of the Cambridge Philosophical Society* 8 (1849): 139–42; Augustus De Morgan, "On the Foundation of Algebra, No. IV, on Triple Algebra" (read Oct 28, 1844), *Transactions of the Cambridge Philosophical Society* 8 (1849): 241–54. (In the world of the mid-century Cambridge Philosophical Society, publication was haphazard enough that the "read" dates I'm here including after each article title are generally considered the salient ones.)

56. Augustus De Morgan, "On the Structure of the Syllogism, and on the Application of the Theory of Probabilities to Questions of Argument and Authority" (read Nov 9, 1846), *Transactions of the Cambridge Philosophical Society* 8 (1849) ("On the Syllogism I"); Augustus De Morgan, "On the Symbols of Logic, the Theory of the Syllogism, and in Particular of the Copula, and the Application of the Theory of Probabilities to Some Questions of Evidence" (read Feb 25, 1850), *Transactions of the Cambridge Philosophical Society* 9 (1851): 79–127 ("On the Syllogism II"); Augustus De Morgan, "On the Syllogism, No. III, and on Logic in General" (read Feb 8, 1858), *Transactions of the Cambridge Philosophical Society* 10 (1864): 173–230 ("On the Syllogism III"); Augustus De Morgan, "On the Syllogism, No. IV, and on the Logic of Relations" (read Apr 23, 1860), *Transactions of the Cambridge Philosophical Society* (1864): 331–58 ("On the Syllogism IV"); Augustus De Morgan, "On the Syllogism, No. V. and on Various Points of the Onymatic System" (read May 4, 1863), *Transactions of the Cambridge Philosophical Society* (1864): 428–87 ("On the Syllogism V"). All of these articles are reprinted in *On the Syllogism and Other Logical Writings,* ed. Peter Heath (New Haven: Yale University Press, 1966).

57. I took this image of the square of opposition from "The Traditional Square of Opposition," in *Stanford Encyclopedia of Philosophy* (2012) The same source offers an explanation of the various relations that exist among them.

58. Augustus De Morgan, *Formal Logic,* ed. A. E. Taylor (London: Open Court Company, 1926), 40–41.

59. Augustus De Morgan, "On the Syllogism I," 380. De Morgan is widely credited with inventing the phrase "universe of discourse," which is in current use. However, it was Boole, not De Morgan, who used that phrase, while De Morgan spoke of the "universe of the proposition." The difference is significant. Boole's "universe of discourse" focuses attention on the words; De Morgan's "universe of the proposition," or "universe under discussion," focuses attention on the universe beyond the words to which a proposition refers. For further discussion of this point, see Hobart and Richards, "De Morgan's Logic," 307.

60. Augustus De Morgan, "On the Syllogism I," 380.

61. De Morgan referred to such divisions as "contraries," but the modern term "complements" is preferable because it points to the way the terms divide the universe of their propositions.

62. Augustus De Morgan, *On the Syllogism and Other Logical Writings,* ed. and comp. Peter Heath (New Haven: Yale University Press, 1966), xiii. For a detailed presentation of this long and convoluted controversy, see Heath's introduction. Augustus published his version of the quarrel as an appendix to Augustus De Morgan, *Formal Logic,* 345–76. De Morgan and Hamilton were neither the only nor the first nineteenth-century

Englishmen to suggest increasing the number of syllogisms to eight. George Bentham, a nephew of the reformer Jeremy Bentham, had considered the idea in a book published in 1828. One of the effects on De Morgan of his priority dispute with Hamilton was that he bound dated lists of his logical readings into his personal copy of *Formal Logic.* At the end of 1850, De Morgan responded to a report of Bentham's work with the comment: "1850 Dec 26 Examined Bentham at Brit. Mus. Found that he has distinct unev. [idea of?] quantif[ication] of predicate . . . but that he has blundered the propositions." For Bentham's role in the development of a logic with eight propositions, see Gordon R. McOuat and Charissa S Varma, "Bentham's Logic," in *British Logic in the Nineteenth Century,* vol. 4 of *Handbook of the History of Logic,* ed. Dov M. Gabbay and John Woods (Boston: Elsevier, 2008), 1–32.

63. Augustus De Morgan, "On the Syllogism II," 99 (emphasis added).

64. Sophia Elizabeth De Morgan, *Memoir,* 96.

65. "Subject, Subjective" *Penny Cyclopædia* 23 (1842) Stearic Acid—Tagus.

66. ADM to WW, Mar 1, 1849. TCL, Add. MS a 202[113].

67. ADM to WW, Oct 5, 1846. TCL, Add. MS a. 202[105].

68. ADM to WW, Oct 21, 1846. TCL, Add. MS a. 202[106].

69. ADM to WW, Oct 5, 1846. TCL, Add. MS a. 202[105].

70. William Whewell, "On the Fundamental Antithesis of Philosophy" (read Feb 5, 1844), *Transactions of the Cambridge Philosophical Society* 8, pt. 2, p. 172. Italics added.

71. Augustus De Morgan, *Formal Logic,* 33n.

72. Augustus De Morgan, *Formal Logic,* 32.

73. This last is Sophia's wonderfully preceptive phrase. Sophia Elizabeth De Morgan, *Memoir,* 115–16n.

74. Augustus De Morgan, *Formal Logic,* 32.

75. ADM to WRH, Oct 11, 1849, in Graves, *Hamilton,* III: 283.

76. ADM to WRH, Dec 31, 1863, in Graves, *Hamilton,* III: 603.

*19. EXPANDING REASON*

1. Sophia lists some of this group in Sophia Elizabeth De Morgan, *Memoir of Augustus De Morgan* (London: Longmans, Green, 1882), 173.

2. ADM quoted in Adrian Rice, "Vindicating Leibniz in the Calculus Priority Dispute: The Role of Augustus De Morgan," in *The History of the History of Mathematics: Case Studies for the Seventeenth, Eighteenth and Nineteenth Centuries,* ed. Benjamin Wardhaugh (Peter Lang AG, Internationaler Verlag der Wissenschaften, 2012), 101.

3. Rice, "Vindicating Leibniz," 104.

4. Rebekah Higgitt, "Why I Don't FRS My Tail: Augustus De Morgan and the Royal Society," *Notes and Records of the Royal Society* 60 (2006): 253–59.

5. ADM to WW, Oct 21, 1846. TCL, Add. MS a.202[106]. Reprinted in Sophia Elizabeth De Morgan, *Memoir,* 197–98.

6. Sophia Elizabeth De Morgan, *Memoir,* 94.

7. A. M. W. Stirling, *William De Morgan and His Wife* (New York: H. Holt, 1922), 32.

8. Sophia Elizabeth De Morgan, *Memoir,* 94.

9. Margaret J. Tuke, *A History of Bedford College for Women* (New York: Oxford University Press, 1939), 3.

10. RHUL, letter included in a packet that Edward De Morgan's granddaughter, Joan M. Antrobus, gave to Bedford College in 1961.

11. For the various intermediate names of this institution, see appendix 8 of Tuke, *Bedford College,* 331. The Bedford College for Women was granted a royal charter in 1909 that allowed it to give college degrees under the University of London. In 1985 it merged with Royal Holloway College to form Royal Holloway and Bedford New College under the University of London.

12. RHUL.

13. "Prospectus for a Ladies' College in Bedford Square," appendix 2, Tuke, *Bedford College,* 319.

14. RHUL, Report of the subcommittee appointed to draw up a Plan of the future Constitution. Jul 16, 1849.

15. ADM to WRH, May 6, 1852. Robert Perceval Graves, *Life of Sir William Rowan Hamilton* (1889), III: 357.

16. As an entrée into the literature on Harriet Becher Stowe, I would recommend Joan D. Hedrick, *Harriet Beecher Stowe: A Life* (New York: Oxford University Press, 1994).

17. Draft proposal on slavery. Sophia autograph. SHL, MS 913B/2/3.

18. J. S. Nicolay to SEDM, Nov 4, 1852. SHL, MS 913B/1/2.

19. Draft proposal on slavery. Alice autograph. SHL, MS 913A/3/2.

20. SEDM to Edwin Chadwick, [Nov 1852]. SHL, MS 913B/1/5.

21. J. S. Nicolay to SEDM, Wednesday, [Nov 12, 1852]. SHL, MS 913B/1/2.

22. J. S. Nicolay to SEDM, Nov 13, [1852]. SHL, MS 913B/1/2.

23. Hedrick, *Stowe,* 232.

24. Sophia Elizabeth De Morgan, *Memoir,* 183.

25. The volumes are housed in the Harriet Beecher Stowe Center in Hartford, Connecticut. https://www.harrietbeecherstowecenter.org/.

26. Sophia Elizabeth De Morgan, *Threescore Years and Ten: Reminiscences of the Late Sophia Elizabeth De Morgan to Which Are Added Letters to and from Her Husband the Late Augustus De Morgan, and Others,* ed. Mary De Morgan (London: Richards Bentley and Son, 1895), 223.

27. Sophia Elizabeth De Morgan, *Reminiscences,* 224.

28. WRH to ADM, Apr 15, 1852. Graves, *Hamilton,* III: 349.

29. ADM to WRH, May 27, 1852. Graves, *Hamilton,* III: 370.

30. Augustus's version of this experience is in a letter to Heald [1849] in Sophia Elizabeth De Morgan, *Memoir,* 206–8. Sophia's version is in C. D. [Sophia De Morgan], *From Matter to Spirit. The Result of Ten Years' Experience in Spirit Manifestations. Intended as a Guide to Enquirers,* with an introduction by A. B. [Augustus De Morgan] (London: Longman, Green, Longman, Roberts & Green, 1863), 47–49.

31. Michael Faraday, "Letter on Table-Turning" (1853); Faraday, "Professor Faraday on Table-Moving," *Athenaeum,* Jul 2, 1853, 801–3.

32. For an overview of the English experience of table-turning and mesmerism in the 1850s and beyond, see Alison Winter, *Mesmerized: Powers of Mind in Victorian Britain* (Chicago: University of Chicago Press, 1998), 276–306.

33. LNB to SEDM, Jul 3, 1853. OBL, box 67, folio 143.

34. *Athenaeum,* Jul 16, 1853, 866.

35. C. D. [Sophia De Morgan], *Matter to Spirit,* 13–14.

36. ADM to Rev. W. Heald, Jul 1853, in Sophia Elizabeth De Morgan, *Memoir,* 221–22.

37. ADM to George Boole, May 31, 1847, in G. C. Smith, *The Boole-De Morgan Correspondence, 1842–1864* (Oxford: Clarendon Press, 1982), 22.

38. Unsent draft letter from ADM to George Boole, [Nov 1847], in Smith, *Boole-De Morgan,* 24.

39. John Corcoran, "Correspondence without Communication," *History and Philosophy of Logic* 7 (1986): 65–75.

40. I've taken these examples from Smith, *Boole-De Morgan,* 23.

41. Unsent draft letter from ADM to George Boole, [Nov 1847], in Smith, *Boole-De Morgan,* 24.

42. Augustus De Morgan, "On the Syllogism I," 381.

43. De Morgan used the term "convertible."

44. Michael E. Hobart and Joan L. Richards, "De Morgan's Logic," in *British Logic in the Nineteenth Century,* vol. 4 of *Handbook of the History of Logic,* ed. Dov M Gabbay and John Woods (Boston: Elsevier, 2008), 299.

45. For a fuller description of the workings of De Morgan's logical system, see Hobart and Richards, "De Morgan's Logic."

46. ADM to WW, Dec 31, 1849. TCL, Add. MS a. 202[118].

47. George Boole to ADM, Oct 17, 1850, in Smith, *Boole-De Morgan,* 38.

48. [Henry Longueville Mansel], "Recent Extensions of Formal Logic," *North British Review* 15 (1851): 94.

49. [ADM], "Syllogism," s.v. *Penny Cyclopædia* 23 (1842).

50. [Mansel], "Formal Logic," 54.

51. [Mansel], "Formal Logic," 54–55.

52. [Mansel], "Formal Logic," 59.

53. [Mansel], "Formal Logic," 57.

54. [Mansel], "Formal Logic," 93.

55. Sophia Elizabeth De Morgan, *Memoir,* 4.

56. Augustus De Morgan, "On the Syllogism II," 114.

57. For a more detailed analysis of the meaning of De Morgan's symbols, see Hobart and Richards, "De Morgan's Logic."

58. I owe this insight to Kyna Leski.

59. ADM to Michael Foster, Nov 15, 1853. Sophia Elizabeth De Morgan, *Memoir,* 223.


*20. BEYOND MATTER*

1. ADM to John Herschel, Aug 10, 1853. RS:HS 6.265.

2. Robin Wilson, *Four Colors Suffice: How the Map Problem Was Solved* (Princeton: Prince-

ton University Press, 2002), 19. This is a major source for the history of the four-color problem; another is Rudolf Fritsch and Gerda Fritsch, *The Four-Color Theorem: History, Topological Foundations, and Idea of Proof,* trans. Julie Peschke (New York: Springer, 1998).

3. The direct quotation from the letter De Morgan wrote to Hamilton is in first person and the present tense: "The more I think of it, the more evident it seems." Wilson, *Four Colors,* 18.

4. Wilson, *Four Colors,* 18.

5. Wilson, *Four Colors,* 23.

6. For an overview of Mansel's views, see Bernard Lightman, *The Origins of Agnosticism: Victorian Unbelief and the Limits of Knowledge* (Baltimore: Johns Hopkins University Press, 1987), 7–32. For a specific focus on the Whewell-Mansel debate, see John Wettersten, *Whewell's Critics: Have They Prevented Him from Doing Good?* (New York: Rudopi, 2005), 77–84.

7. ADM to WW, Dec 9, 1853. TCL, Add. MS a. 202$^{125}$.

8. ADM to WRH, Dec 15, 1853. Trinity College Archives, Dublin, MS 1493, 767.

9. ADM to WRH, Jan 10, 1854. Trinity College Archives, Dublin, MS 1493, 772.

10. Sophia Elizabeth De Morgan, *Memoir of Augustus De Morgan* (London: Longmans, Green, 1882), 190.

11. William De Morgan, *The Old Man's Youth and the Young Man's Old Age* (New York: Henry Holt, 1920), 101.

12. ADM to WRH, Jan 10, 1854. Robert Perceval Graves, *Life of Sir William Rowan Hamilton* (1889), III: 470. ADM wrote this in the first person, so was "set down at my desk."

13. ADM to WRH, Feb 13, 1854. Graves, *Hamilton,* III: 470.

14. William Whewell, *Of the Plurality of Worlds: An Essay,* reprinted from first edition, ed. and comp. Michael Ruse (Chicago: University of Chicago Press, 2001), 253.

15. ADM to WW, Jan 24, 1854. Sophia Elizabeth De Morgan, *Memoir,* 228.

16. Sophia Elizabeth De Morgan, *Memoir,* 229.

17. ADM to WW, May 21, 1854. Sophia Elizabeth De Morgan, *Memoir,* 230.

18. SHL, MS 913B/2/2.

19. Artifacts from the later Playground Society have been displayed on the basement level of the British Library.

20. Augustus De Morgan, "On the Syllogism III," quoted in Daniel D Merrill, *Augustus De Morgan and the Logic of Relations* (Boston: Kluwer Academic Publishers, 1990), 100.

21. Augustus De Morgan, "On the Syllogism III," 74.

22. Augustus De Morgan, "On the Syllogism III," 175.

23. Augustus De Morgan, "On the Foundation of Algebra, No. II" (read Nov 29, 1841), *Transactions of the Cambridge Philosophical Society* 7 (1842): 289–90.

24. ADM to WRH, Apr 29, 1854. Graves, *Hamilton,* III: 479.

25. Augustus De Morgan, "On the Syllogism III," 179.

26. Augustus De Morgan, "On the Syllogism III," 174–75.

27. ADM to WRH, Jul 27, 1858. Graves, *Hamilton,* 553.

28. Augustus De Morgan, "On the Syllogism II," 106–7.

29. Henry Longueville Mansel], "Recent Extensions of Formal Logic," *North British Review* 15 (1851).

30. Augustus De Morgan, "On the Syllogism II," 126–27.

31. Augustus De Morgan, "On the Syllogism III," 175.

32. De Morgan "Logic" from the *English Encyclopedia,* in *On the Syllogism and Other Logical Writings,* ed. Peter Heath (New Haven: Yale University Press, 1966), 252.

33. W. Stanley Jevons, *Studies in Deductive Logic, a Manual for Students* (New York: Macmillan, 1880), xii–xiii.

34. Augustus De Morgan, "On the Syllogism IV," 341.

35. UCL, MS Add. 7.

36. Augustus De Morgan, *Syllabus of a Proposed System of Logic* (London: Walton and Maberly, 1860), 21. He also briefly alluded to it in Augustus De Morgan, "On the Syllogism III," 219n.

37. Michael Faraday, "On the Education of the Judgment," in *The Culture Demanded by Modern Life,* ed. E. L. Youmans (New York: D. Appleton, 1867), 188.

38. Faraday, "On Education," 205.

39. Faraday, "On Education," 207.

40. [Augustus De Morgan], "Review of 'On the Conservation of Force. A Lecture Delivered by Prof. Faraday at the Royal Institution, February 27, 1857,'" *Athenaeum,* no. 1535 (Mar. 28, 1857): 397.

41. AMD to WRH, Apr 23, 1852. Graves, *Hamilton,* III: 404.

42. [De Morgan], "Review of Faraday," 399.

43. C. D. [Sophia De Morgan], *From Matter to Spirit. The Result of Ten Years' Experience in Spirit Manifestations. Intended as a Guide to Enquirers,* with an introduction by A. B. [Augustus De Morgan] (London: Longman, Green, Longman, Roberts & Green, 1863), xi.

44. C. D. [Sophia De Morgan], *Matter to Spirit,* x–xi.

45. C. D. [Sophia De Morgan], *Matter to Spirit,* xviii–xix.

46. C. D. [Sophia De Morgan], *Matter to Spirit,* xviii–xix.

47. Sophia explained that sometimes spirit writers worked better in pairs.

48. C. D. [Sophia De Morgan], *Matter to Spirit,* 268.

49. C. D. [Sophia De Morgan], *Matter to Spirit,* 274. I am capitalizing Sophia's Principle of Correspondence to make it comparable to the Principle of Equivalent Forms, which Peacock capitalized.

50. C. D. [Sophia De Morgan], *Matter to Spirit,* 267.

51. C. D. [Sophia De Morgan], *Matter to Spirit,* 15.

52. C. D. [Sophia De Morgan], *Matter to Spirit,* 23.

53. C. D. [Sophia De Morgan], *Matter to Spirit,* 110.

54. C. D. [Sophia De Morgan], *Matter to Spirit,* 273.

55. C. D. [Sophia De Morgan], *Matter to Spirit,* 268.

56. C. D. [Sophia De Morgan], *Matter to Spirit,* 269.

57. C. D. [Sophia De Morgan], *Matter to Spirit,* v.

58. C. D. [Sophia De Morgan], *Matter to Spirit,* xlv.

59. Sophia Elizabeth De Morgan, *Memoir,* 15.

## 21. BEYOND REASON

1. The initial address was 41 Chalcot Villas, but it was soon changed to 91 Adelaide Road.
2. ADM to a Friend, Jan 19, 1861. Sophia Elizabeth De Morgan, *Memoir of Augustus De Morgan* (London: Longmans, Green, 1882), 304.
3. A. M. W. Stirling, *William De Morgan and His Wife* (New York: H. Holt, 1922), 61.
4. Stirling, *William De Morgan,* 64–65.
5. ADM to WRH, Apr 29, 1854. Robert Perceval Graves, *Life of Sir William Rowan Hamilton* (1889), III: 479.
6. Augustus De Morgan, "Speech of Professor De Morgan," *Proceedings of the London Mathematical Society* 1 (1865): 9.
7. ADM to WRH, Feb 2, 1852. Graves, *Hamilton,* III: 220.
8. [ADM], "Viga Ganita," *Penny Cyclopædia* 26 (1843).
9. Sophia Elizabeth De Morgan, *Memoir,* 268–69; Druv Raina and S. Irfan Habib, *Domesticating Modern Science: A Social History of Science and Culture in Colonial India* (New Delhi: Tulika Books, 2004), 1–59. I am following Raina and Habib in spelling the name Ramchandra; De Morgan spelled it Ramchundra.
10. Sophia Elizabeth De Morgan, *Memoir,* 269.
11. Augustus De Morgan, "On the Beats of Imperfect Consonances" (read, Nov 9, 1857), *Transactions of the Cambridge Philosophical Society* 10 (1864): 129–45.
12. ADM to WRH, Apr 26, 1863. Graves, *Hamilton,* III: 281.
13. ADM to WRH, Apr 13, 1863. Graves, *Hamilton,* III: 380.
14. ADM to WRH, Apr 26, 1863. Graves, *Hamilton,* III: 281.
15. This quotation is from the first column, published in the *Athenaeum,* no.1867 (Oct 10, 1863): 466–68. It would be far too much work to locate all of the other columns in that form and so future references will refer to: Augustus De Morgan, *A Budget of Paradoxes,* ed. David Eugene Smith (Chicago: Open Court, 1915), I: 2.
16. Augustus De Morgan, *Budget,* I: vi.
17. Graves, *Hamilton,* 287.
18. For the founding of the society, see Sophia Elizabeth De Morgan, *Memoir,* 281–86; Adrian Rice, "From Student Club to National Society: The Founding of the London Mathematical Society in 1865," *Historia Mathematica* 22 (1995): 402–21.
19. Thomas Hirst's diary, Jun 15, 1862. http://www-groups.dcs.st-and.ac.uk/~history/HistTopics/Hirst_comments.html.
20. Augustus De Morgan, "Speech," 2–3.
21. Augustus De Morgan, "Speech," 3.
22. Augustus De Morgan, "Speech," 1.
23. Augustus De Morgan, "Speech," 6.
24. Augustus De Morgan, "Speech," 9.
25. Adrian Rice, "Introduction," in *The Book of Presidents, 1865–1965,* ed. Susan Oakes, Alan Pears, and Adrian Rice (London: London Mathematical Society, 2005), 4.
26. [Augustus De Morgan], "Philosophy of Discovery, Chapters Historical and Critical. By W. Whewell," *Athenaeum,* no. 1694 (Apr 14, 1860): 502.

27. [De Morgan], "Philosophy of Discovery," 501–2.

28. Robin Wilson, *Four Colors Suffice: How the Map Problem Was Solved* (Princeton: Princeton University Press, 2002), 62.

29. George Bruce Halstedt, "De Morgan to Sylvester," *Monist* 10 (1900): 188.

30. ADM to J. S. Mill, Feb 5, 1865. Sophia Elizabeth De Morgan, *Memoir*, 329.

31. Stirling, *William De Morgan*, 57.

32. Sophia Elizabeth De Morgan, *Memoir*, 358.

33. For the details of these developments, see Sophia Elizabeth De Morgan, *Memoir*, 336–62.

34. ADM to John Herschel, Oct 18, 1867. Sophia Elizabeth De Morgan, *Memoir*, 377.

35. Sophia Elizabeth De Morgan, *Threescore Years and Ten: Reminiscences of the Late Sophia Elizabeth De Morgan to Which Are Added Letters to and from Her Husband the Late Augustus De Morgan, and Others,* ed. Mary De Morgan (London: Richards Bentley and Son, 1895), 364.

36. Sophia Elizabeth De Morgan, *Memoir*, 364.

37. Sophia Elizabeth De Morgan, *Memoir*, 364.

38. ADM to SEDM, Aug 19, 1870. SHL, MS 913A/1/2.

39. Sophia Elizabeth De Morgan, *Reminiscences*, 367.

40. Technically, Sophia sold her husband's books to Samuel Jones-Lloyd, 1st Baron Overstone, who then donated them to the University of London. The Senate House Library has now created a virtual library of De Morgan's books, which can be found at http://www.senatehouselibrary.ac.uk/our-collections/special-collections/printed-special-collections/de-morgan-library.

41. Joan L. Richards, *Mathematical Visions: The Pursuit of Geometry in Victorian England* (Boston: Academic Press, 1988), 61–114.

42. William Kingdon Clifford, "The Postulates of the Science of Space," *Contemporary Review* 25 (1875): 363.

43. June Barrow-Green, "Euclid, William De Morgan and Charles Dodgson," in *The London Mathematical Society and Sublime Symmetry* (London: London Mathematical Society, 2018), 17.

## EPILOGUE

1. Thomas Kuhn, "Mathematical versus Experimental Traditions in the Development of Physical Science," in *The Essential Tension* (Chicago: University of Chicago Press, 1976), 31–65.

# Index

Note: An italicized *f* or *n* following a page number indicates a figure or an endnote, respectively.

Frend, William (continued)
107–11, 199; J. Herschel and, 178;
imaginary numbers and, 221; Jewish community and, 77–78, 106, 139,
354, 366*n*16; Lady Byron and, 1, 122,
189–91, 204–5, 226, 232; law career
and, 113; T. Lindsey and, x, 3, 4, 46,
73, 82, 101, 121–22, 130, 241; Locke
and Newton and, 52–56, 61; London
and, 57–59, 101, 194; London Corresponding Society (LCS) and, 104;
London radicals and, 101–2; Louis
XVI's execution and, 103; Maddingly
Sunday School and, 61–62, 68, 71,
89, 106, 363*n*5; marriage and, 58–59,
98, 106, 111, 132–33, 371*n*20; Maseres
and, 116–17, 120, 122, 340; mathematics and, 5, 106, 187; Mechanics'
Institute and, 193; *Memoirs of Emma
Courtney* (Hays) and, 109–11; Napoleon and, 127–29; natural theology
and, 122–23; negative numbers and,
115–16, 117–20, 186, 221; openness
of, 271; Paine and, 90; paper money
and, 120–21; poor people and, 71,
189; portrait, 42, 61*f*, 362*n*22; Priestley and, 352; prophecy and, 128;
*Rachovian Catechism* and, 67; reason
and, 46, 65, 67, 354; reform and, 134;
religious tolerance and, 232; right to
free speech and, 72; Robinson and,
76–77, 89; Roy and, 203; secularism
and, 47–48, 113; spiritualism and,
333–34; students and, 111–12; subscription requirement and, 68–70;
theology and, 75, 163–64, 187; tolerance and, 129; travel to Continent (1786 & 1789), 63–67, 78–80,
366*n*16; Trinitarian controversy and,
95; Trinitarianism and, 67–72; truth
and freedom and, 182; "Two Acts"
and, 105; Unitarianism and, 4, 5, 125,
204–5; Watt and, 266–67; wedding
of, 132–33; women's education and,

106–7; workers' cooperatives and,
193–94; working classes and, 71, 189.
*See also* Hays, Mary (Eusebia); London University; Long Stanton; Stoke
Newington (Frend's home)

Frend, William, **writings of**: *An Account
of the Proceedings in the University of
Cambridge, against William Frend
M.A.*, 97; *An Address to the Members of the Church of England, and
to Protestant Trinitarians in General,
Exhorting Them to Turn from the
False Worship of Three Persons to the
Worship of One True God*, 70; *Considerations on the Oaths Required by
the University at the Time of Taking
Degrees, and on other Subjects Which
Relate to the Discipline of That Seminary*, 69, 70; "The Effect of the War
on the Poor," 89–92, 95, 104, 269*n*11;
*Evening Amusements: Or, The Beauty
of the Heavens Displayed*, 123–24,
125, 127, 130, 132, 175; "The London
University," 382*n*32; *Monthly Repository of Theology and General Literature*, 128; "A Monthly Retrospect of
Public Affairs; or A Christian's Survey of the Political World," 128, 129,
134–35, 166, 201; "Paley," 361*n*14;
"Patriotism," 127; *Peace and Union
Recommended to the Associated Bodies
of Republicans and Anti-Republicans*,
89–92, 93, 95; on plurality of worlds,
320; *The Principles of Algebra*, 113, 119,
120, 127, 182, 187; *Scarcity of Bread
A Plan for Reducing the High Price
of This Article*, 104–5; *A Second Address*, 71–72; *A Sequel to the Account
of the Proceedings in the University
of Cambridge against the Author of a
Pamphlet, Entitled Peace and Union*,
22, 112, 365*n*56; *Tangible Arithmetic*,
123; theological lectures (private),
75; *Thoughts on Subscription to Reli-*